TRANSMISSION LINES ANTENNAS AND WAVE GUIDES

BY

RONOLD W. P. KING, A.B., Ph.D.
Gordon Mackay Professor of Applied Physics
Harvard University

HARRY ROWE MIMNO, E.E., Ph.D.
Gordon Mackay Professor of Applied Physics
Harvard University

ALEXANDER H. WING, E.E., Ph.D.
Professor of Electrical Engineering
University of Florida

DOVER PUBLICATIONS, INC.,
NEW YORK

Published in Canada by General Publishing Company, Ltd., 30 Lesmill Road, Don Mills, Toronto, Ontario.
Published in the United Kingdom by Constable and Company, Ltd., 10 Orange Street, London W.C.2.

This Dover edition, first published in 1965, is a corrected and enlarged version of the work first published by McGraw-Hill Book Company, Inc., in 1945.

Library of Congress Catalog Card Number 65-15512

Manufactured in the United States of America

Dover Publications, Inc.
180 Varick Street
New York 14, N.Y.

To

the men and women of the
Armed Forces for whom
this work was written

FOREWORD

This book has developed from the lecture notes of a special wartime training course given in the Graduate School of Engineering, Harvard University.

The need for highly trained officers in the new uses of electronics was definitely appreciated even in 1941. In July of that year, a course for officers of the Signal Corps who were graduate electrical engineers was established at Harvard University in the Graduate School of Engineering to give intensive training in the fundamentals of electronics and high-frequency circuits. This course served as preparation for further training. Immediately after the United States declared war, the Navy also sent officers to Harvard for preradar training.

The rapid expansion of the course necessitated a greatly increased instructional staff. Professors and instructors from many educational institutions were invited to Cambridge to aid the regular staff at Harvard in the field of communication engineering.

Although the war-training course was distinct from the regular graduate courses in communication engineering which have been given for more than two decades in the Cruft Laboratory, it was planned originally and patterned to a considerable degree on the lines of some of the Cruft Laboratory courses. Since the scope and the method of presentation of the material in this volume are not on a graduate level, they differ greatly from corresponding course work in the Cruft Laboratory. The members of the wartime staff have brought to the work many excellent ideas from their teaching experience. Frequent repetition of the course has given opportunity to improve its character both as to lecture presentations and laboratory experiments. Much new material has been injected into the course content. In short, this war-training program has presented a rare opportunity to develop and improve teaching methods in electronics.

During the progress of the course, lecture notes were prepared mainly by twelve of the members of the lecturing staff. It was the original intention to publish the entire lecture material in a single volume. Since the portion of the manuscript included in the present book was completed in advance of some of the remainder of the manuscript, it seemed best to make it available in published form at this time even though it precedes the publication of the first part of the lecture material.

The chapters on Antennas and Ultra-high-frequency Circuits were written by Dr. R. W. P. King, Associate Professor at Harvard, and include much original material not published elsewhere. The chapter on Wave Propagation was written by Dr. H. R. Mimno, Associate Professor at Harvard, who has devoted many years to original research in this field. The chapter on Lines was written by Dr. A. H. Wing, on leave of absence from the School of Technology of the College of the City of New York.

As editor, Dr. Wing from the beginning has devoted himself to the technical aspects of the manuscript.

Professor Harry E. Clifford, consulting editor of the McGraw-Hill Series of Electrical Engineering Texts, as general editor of the manuscript has rendered a most valuable service by giving the staff the benefit of his long editorial experience; and for his painstaking and constructively critical assistance with both the manuscript and proof, the authors have expressed their gratitude.

E. L. CHAFFEE
Director of Cruft Laboratory

PREFACE TO THE DOVER EDITION

This book provides an introduction to the basic principles of transmission lines, antennas, wave guides, and radio-wave propagation in the ionosphere. In accordance with its objective, the presentation avoids complicated mathematics but is nevertheless consistent with the more advanced mathematical theory. It seeks to give a thought-provoking insight into the fundamentals of radio transmission and radiation for the practical engineer, the radio astronomer, the radio amateur, and the college student.

Those who are stimulated to more advanced study are referred especially to *Fundamental Electromagnetic Theory* (Dover, 1963), *Transmission-line Theory* (McGraw-Hill, 1955; Dover reprint, 1964), and *Theory of Linear Antennas* (Harvard, 1956) by R. W. P. King.

In addition to a number of corrections, this edition has been provided with a completely new set of graphs for distributions of current, impedances, and effective lengths of dipole antennas and for self- and mutual impedances of coupled antennas. These are based on the King-Middleton second-order results which are known to be in good agreement with experiment. They have been selected from an extensive set of theoretical and experimental graphs and tables in *Theory of Linear Antennas*. The quantitative treatment of end and coupling effects between transmission lines and antennas, which were not understood in 1945 and to which brief references have been inserted at appropriate points in this new edition, is also contained in *Theory of Linear Antennas*.

In Chapters I and III, the definition of susceptance has been changed to agree with the equation $Y = G + jB$, where Y is the admittance, G the conductance, and B the susceptance. Accordingly, the necessary sign changes, and certain deletions, have been made.

The following figures in Chapter II of this edition are new:

9.1 (II.22.2), 9.2 (II.22.2), 9.3 (II.22.5), 9.4 (II.22.5); 10.3 (II.20.7), 10.4 (II.30.5a), 10.5 (II.30.5b), 10.6 (II.30.6a), 10.7 (II.30.9), 10.8 (II.30.10a), 10.10 (II.30.7a), 10.11 (II.30.7b), 10.12 (II.30.4b), 10.13 (II.33.2a), 10.14 (II.33.3a); 19.2 (III.8.13), 19.3 (III.8.5), 19.4 (III.8.6a), 19.5 (III.8.6b); 31.2 (IV.9.6a), 31.4 (IV.9.2c), 31.5 (IV.9.2a), 31.6 (IV.9.2b).

They correspond to the figures with numbers in parentheses from *Theory of Linear Antennas*. They are reproduced by permission of the Harvard University Press which retains the copyright dated 1956.

CAMBRIDGE, MASS.
GAINESVILLE, FLA.
September 1964.

RONOLD W. P. KING,
HARRY ROWE MIMNO,
ALEXANDER H. WING.

PREFACE TO THE FIRST EDITION

The material of Chap. I comprises the work given originally in a series of approximately 15 one-hour lectures, devoted mostly to the dissipationless line and impedance-matching devices. Supplementary material has been added to meet requirements of a more general nature. Many helpful suggestions have been made by the staff at Cruft Laboratory. In particular, thanks are due to Dr. Philippe E. LeCorbeiller for some of the material of Sec. 42 and for Sec. 53 of Chap. I.

Chapters II and III are the result of extensive study of the educational problem of presenting general electromagnetic phenomena without the use of higher mathematics. The two chapters are largely self-contained because the study of antennas, ultra-high-frequency circuits, and wave guides followed a curriculum based on conventional electric-circuit theory. The attempt is made throughout not only to provide necessary factual material, but to develop a real understanding for the electromagnetic point of view. The greater part of the material in Chap. II together with a discussion of impedance matching, including single- and double-stub tuners, was contained originally in mimeographed form under the title "Notes on Antennas, Engineering 270" (copyright 1943). Acknowledgment is due Lt. Comdr. Charles W. Harrison, Jr., for assisting in extensive research carried out to provide many of the curves and charts, and for preparing numerous problems for Chap. II. Valuable suggestions, contributions, and corrections were made by student officers and by members of the staff.

The section on wave propagation is brief and is intended primarily for students with a general knowledge of physics who have had no previous opportunity to study the basic principles of radio transmission. The essential facts governing ground-wave and sky-wave propagation are brought out and related to well-known scientific principles in order to show that the newly described phenomena are in reality wholly natural and easily understandable. In order to focus attention chiefly upon the simple factors controlling practical long-distance sky-wave communication, all refinements related to the effect of the earth's magnetic field are deliberately minimized. Within the limits thus determined the treatment is concise and largely nonmathematical. The material was selected in an attempt to make the most effective use of four one-hour lectures allocated to radio-wave propagation in the Cruft Army-Navy course. Acknowledgment is due Dr. J. H. Dellinger and Dr. Lyman J. Briggs for permission to republish three diagrams issued originally by the Bureau of Standards.

Ronold W. P. King,
Harry Rowe Mimno,
Alexander H. Wing.

Cambridge, Mass.,
May, 1945.

CONTENTS

Chapter I

TRANSMISSION LINES

ALEXANDER H. WING

TRANSMISSION LINES, ANTENNAS, AND WAVE GUIDES

CHAPTER I

TRANSMISSION LINES

1. Introduction.—The transmission of power over long distances by means of wires or "lines" is well known. There are in use transmission lines several hundred miles long, transmitting power from generators of commercial power frequencies, 25 to 60 cps. Long as these lines are, they are relatively short when compared with the electrical wavelength. Since electromagnetic disturbances on open-wire transmission lines travel approximately with the speed of light (186,000 miles per second), the wavelength at 60 cps is 3,100 miles. Thus a 310-mile 60-cycle transmission line is only 0.1 wavelength long. At radio frequencies, the wavelength may be measured in a few inches or in a few centimeters.

Each wavelength of line provides for a possible phase change of 360°, to speak of no other changes. For such a length, the rms current along a given wire is not everywhere the same, nor is the rms voltage between the wires everywhere the same. Sometimes these variations are objectionable; sometimes they are extremely useful. Since these changes occur on any line that is not short compared to a wavelength, such a line must be considered electrically "long" regardless of its physical length.

The simplest transmission line to construct is the familiar two-wire line consisting of two round wires separated by a uniform distance. Certain properties of such a line are common to all types of transmission line. The motion of charges in the wires constitutes a current. A description of the electrical behavior of the line is really a description of the motion of charges in terms of the currents in the wires and the voltage between the wires. This description is simplified if, as is usual, the distance between wires is very small compared with the wavelength, and the length of the wires is large compared with their separation. If the distance between wires is not small compared with the wavelength, transmission-line theory as here formulated does not apply. Where the spacing of circuit elements is an appreciable fraction of a wavelength, as in antennas and hollow wave guides, the time delay involved between a cause in one element of the circuit and its effect in another element of the circuit is an appreciable fraction of a period. This situation requires a different treatment from that used in circuit theory, in which the action of one circuit on a neighboring circuit is considered to be instantaneous. Transmission lines can be described in terms of circuit theory only because one wire is considered to be so close to the other that the effect on one wire of a change of current in the other wire is instantaneous. The necessity for this restriction will be clearer after the study of antennas and wave guides (Chaps. II and III).

This chapter is concerned with the behavior of lines at radio frequencies, their use for the transmission of power, the use of short sections of line as reactances and for impedance matching, their use in the measurement of frequency and the measurement of impedances at radio frequencies.

2. Construction.—Radio-frequency transmission lines are constructed commonly in three forms: the two-wire line, the four-wire line, and the coaxial line.

The two-wire line in Fig. 2.1 consists of two parallel wires whose uniform spacing is small compared with the electrical wavelength. If the line is far removed from other objects or is symmetrically disposed with respect to near-by circuits including its own terminations, the current in one wire is equal and opposite to the current in the other wire at any point along the line. This is a fundamental assumption in the theory formulated in this chapter.

FIG. 2.1.—Two-wire transmission line.

The four-wire line as ordinarily constructed employs four wires placed at the corners of a small square, Fig. 2.2. Diagonally opposite wires are connected in parallel. This type of line has a weaker external field than the two-wire line. Neighboring four-wire lines have less mutual effect than similarly situated two-wire lines.

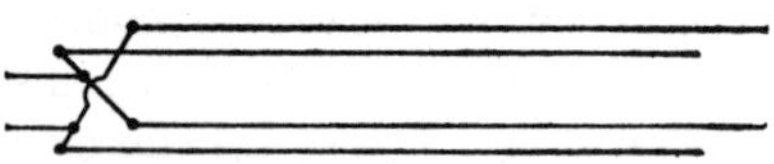

FIG. 2.2—Four-wire transmission line.

FIG. 2.3.—Coaxial transmission line.

The coaxial line consists of a round conductor supported coaxially (by insulators) within a round tube which serves as the second conductor, Fig. 2.3. This type of line has practically no external field when used with shielded terminations.

3. Skin Effect.—Owing to skin effect, the current in an open-wire line is confined to a thin layer adjacent to the outer surface. The surface area rather than the cross-sectional area therefore determines the resistance per unit length. In the coaxial line, the current flows on the outside surface of the inner conductor and on the inside surface of the outer conductor. Owing to the smaller surface area of the inner conductor, more of the I^2R loss occurs in the inner conductor. Since the electromagnetic field due to transmission-line currents is confined within the tube, the outside surface of the outer conductor of a coaxial line normally carries no current and normally requires no external insulation.

4. "Antenna" Currents.—The presence of antennas or other circuits may cause currents to flow on the *outside* surface of the outer conductor of a coaxial line, which may be quite independent of the "transmission-line" current that flows on the *inside* surface. The currents on the outside surface cause the outside surface to act as an antenna. Similarly, there may be "antenna" currents flowing on the two or more wires of an open-wire line, in addition to the "transmission-line" currents, causing the resultant current distribution to be different from that which would exist if the line were properly isolated or balanced with respect to near-by objects. The cause and effect of these "antenna" currents and the means of suppressing them are discussed more completely in Chap. II.

5. Distributed Constants.—The transmission line cannot be analyzed as a simple series circuit, because the current in the wires is not everywhere the same. The variations in current and voltage must be examined inch by inch, centimeter by centimeter, meter by meter, etc., depending upon the wavelength and the unit of length chosen. The factors of fundamental importance are the series resistance and inductance per unit length, and the shunt conductance and capacitance per unit length. These parameters are denoted by r, l, g, c. A line is said to be uniform when these "distributed constants" are the same over the whole line.

These constants are said to be "distributed" rather than "lumped" since, for example, the total capacitance of the line is spread out along the line, and the effects of this capacitance are not the same as would be obtained if all the capacitance were concentrated at one point. The values of the distributed constants r, l, g, c will be given in Sec. 12.

6. Differential Equations of a Line.—The analysis will be made in terms of the two-wire line, since the analysis leads to the same equations for the other types of construction. Figure 6.1 illustrates the terminology that will be used. The line is assumed to be fed by a generator of voltage E_g and internal impedance Z_g. The total length of the line is s (s meaning space). The load, located at the receiving end of the line, has an impedance Z_R. The distance of a point M on the line from the generator or sending end is denoted by x, and the distance of the same point from the load or receiving end is d. The current and voltage at point M are denoted by I and E; at the load, the current and voltage are denoted by I_R and E_R (subscript R for "receiving"); and at the sending end ($x = 0$),

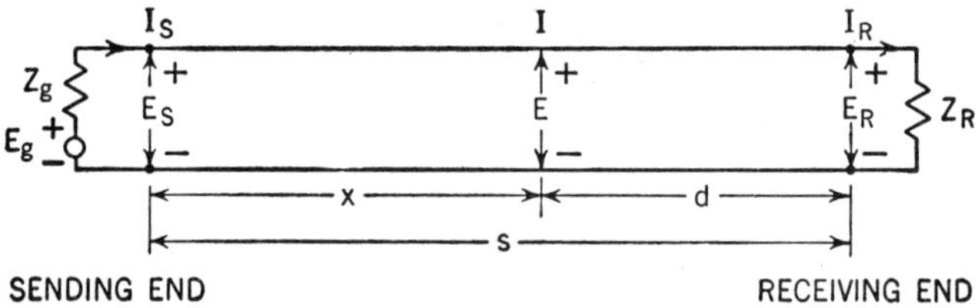

FIG. 6.1.—Diagram showing notations used in the text.

the current and voltage are denoted by E_S and I_S (subscript S for "sending"). The currents and voltages are the complex rms values in the steady state.

For purposes of calculation, a short length Δx of the line may be represented closely by an equivalent T section in which the series impedance is $(r + j\omega l)\,\Delta x$ and the shunt admittance is $(g + j\omega c)\,\Delta x$. If z and y be used to represent the series impedance and shunt admittance per unit length

$$z = r + j\omega l \tag{6.1}$$
$$y = g + j\omega c \tag{6.2}$$

with the arms of the T section as indicated in Fig. 6.2. The input and output voltages of the section are denoted by E and $E + \Delta E$, where ΔE is the change. Similarly, the currents are denoted by I and $I + \Delta I$. Since the output voltage is equal to the input voltage less the voltage drop in the series arms,

$$E - I\frac{z\,\Delta x}{2} - (I + \Delta I)\frac{z\,\Delta x}{2} = E + \Delta E \tag{6.3}$$

whence

$$\frac{\Delta E}{\Delta x} = -Iz - \Delta I \frac{z}{2} \tag{6.4}$$

As Δx is taken smaller and smaller, the ratio $\Delta E/\Delta x$ approaches the derivative dE/dx, and the term containing ΔI becomes zero. Then

$$\frac{dE}{dx} = -Iz \tag{6.5}$$

Similarly in the limit as Δx approaches zero

$$\frac{dI}{dx} = -Ey \tag{6.6}$$

Equations (6.5) and (6.6) are similar, and equations identical in form

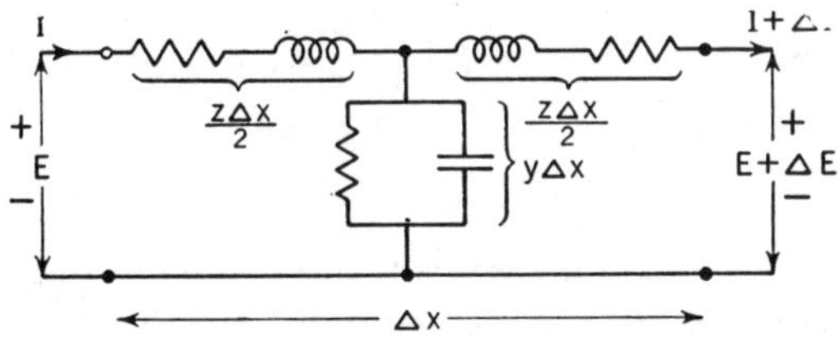

FIG. 6.2.—Nominal T section of a short length of transmission line.

can be obtained by differentiating once. The derivative of (6.5), after substitution from (6.6), is

$$\frac{d^2E}{dx^2} = zyE \tag{6.7}$$

and similarly the derivative of (6.6) is

$$\frac{d^2I}{dx^2} = zyI \tag{6.8}$$

Equations (6.5) to (6.8) are the equations upon which the analysis of transmission lines is based.

The solution of (6.7) is, in general,

$$E = A\epsilon^{-\sqrt{zy}\,x} + B\epsilon^{+\sqrt{zy}\,x} \tag{6.9}$$

where A and B are constants depending on the termination and the length of the line. The solution of (6.8) also involves two constants, which are not independent of A and B because the current in a line is not independent of the voltage. The values of these constants may be obtained in terms of A and B by differentiating (6.9) and equating the result to (6.5), so that

$$I = \frac{A}{\sqrt{z/y}}\epsilon^{-\sqrt{zy}\,x} - \frac{B}{\sqrt{z/y}}\epsilon^{+\sqrt{zy}\,x} \tag{6.10}$$

The quantity $\sqrt{zy}$ is a numeric per unit length and will be denoted by γ. The quantity $\sqrt{z/y}$ has the dimensions of impedance and will be denoted by Z_c. Then (6.9) and (6.10) become

$$E = A\epsilon^{-\gamma x} + B\epsilon^{+\gamma x} \tag{6.11}$$

$$I = \frac{A}{Z_c}\epsilon^{-\gamma x} - \frac{B}{Z_c}\epsilon^{+\gamma x} \tag{6.12}$$

7. Propagation Constant.—The quantity γ is called the *propagation constant* and is defined by

$$\gamma = \sqrt{zy} = \sqrt{(r + j\omega l)(g + j\omega c)} \tag{7.1}$$

It has a real and an imaginary part given by

$$\gamma = \alpha + j\beta = \sqrt{(r + j\omega l)(g + j\omega c)} \tag{7.2}$$

wherein α is the *attenuation constant* in nepers per unit length and β is the *phase constant* in radians per unit length.

It is shown in Sec. 10 that in a line to which Z_c is connected as a load, the current is attenuated α nepers by each unit length of line, and at the end of each unit length the voltage and current will lag the voltage and current at the beginning of that unit length by an angle of β radians. Thus γ is the constant that determines the current and voltage variations in the line.

8. Characteristic Impedance.—The quantity

$$Z_c = \sqrt{\frac{z}{y}} = \frac{\sqrt{r + j\omega l}}{\sqrt{g + j\omega c}} \tag{8.1}$$

is called the *characteristic impedance* (sometimes surge impedance). It is shown in Sec. 9 that any line, however long, when terminated by an impedance Z_c has an input impedance Z_c. Also, on a line terminated in Z_c there are no "standing waves." These statements will be clearer after a study of the variations in current and voltage along a transmission line.

I. NONRESONANT LINES

9. Line Terminated in Z_c.—A line terminated in its characteristic impedance Z_c is sometimes called a nonresonant line. To determine the value of the constants A and B, the terminal conditions must be investigated. At the load, $x = s$ and (6.11) and (6.12) take on the particular values

$$E_R = A\epsilon^{-\gamma s} + B\epsilon^{+\gamma s} \tag{9.1}$$

$$I_R = \frac{A}{Z_c}\epsilon^{-\gamma s} - \frac{B}{Z_c}\epsilon^{+\gamma s} \tag{9.2}$$

Since the load is equal to Z_c

$$E_R = Z_c I_R \tag{9.3}$$

After substitution from (9.1) and (9.2)

$$A\epsilon^{-\gamma s} + B\epsilon^{+\gamma s} = A\epsilon^{-\gamma s} - B\epsilon^{+\gamma s}$$

therefore,

$$B = 0 \tag{9.4}$$

At the generator end, $x = 0$, and (6.11) and (6.12) reduce to

$$E_S = A \tag{9.5}$$

$$I_S = \frac{A}{Z_c} \tag{9.6}$$

which are the input voltage and current.

The input impedance of the line is the ratio of input voltage and current, or

$$Z_S = \frac{E_S}{I_S} = Z_c \tag{9.7}$$

This is a remarkable result and indicates that a transmission line of any length, *connected to a load impedance* Z_c, has an input impedance Z_c. Z_c is a function of r, l, g, c, and the frequency, and therefore is a function of the size of the wires and their spacing but is independent of the length of the line.

The result of (9.7) also would be obtained if the line had been considered infinitely long and E_R and I_R set equal to zero as they would be on an infinitely long line having losses. Thus Z_c is also the input impedance of an infinitely long line.

The steady-state input current and voltage now may be calculated, since the generator "sees" an impedance Z_c connected in series with its internal impedance Z_g, Fig. 6.1. Then

$$E_S = E_g \frac{Z_c}{Z_g + Z_c} \tag{9.8}$$

and

$$I_S = \frac{E_g}{Z_g + Z_c} \tag{9.9}$$

With $B = 0$ from (9.4) and $A = E_S$ from (9.5), the general equations (6.11) and (6.12) become

$$E = E_g \frac{Z_c}{Z_g + Z_c} \epsilon^{-\gamma x} \tag{9.10}$$

$$I = \frac{E_g}{Z_g + Z_c} \epsilon^{-\gamma x} \tag{9.11}$$

Thus at any point on a nonresonant line

$$E = Z_c I \tag{9.12}$$

10. Current and Voltage Distribution on a Line Terminated in Z_c The voltage variation along the line is given by (9.10), which, replacing γ by its real and imaginary parts, becomes

$$E = E_S\epsilon^{-(\alpha+j\beta)x} = E_S\epsilon^{-\alpha x}\epsilon^{-j\beta x} \tag{10.1}$$

indicating that, as x increases, the magnitude of E is decreased by the factor $\epsilon^{-\alpha x}$ and the phase is retarded by βx radians. If the

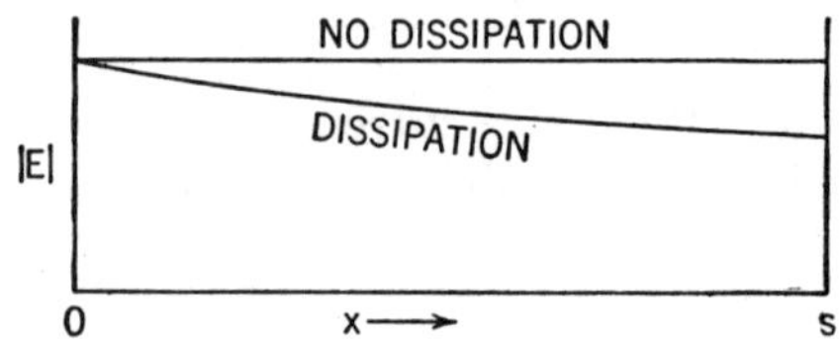

FIG. 10.1.—Variation in the magnitude of the voltage along a line terminated in Z_c. The current varies in the same manner.

magnitude of E is plotted vertically and distance along the line horizontally, the result is Fig. 10.1, showing the magnitude decreasing exponentially for the case marked "Dissipation." If there were no dissipation, r and g would be zero and α would be zero. There would be no attenuation, but the phase retardation would remain. The magnitude of the voltage would be unchanged, as plotted in Fig. 10.1 for the case marked "No Dissipation." In many cases the dissipation may be so small that the line marked "No Dissipation" is a close approximation. A line in which the dissipation is zero or is very small and which is terminated in Z_c is said to be "flat," *i.e.*, the *magnitude* of the voltage is practically the same everywhere along the line.

For a line terminated in Z_c, the variations in current and voltage are exactly the same, and Fig. 10.1 also suffices as a diagram of the current distribution, the vertical scale being changed in accordance with the fact that the ratio of voltage to current is Z_c.

If the sending end voltage E_S is taken as the reference vector, and the results of (10.1) are plotted for each value of x, the locus of the tips of all the vectors representing E is a logarithmic spiral for the dissipative line, or a circle for the nondissipative line, Fig. 10.2. Each unit length of line causes the voltage to lag by

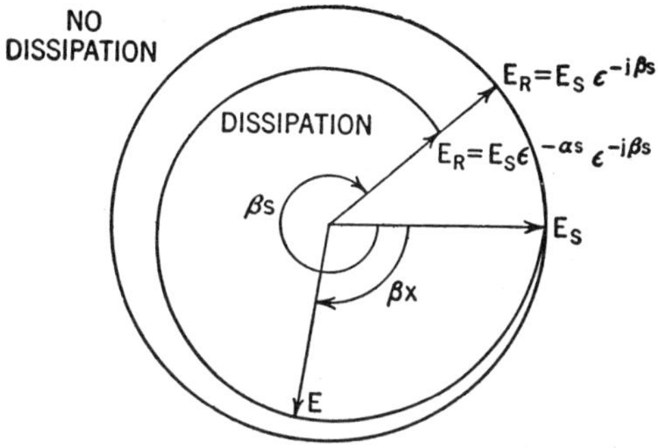

FIG. 10.2.—Polar diagram showing the variation in magnitude and phase of the voltage along a line terminated in Z_c.

an angle β, so that the voltage at the receiving end lags the voltage at the sending end by βs radians. The same is true for the current.

11. Wavelength.—If the length s of the line of Fig. 10.2 were made a little greater, the total phase difference between E_R and E_S could be made equal to 2π or 360°. The line would then be one wavelength long. Thus the wavelength in a transmission line is the length of line which, when terminated in Z_c, causes a phase retardation of 360° or one complete cycle. Since the phase change is β radians per unit length, the wavelength, in the units of length in which β is expressed, is given by

$$\lambda = \frac{2\pi}{\beta} \tag{11.1}$$

12. Phase Velocity and Time Delay.—Each of (9.10) and (9.11) represents a wave of wavelength λ traveling from the generator toward the load. In general, wave motion involves variations in both time and space. Equations (9.10) and (9.11) indicate only the variations in space, since these equations are intended to represent only the steady rms values.

Equation (10.1) states that at a distance x from the sending end the voltage has an rms magnitude $\epsilon^{-\alpha x}$ times that of E_S, and lags E_S by an angle of βx radians. If the instantaneous value of the sending-end voltage is

$$e_s = \hat{E}_S \cos \omega t \tag{12.1}$$

where $\hat{E}_S$ is a real positive number, then the equation for this voltage after being changed in magnitude by $\epsilon^{-\alpha x}$ and retarded in phase by βx radians is

$$e = \hat{E}_S \epsilon^{-\alpha x} \cos (\omega t - \beta x) \tag{12.2}$$

If the attenuation is so small that $\epsilon^{-\alpha x} \doteq 1$, (12.2) reduces to

$$e = \hat{E}_S \cos (\omega t - \beta x) \tag{12.3}$$

This equation represents a wave, periodic in time *and space,* moving to the right with a certain velocity called the *phase velocity.* In Fig. 12.1, the value of e given by (12.3) is plotted as a function of the distance x, for a line one wavelength long, for values of t equal to 0, $T/4$, and $T/2$, where T is the period of one cycle. The

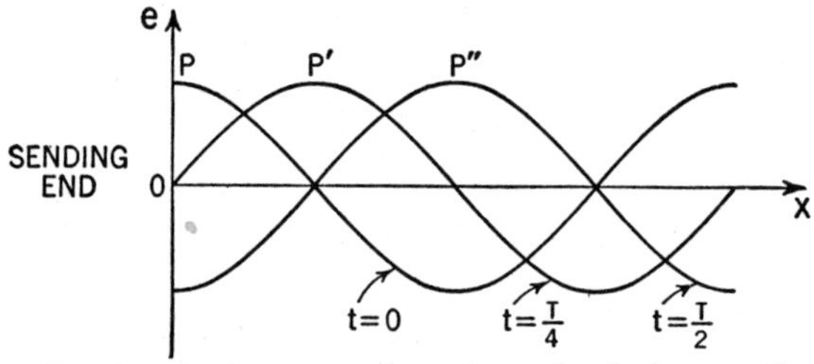

FIG. 12.1.—Showing the instantaneous voltage (at the instants indicated) along a transmission line having no attenuation and terminated in its characteristic impedance.

curve for $t = T/4$ is the same as that for $t = 0$ but moved to the right by a distance $\lambda/4$. The curve for $t = T/2$ is the same as that for $t = T/4$ but moved to the right by $\lambda/4$. If a point at which the instantaneous voltage has a definite phase, such as positive maximum, be denoted by P, Fig. 12.1, this point moves to the right a distance $\lambda/4$ in a time $T/4$. The *velocity of a point* denoting the location *in space* of a definite phase of the periodic disturbance is the *phase velocity.* Since any such point, like the point P, Fig. 12.1, moves a distance $\lambda/4$ in a time $T/4$, the phase velocity is

$$v_p = \frac{\lambda/4}{T/4} = \frac{\lambda}{T} = \lambda f \tag{12.4}$$

This is an important equation, since it shows the relation between wavelength and frequency. From (12.4) and (11.1),

$$v_p = \frac{2\pi f}{\beta} = \frac{\omega}{\beta} \tag{12.5}$$

In an open-wire transmission line, the phase velocity is approximately equal to the speed of light, 300,000,000 or $3 \cdot 10^8$ meters per second.

This does not mean that the electrons in the wire travel with this speed. The actual velocity of the electrons in the wire is of the order of centimeters per second. A perfect analogy to this apparently contradictory situation is found in acoustics. Sound "travels" with a speed of approximately 1,100 feet per second or about 700 miles per hour. Yet, when a note is sounded (a whistle produces a practically pure tone), there is no 700 mile per hour gale in the surrounding air. The actual motion of the air particles is only a fraction of an inch. The fact that sound "travels" at about 700 miles per hour does not mean that the air travels that fast. It means that, if P is a point in space at which the air pressure is a maximum, this point travels with a velocity of 700 miles per hour. Note that it is the descriptive point that travels, not the air. In the transmission line, the phase velocity is very high, but the motion of the electrons in the conductors is very slow. The electrons could not move with the speed of light, because their mass would become infinite at this speed. In conductors, the maximum velocity of electrons is always small. In evacuated spaces, as in cathode-ray tubes or in other high-voltage tubes, the electrons may acquire a velocity that is an appreciable fraction of the velocity of light, but is still a *fraction*. In hollow wave guides, the phase velocity is even greater than the velocity of light, and again the velocity of the electrons in the conducting walls is extremely low and of the same order of magnitude as it is in transmission lines.

The time delay t_d of a transmission line is the time it takes for a point such as P, Fig. 12.1, to travel the length of the line. If the phase velocity is independent of frequency, this time is the same as the time it takes for a pulse or signal to travel the length of the line. From (12.5) and (12.4),

$$t_d = \frac{s}{v_p} = \frac{\beta s}{\omega} = \frac{s}{\lambda} T \tag{12.6}$$

so that the time delay in terms of the period T is equal to the number of wavelengths in the line.

II. TRANSMISSION-LINE CONSTANTS

The "constants" r, l, g, c depend upon the size of the conductors, their spacing, and the nature of the insulation. At radio frequencies, the skin effect is pronounced, and r, l, c may therefore be calculated on the basis of a very thin surface layer of current. The constant g is not easily evaluated except for a coaxial line having a solid dielectric whose properties are known. It is usually determined by test. The effect of spacers or insulators spaced at intervals may be determined sometimes in terms of an equivalent continuous insulation if there are many insulators per wavelength. More will be said later about the spacing of insulators. At extremely high frequencies the loss in the insulation may become excessive. The weather has considerable effect on the shunt conductance of exposed open-wire lines. The insulators must be chosen to resist dampness and to have a minimum surface leakage, in addition to having small dielectric losses.

13. Skin Depth and Proximity Effect.—At high frequencies, current does not penetrate deeply into metal. The current density is greatest at the surface and decreases in magnitude and shifts in phase as the distance from the surface increases. If the radius of curvature of the conductor surface is large compared with what is known as the "skin depth," the resistance of the conductor may be calculated by assuming the current density to be uniform and confined to a surface layer of thickness

$$\delta = \frac{1}{\sqrt{f}\,\sqrt{\pi\mu\sigma}} \text{ meters} \tag{13.1}$$

where δ is the "skin depth," f is the frequency in cycles per second, and μ and σ are the absolute permeability and the conductivity in mks units. For nonmagnetic metals, $\mu = 4\pi \cdot 10^{-7}$, so that

$$\delta = \frac{504}{\sqrt{f\sigma}} \text{ meters} \tag{13.2}$$

The conductivity for copper at 20° C is $5.8 \cdot 10^7$ mhos/m (corresponding to a resistivity of $1.724 \cdot 10^{-6}$ ohm-cm), so that for copper

$$\delta = \frac{6.62}{\sqrt{f}} \text{ cm} = \frac{2.61}{\sqrt{f}} \text{ in.} \tag{13.3}$$

Practically every conductor used for transmission lines has a radius that is very much greater than the skin depth, and for currents traveling along the length of the conductor the effective cross-sectional area is therefore equal to the skin depth multiplied by the perimeter.

There is some current within the metal at a depth greater than the skin depth. The current density is greatest at the surface. As the distance from the surface increases, the current density decreases exponentially. For example, at a depth equal to 3 times the skin depth, the current density is ϵ^{-3} or 5 per cent of the value at the surface. The total I^2R loss, however, is the same as if the current were uniformly spread over a surface layer of depth equal to the skin depth.

If the wires of a two- or four-wire transmission line are very close together, an effect called the "proximity effect" causes a further change in the current distribution. The surface of the wires nearest the wires of opposite polarity has a higher current density than the mutually more distant parts of the surface. For spacings greater than 8 times the wire diameter, the effect causes less than 1 per cent increase in the resistance. In coaxial cables there is no proximity effect except when the inside conductor is displaced from its coaxial position.

14. Approximations for Low-loss Lines.—The rigorous formulas for γ and Z_c are

$$\gamma = \alpha + j\beta = \sqrt{(r + j\omega l)(g + j\omega c)} \tag{14.1}$$

$$Z_c = R_c + jX_c = \sqrt{\frac{r + j\omega l}{g + j\omega c}} \tag{14.2}$$

If $g = 0$ and $r << \omega l$, the following approximations are valid:

$$Z_c = R_c = \sqrt{\frac{l}{c}} \text{ ohms} \tag{14.3}$$

$$\alpha = \frac{r}{2R_c} \text{ nepers/unit length} \tag{14.4}$$

$$\beta = \omega\sqrt{lc} \text{ radians/unit length} \tag{14.5}$$

The characteristic impedance is practically a pure resistance and is practically independent of frequency. The phase velocity, from (12.5), is

$$v_p = \frac{\omega}{\beta} = \frac{1}{\sqrt{lc}} \tag{14.6}$$

and is also practically independent of frequency.

It is a property of the electromagnetic field that, for a dissipationless line in free space, l and c are so related that their product is constant and the phase velocity is equal to the velocity of light, 300,000,000 m/sec. The effect of a dielectric is to increase c. In lines having a solid dielectric, the phase velocity may be considerably less than the free-space value; and since

$$\lambda = \frac{2\pi}{\beta} = \frac{v_p}{f} = \frac{1}{f\sqrt{lc}} \tag{14.7}$$

the wavelength on the line is shortened from its free-space value.

The effect of losses in the line is to cause a slight increase in β and a slight decrease in λ below that indicated by (14.7). In practice this effect is usually small, and calculations based upon the free-space value of λ are quite accurate for lines having air or gaseous dielectric. (Coaxial lines are often filled with dry nitrogen under slight pressure to keep them from "sweating" and "breathing" owing to changes in atmospheric temperature, humidity, and pressure.)

15. Constants for the Two-wire Line.—The following formulas neglect g and are based upon a well-developed skin effect. For round wires of radius a, spaced at a center-to-center distance D, Fig. 15.1, the inductance per unit length of line is

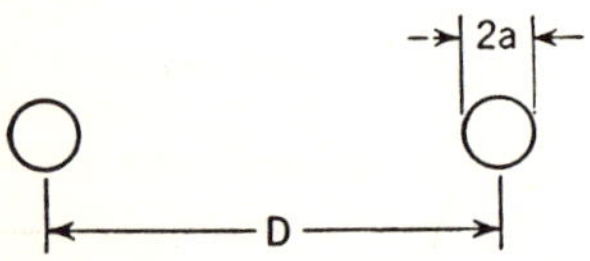

FIG. 15.1.—Dimensions of two-wire line.

$$l = 0.921 \log_{10} \frac{D}{a} \ \mu\text{h/m} \tag{15.1}$$

The capacitance per unit length of line is

$$c = \frac{12.06\epsilon_r}{\log_{10} \frac{D}{a}} \ \mu\mu\text{f/m} \tag{15.2}$$

where ϵ_r is the relative dielectric constant of the medium in which the wires are embedded. For air, $\epsilon_r \doteq 1$. The resistance per unit length of line for standard annealed copper at 20° C is

$$r = \frac{8.3\sqrt{f}}{a} \text{ microhms/m} \tag{15.3}$$

where a is in centimeters. For the open-wire line

$$R_c = 276 \log_{10} \frac{D}{a} \tag{15.4}$$

For hard-drawn copper the conductivity is slightly lower (that is, $\sigma = 5.65 \cdot 10^7$ mhos per meter, corresponding to a resistivity of $1.77 \cdot 10^{-6}$ ohm-cm) and the constant 8.3 in (15.3) becomes 8.4. The resistance of thick wires at radio frequencies varies as the square root of the resistivity.

16. Constants for the Four-wire Line.—The resistance per unit length of line is half that for the two-wire line, or

$$r = \frac{4.2\sqrt{f}}{a} \text{ microhms/m}$$

where a is in centimeters. For the square spacing, Fig. 16.1, the inductance per unit length of line is

$$l = 0.460 \log_{10} \frac{D}{a\sqrt{2}} \ \mu\text{h/m}$$

The capacitance per unit length of line is

$$c = \frac{24.1\epsilon_r}{\log_{10} \dfrac{D}{a\sqrt{2}}} \ \mu\mu\text{f/m}$$

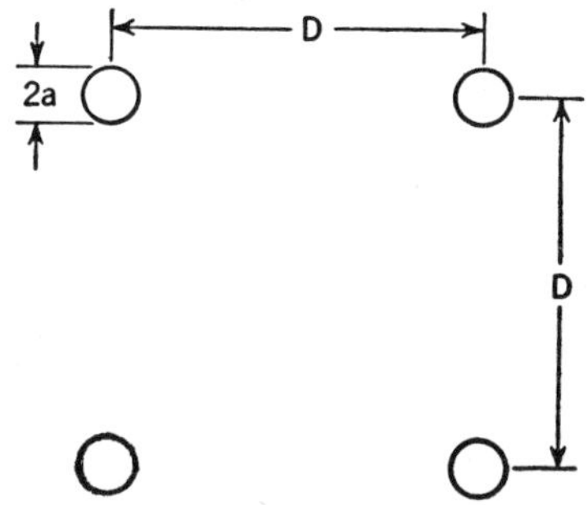

FIG. 16.1.—Dimensions of four-wire line with square spacing.

where ϵ_r is the relative dielectric constant of the medium. For the open-wire line, neglecting losses, the characteristic impedance is

$$R_c = 138 \log_{10} \frac{D}{a\sqrt{2}} \text{ ohms}$$
$$= 138 \log_{10} \frac{D}{a} - 20.8 \text{ ohms}$$

The attenuation on a four-wire line is approximately the same as on a two-wire line for the same ratio D/a.

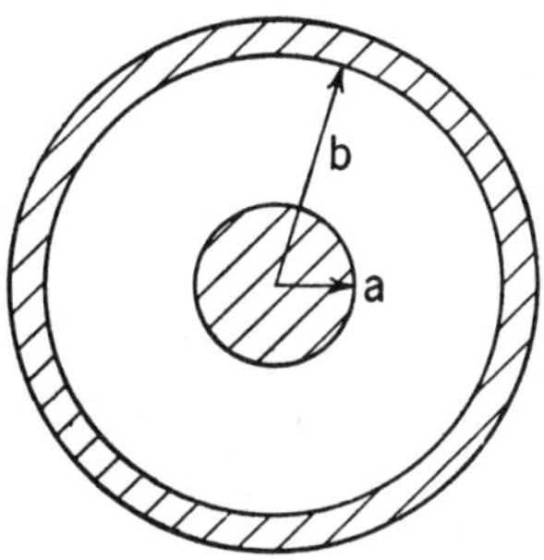

FIG. 17.1.—Dimensions of coaxial cable.

17. Constants for the Coaxial Cable. The following formulas again assume a well-developed skin effect. The resistance per unit length for copper conductors at 20° C is

$$r = 4.2\sqrt{f}\left(\frac{1}{a} + \frac{1}{b}\right) \tag{17.1}$$
$$= \frac{4.2\sqrt{f}}{b}\left(\frac{b}{a} + 1\right) \text{ microhms/m} \tag{17.2}$$

where a is the outside radius of the inner conductor and b is the inside radius of the outer conductor, both in centimeters, Fig. 17.1. The inductance per unit length is

$$l = 0.46 \log_{10} \frac{b}{a} \ \mu\text{h/m} \tag{17.3}$$

and the capacitance per unit length is

$$c = \frac{24.1\epsilon_r}{\log_{10} \dfrac{b}{a}} \ \mu\mu\text{f/m} \tag{17.4}$$

where ϵ_r is the relative dielectric constant, 1 for air. The characteristic resistance is

$$R_c = \frac{138}{\sqrt{\epsilon_r}} \log_{10} \frac{b}{a} \text{ ohms} \tag{17.5}$$

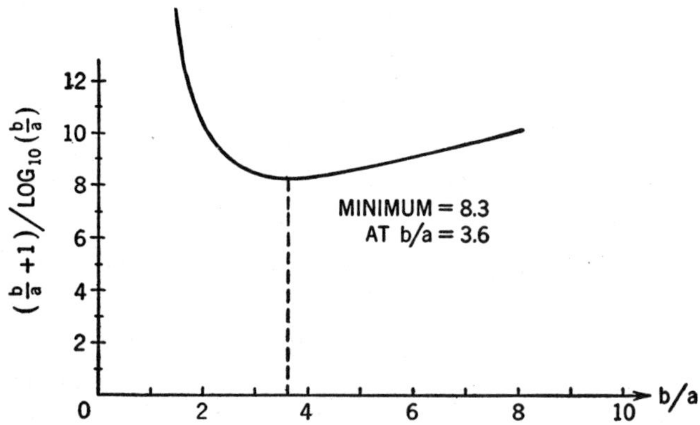

FIG. 17.2.—Variation in the factor appearing in the expression for the attenuation of coaxial cable.

The attenuation constant is $r/2R_c$, (14.4), or

$$\alpha = \frac{\sqrt{f\epsilon_r}\left(\dfrac{b}{a} + 1\right)}{65.7b \log_{10} \dfrac{b}{a}} \cdot 10^{-6} \text{ nepers/m} \tag{17.6}$$

For a constant outer radius b, the attenuation constant is proportional to the factor

$$\frac{(b/a) + 1}{\log_{10} \dfrac{b}{a}}$$

which has a rather broad minimum, Fig. 17.2, for $b/a = 3.6$. The characteristic resistance of a line with $b/a = 3.6$ and $\epsilon_r = 1$ is 77 ohms.

The minimum attenuation on copper coaxial lines, neglecting loss in the dielectric, is

$$\alpha_{\min} = 0.13 \frac{\sqrt{f\epsilon_r}}{b} \cdot 10^{-6} \text{ nepers/m} \tag{17.7}$$

The attenuation varies little in the vicinity of this minimum. The ratio b/a may be as low as 2.5 or as high as 6 without causing the attenuation to increase above the minimum by more than 10 per cent.

It is worthy of note that a dissipationless solid dielectric increases the attenuation. The dielectric decreases R_c, and from (14.4) the attenuation increases. For a given power transmitted by a line, a lower R_c requires a higher current, and the copper loss in the conductor increases. Therefore a dielectric with a dielectric constant greater than unity increases the attenuation on the line even when there is no energy loss in the dielectric itself.

III. DISSIPATIONLESS TRANSMISSION LINES

In many applications, the losses in the transmission line can be neglected. Calculations often may be based upon the dissipationless line and corrections or adjustments made to allow for the dissipation. This is especially true where short lengths of line are used for impedance matching.

Of course, if an impossible assumption is made, an impossible consequence is often encountered. For example, it is shown later that, for an assumed dissipationless line, the input impedance may be infinite under certain conditions. Actually the impedance under these conditions is a high resistance. Likewise, when the dissipationless case indicates zero impedance, the actual impedance is a very low resistance. When the impedances encountered are very low or very high compared with the R_c of the line, or when the dissipation in the line is appreciable compared with that in the load, the attenuation cannot be neglected. But for short lengths, or when the dissipation in the load is very large compared with the dissipation in the line, calculations based upon the dissipationless line are sufficiently accurate to be useful.

18. Dissipationless Line Terminated in Any Impedance Z_R.—With $r = g = 0$, the propagation constant becomes, from (7.2),

$$\gamma = j\beta = j\omega\sqrt{lc}$$

and from (8.1),

$$Z_c = R_c = \sqrt{\frac{l}{c}}$$

The general equations (6.11) and (6.12) become

$$E = A\epsilon^{-j\beta x} + B\epsilon^{+j\beta x} \tag{18.1}$$

$$I = \frac{A}{R_c}\epsilon^{-j\beta x} - \frac{B}{R_c}\epsilon^{+j\beta x} \tag{18.2}$$

The constants A and B are determined by the terminal conditions.

In terms of the conditions at the load

$$E_R = A\epsilon^{-j\beta s} + B\epsilon^{+j\beta s} \tag{18.3}$$

$$I_R = \frac{A}{R_c}\epsilon^{-j\beta s} - \frac{B}{R_c}\epsilon^{+j\beta s} \tag{18.4}$$

Solving for A and B

$$A = \frac{I_R}{2}(Z_R + R_c)\epsilon^{+j\beta s} \tag{18.5}$$

$$B = \frac{I_R}{2}(Z_R - R_c)\epsilon^{-j\beta s} \tag{18.6}$$

which, substituted in (18.1) and (18.2), give

$$E = E_R \cos \beta d + jI_R R_c \sin \beta d \tag{18.7}$$

$$I = I_R \cos \beta d + j\frac{E_R}{R_c}\sin \beta d \tag{18.8}$$

since $d = s - x$, and $\epsilon^{\pm j\beta d} = \cos \beta d \pm j \sin \beta d$. Equations (18.7) and (18.8) are a complete description of the voltage and current at any point on the line in terms of the voltage and current at the load.

In terms of the conditions at the sending end, $x = 0$, and (18.1) and (18.2) become

$$E_S = A + B \tag{18.9}$$

$$I_S = \frac{A}{R_c} - \frac{B}{R_c} \tag{18.10}$$

so that

$$A = \frac{E_S + I_S R_c}{2} \tag{18.11}$$

$$B = \frac{E_S - I_S R_c}{2} \tag{18.12}$$

Then (18.1) and (18.2) become

$$E = E_S \cos \beta x - jI_S R_c \sin \beta x \tag{18.13}$$

$$I = I_S \cos \beta x - j \frac{E_S}{R_c} \sin \beta x \tag{18.14}$$

If the voltage of the generator is the only known voltage, E_S and I_S may be evaluated by first determining the input impedance to the line, which may be done by setting $d = s$ in (18.7) and (18.8), whereupon

$$E_S = E_R \cos \beta s + jI_R R_c \sin \beta s \tag{18.15}$$

$$I_S = I_R \cos \beta s + j \frac{E_R}{R_c} \sin \beta s \tag{18.16}$$

The input impedance is

$$Z_S = \frac{E_S}{I_S} = \frac{E_R \cos \beta s + jI_R R_c \sin \beta s}{I_R \cos \beta s + j \dfrac{E_R}{R_C} \sin \beta s} \tag{18.17}$$

which may be simplified by dividing numerator and denominator by $I_R \cos \beta s$, giving

$$Z_S = R_c \frac{Z_R + jR_c \tan \beta s}{R_c + jZ_R \tan \beta s} \tag{18.18}$$

The generator of Fig. 6.1 "sees" this impedance in series with its own Z_g, so that

$$E_S = E_g \frac{Z_S}{Z_g + Z_S} \tag{18.19}$$

and

$$I_S = \frac{E_g}{Z_g + Z_S} \tag{18.20}$$

The voltage and current at the load now may be determined from (18.13) and (18.14) by setting $x = s$. Naturally,

$$I_R = \frac{E_R}{Z_R} \tag{18.21}$$

An alternative expression for I_R not involving E_R or E_S is

$$I_R = I_S \frac{R_c}{R_c \cos \beta s + jZ_R \sin \beta s} \tag{18.22}$$

obtained by dividing (18.16) by I_R.

If the phase of I_R is not of importance, work may be saved by calculating the magnitude of I_R from the power relationships. With the line assumed dissipationless, the power input is equal to the power output, *i.e.*,

$$P_S = P_R \tag{18.23}$$

or

$$|I_S|^2 R_S = |I_R|^2 R_R \tag{18.24}$$

where R_S and R_R are the real parts of Z_S and Z_R. Then

$$|I_R| = |I_S| \sqrt{\frac{R_S}{R_R}} \tag{18.25}$$

19. Dissipationless Line Terminated in R_c.—The relationships of Sec. 9 apply directly with Z_c set equal to R_c and γ set equal to $j\beta$. Then

$$Z_S = R_c \tag{19.1}$$

$$E_S = E_g \frac{R_c}{R_c + Z_g} \tag{19.2}$$

$$I_S = \frac{E_g}{R_c + Z_g} \tag{19.3}$$

$$E_R = E_S \epsilon^{-j\beta s} \tag{19.4}$$

$$I_R = I_S \epsilon^{-j\beta s} \tag{19.5}$$

The magnitudes of the current and voltage do not change along the line, and the load voltage and current lag the input voltage and current by βs radians.

20. Incident and Reflected Waves.—The voltages and currents at the ends of the transmission line are not the only voltages and currents of interest. There may be places on the line where the voltages and currents are either larger or smaller than the input and output values. To determine the variations in current and voltage along the line, (18.1) and (18.2) must be examined in greater detail. Each of these equations has terms involving the coefficients A and B. It is shown in Sec. 9 that, for a line terminated in Z_c, the B term vanishes, leaving only the A term. In Sec. 12 it is shown, in connection with Fig. 12.1, that the voltage expressed by the A term represents a wave moving to the right. Similarly, the B term, which is no longer zero when the line is terminated in an impedance other than Z_c, represents a wave moving to the left.

Equation (18.1) is repeated here in order to discuss this point.

$$E = A\epsilon^{-j\beta x} + B\epsilon^{+j\beta x} \tag{20.1}$$

If the values of A and B given by (18.5) and (18.6) are substituted in (20.1),

$$E = \frac{I_R}{2}(Z_R + R_c)\epsilon^{+j\beta(s-x)} + \frac{I_R}{2}(Z_R - R_c)\epsilon^{-j\beta(s-x)} \qquad (20.2)$$

or, since $s - x = d$

$$E = \frac{I_R}{2}(Z_R + R_c)\epsilon^{+j\beta d} + \frac{I_R}{2}(Z_R - R_c)\epsilon^{-j\beta d} \qquad (20.3)$$

where the first and second terms on the right are A and B terms. The first term on the right now has $\epsilon^{+j\beta d}$ in place of $\epsilon^{-j\beta x}$, but it is still the A term and represents a wave moving to the right.

The + sign of the exponent denotes wave motion *toward* the point of reference and the − sign of the exponent denotes wave

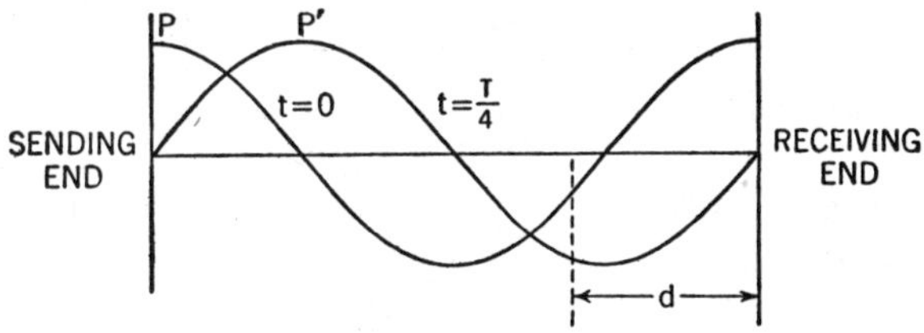

FIG. 20.1.—Plot of cos $(\omega t + \beta d)$ showing wave motion toward the right or toward the load.

motion *away* from the point of reference. From this it is concluded that the $\epsilon^{-j\beta d}$ term in (20.3) indicates a wave moving away from the point from which d is measured, *i.e.*, away from the load and toward the generator.

This may be verified in a manner similar to that used in connection with Fig. 12.1. Assuming, for example, that the instantaneous value of I_R is

$$i_R = \hat{I}_R \cos \omega t$$

and assuming Z_R, for simplicity, to be purely resistive, the instantaneous value of E in (20.3) is

$$e = \frac{\hat{I}_R}{2}(Z_R + R_c)\cos(\omega t + \beta d) + \frac{\hat{I}_R}{2}(Z_R - R_c)\cos(\omega t - \beta d)$$

A plot of cos $(\omega t + \beta d)$ at $t = 0$ and $T/4$ is shown in Fig. 20.1 for a line one wavelength long; the wave motion is to the right or toward the point from which d is measured. A plot of cos $(\omega t - \beta d)$ is shown in Fig. 20.2; the wave motion is to the left or away from the point from which d is measured.

Equation (20.3) and similar preceding equations indicate that the voltage along the line in the steady state may be represented in terms of two components; the first, represented by the first or A term on the right of these equations, is a wave moving toward the load; the second, represented by the second or B term on the right of these equations, represents a wave moving away from the load. The wave moving from the generator *toward the load* is called the *incident* wave, and the wave moving away from the load *toward the generator* is called the *reflected* wave. The incident wave or first

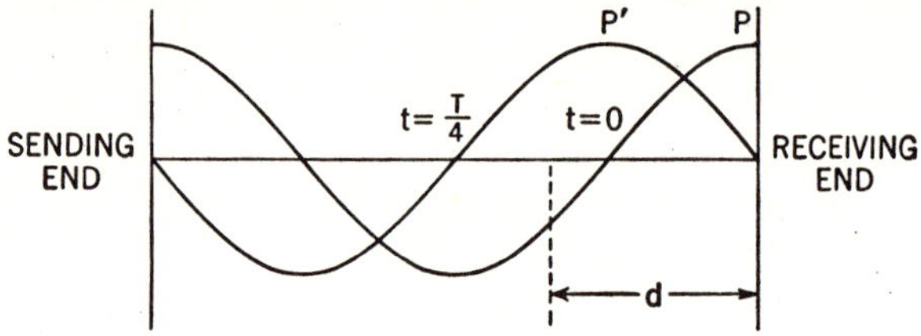

FIG. 20.2.—Plot of cos $(\omega t - \beta d)$ showing wave motion toward the left or away from the load.

term on the right of (20.3) will be denoted by E^+, and the reflected wave by E^-, so that

$$E = E^+ + E^- \tag{20.4}$$

where

$$E^+ = \frac{I_R}{2}(Z_R + R_c)\epsilon^{+j\beta d} = \frac{E_R}{2Z_R}(Z_R + R_c)\epsilon^{+j\beta d} \tag{20.5}$$

and

$$E^- = \frac{I_R}{2}(Z_R - R_c)\epsilon^{-j\beta d} = \frac{E_R}{2Z_R}(Z_R - R_c)\epsilon^{-j\beta d} \tag{20.6}$$

Similarly, from (18.2), (18.5), and (18.6),

$$I = \frac{I_R}{2R_c}(Z_R + R_c)\epsilon^{+j\beta d} - \frac{I_R}{2R_c}(Z_R - R_c)\epsilon^{-j\beta d} \tag{20.7}$$

$$= \frac{E^+}{R_c} - \frac{E^-}{R_c} \tag{20.8}$$

$$= I^+ + I^- \tag{20.9}$$

Thus the incident and reflected waves of voltage are accompanied by incident and reflected waves of current obtained by dividing the respective voltage by R_c, and reversing the sign for the reflected current term. The reversal in sign for the current term is required by the fact that, for a given direction of the electric field, the magnetic field reverses when there is a reversal in the motion of an electromagnetic wave.

21. Reflection Coefficient.—The ratio of the reflected component to the incident component of voltage at the load is denoted by Γ, and from (20.5) and (20.6),

$$\Gamma = \frac{E_R^-}{E_R^+} = \frac{Z_R - R_c}{Z_R + R_c} = |\Gamma| \underline{/\psi} \tag{21.1}$$

The quantity Γ is called the *reflection coefficient.* It is in general a complex number. Whenever it has a value different from zero, the voltage on the line may be described in terms of two components, which from (20.5) and (20.6) have different phases at different points on the line.

In terms of the reflection coefficient and E_R^+ or I_R^+, (20.3) and (20.4) become

$$\begin{aligned} E &= E_R^+\epsilon^{+j\beta d} + E_R^-\epsilon^{-j\beta d} \\ &= E_R^+\epsilon^{+j\beta d}(1 + \Gamma\epsilon^{-j2\beta d}) \\ &= E_R^+\epsilon^{+j\beta d}(1 + |\Gamma|\underline{/\psi - 2\beta d}) & (21.2) \\ &= R_c I_R^+\epsilon^{+j\beta d}(1 + |\Gamma|\underline{/\psi - 2\beta d}) & (21.3) \end{aligned}$$

Similarly, (20.7) becomes

$$\begin{aligned} I &= I_R^+\epsilon^{+j\beta d}(1 - |\Gamma|\underline{/\psi - 2\beta d}) & (21.4) \\ &= \frac{E_R^+}{R_c}\epsilon^{+j\beta d}(1 - |\Gamma|\underline{/\psi - 2\beta d}) & (21.5) \end{aligned}$$

When $\psi - 2\beta d$ has a magnitude of zero or any multiple of 2π radians, the quantity within the parentheses has a maximum magnitude in (21.2) and a minimum magnitude in (21.4). Hence

$$\left.\begin{aligned} |E| &= |E_{\max}| \\ |I| &= |I_{\min}| \end{aligned}\right\} \text{ at } |\psi - 2\beta d| = 0, 2\pi, 4\pi, \ldots \tag{21.6}$$

Similarly,

$$\left.\begin{aligned} |E| &= |E_{\min}| \\ |I| &= |I_{\max}| \end{aligned}\right\} \text{ at } |\psi - 2\beta d| = \pi, 3\pi, 5\pi, \ldots \tag{21.7}$$

At a point where the two components of voltage are in phase, they add numerically, and the rms voltage is a maximum at that point. Similarly, at points $\lambda/4$ distant from the points of $|E_{\max}|$, the two components differ in phase by 180°, and at such points the rms voltage is a minimum. When such conditions obtain, *standing waves* are said to exist on the line. Points of $|E_{\max}|$ are points of $|I_{\min}|$, and points of $|E_{\min}|$ are points of $|I_{\max}|$.

This may be shown by a diagram such as Fig. 21.1, where Γ is $0.5\underline{/45°}$ as an example. The voltage E is proportional to

$1 + |\Gamma|\underline{/\psi - 2\beta d}$; the current I is proportional to $1 - |\Gamma|\underline{/\psi - 2\beta d}$. At $d = 0$, $E = E_R = E_R^+(1 + \Gamma)$ and $I = I_R = (E_R^+/R_c)(1 - \Gamma)$. As d increases, the line representing $\Gamma\underline{/-2\beta d} = |\Gamma|\underline{/\psi - 2\beta d}$ rotates clockwise, so that E and I pass through maximum and minimum values. When E is a maximum, I is a minimum, and vice versa. The phase angle between E and I is as indicated in the figure.

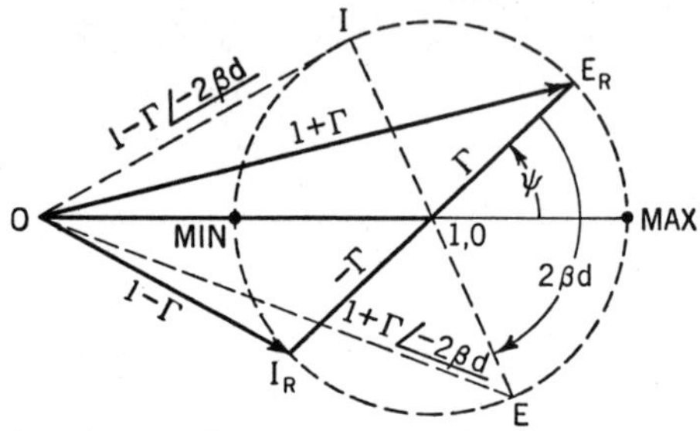

FIG. 21.1.—Variations in E and I in terms of the reflection coefficient Γ and the angle $\psi - 2\beta d$.

22. Short-circuited Dissipationless Line.—Since the reflection coefficient with $Z_R = 0$ is, from (21.1),

$$\Gamma = \frac{-R_c}{+R_c} = 1\underline{/180^\circ}$$

the reflected wave is equal in magnitude to the incident wave. From (21.7) a current maximum and a voltage minimum occur at $d = 0$, *i.e.*, at the closed end. A study of the standing-wave pattern on such a line now will be made, to show in graphical form the effect of the length of the line.

Since the current in the short circuit is usually finite while the voltage is zero, the current I_R in the short circuit will be taken as the reference vector. The current I at any point distant d from the short-circuited end of the line is, from (20.7) and (20.9),

$$I = \frac{I_R}{2}\epsilon^{+j\beta d} + \frac{I_R}{2}\epsilon^{-j\beta d} = I^+ + I^- \tag{22.1}$$

When $d = 0$, the incident and reflected components of current are in phase, and their sum is I_R, as it should be. As d increases, the I^+ vector rotates counterclockwise and the I^- vector rotates clockwise. The rotation is 360° for each vector for every wavelength of line. In Fig. 22.1, the vectors I^+, I^-, and their sum I are plotted at intervals of d equal to $\lambda/8$ for a line $5\lambda/8$ long. As d increases from 0 to

$\lambda/4$, I decreases from a maximum to zero, but the phase does not change. After $d = \lambda/4$, the phase of the current reverses, and the magnitude increases to a maximum at $d = \lambda/2$. The magnitude of the maximum current equals the magnitude of I_R. In Fig. 22.2

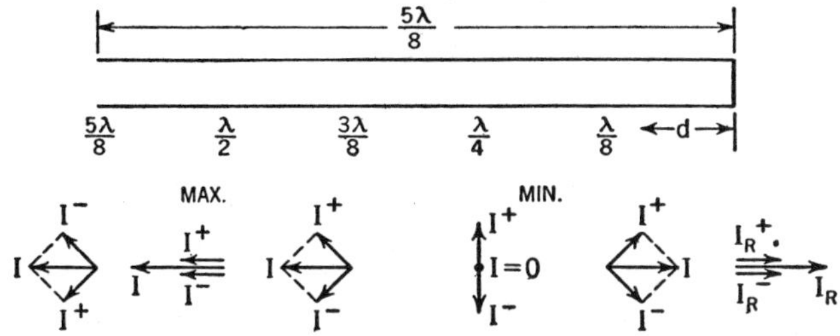

FIG. 22.1.—Current in a short-circuited dissipationless line shown as the sum of incident and reflected components

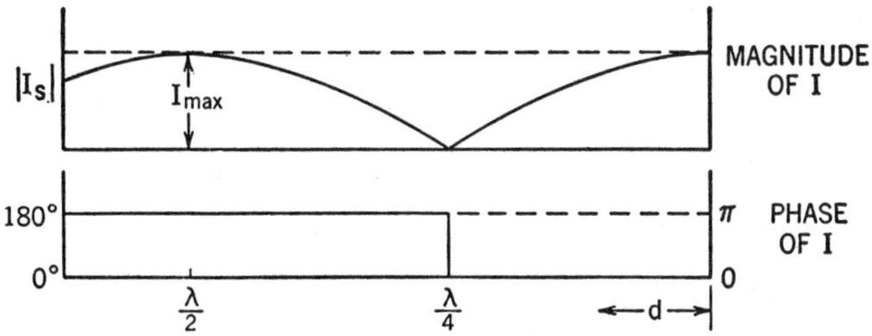

FIG. 22.2.—Magnitude and phase of the current in a short-circuited dissipationless line, I_R taken as reference for phase.

curves are plotted showing the magnitude and phase of I, as functions of the distance d from the short circuit. I_R is taken as the reference for phase.

The voltage distribution on the short-circuited line may be analyzed in a similar manner. Setting $Z_R = 0$ in (20.3),

$$\begin{aligned} E &= \frac{I_R}{2}(R_c)\epsilon^{+j\beta d} + \frac{I_R}{2}(-R_c)\epsilon^{-j\beta d} \\ &= E^+ + E^- \end{aligned} \tag{22.2}$$

The voltages E^+, E^-, and their sum E are plotted in Fig. 22.3. Note that the vector E points upward, has the same phase for all values of d between zero and $\lambda/2$, and reverses its phase after $d = \lambda/2$. In Fig. 22.4, the magnitude and phase of E are plotted as functions of the distance d. Just as for the current, there is a reversal of phase when the voltage passes through zero, but no change of phase between successive zeros.

In Fig. 22.5, the diagrams of Figs. 22.2 and 22.4 have been extended to cover a line $5\lambda/4$ long. The scales for E and I have been chosen so that $|E_{max}|$ and $|I_{max}|$ are plotted at equal heights. Comparison of (22.2) and (22.1) shows that

$$|E^+| = R_c|I^+| \tag{22.3}$$
$$|E^-| = R_c|I^-| \tag{22.4}$$

so that each wave of current is accompanied by a wave of voltage

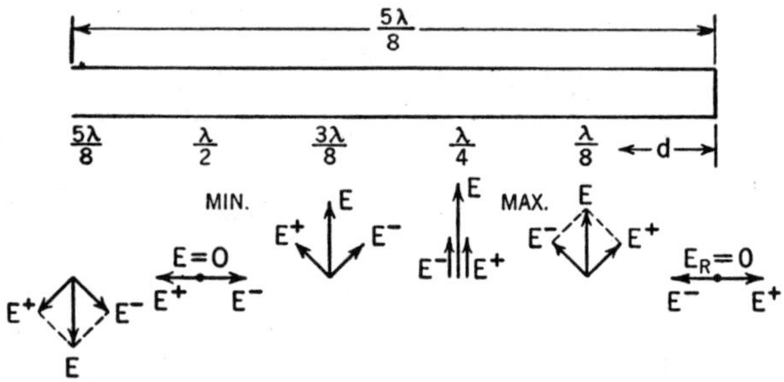

FIG. 22.3.—Voltage across a short-circuited dissipationless line shown as the sum of incident and reflected components.

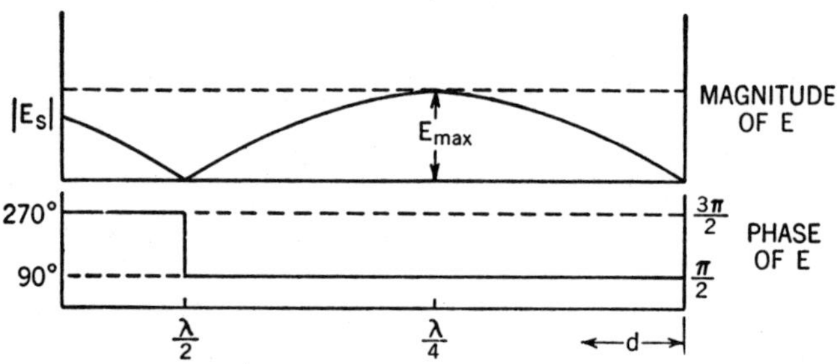

FIG. 22.4.—Magnitude and phase of the voltage across a short-circuited dissipationless line. I_R taken as reference for phase.

R_c times as great. Also, inspection of Figs. 22.3 and 22.4 indicates that

$$|E_{max}| = |E^+| + |E^-| \tag{22.5}$$

and similarly, from Figs. 22.4 and 22.5,

$$|I_{max}| = |I^+| + |I^-| \tag{22.6}$$

so that

$$|E_{max}| = R_c|I_{max}| \tag{22.7}$$

Thus in Fig. 22.5, the current has been plotted to a scale R_c larger than the scale for voltage.

Equations (22.3) to (22.7) are relationships that hold for any dissipationless line and will be used in plotting other "standing-wave" patterns. Note that $|E_{max}|$ and $|I_{max}|$ occur at different points on the line, these points being separated by $\lambda/4$. This fact also holds true for all standing-wave patterns on dissipationless lines.

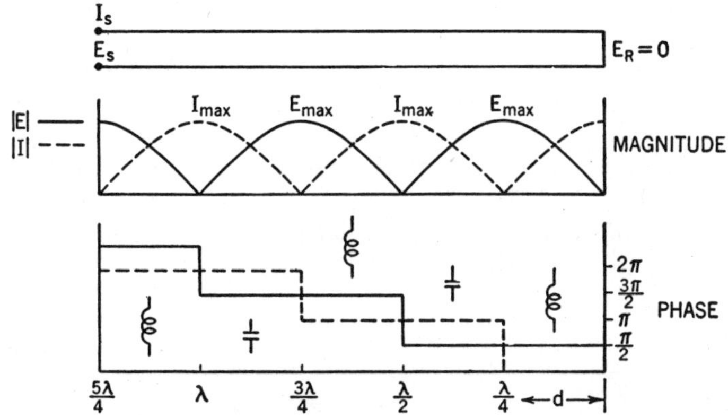

FIG. 22.5.—Voltage and current distribution for a short-circuited dissipationless line. $|E_{max}| = R_c|I_{max}|$; voltage and current magnitudes are plotted to different scales. Inductances and capacitances are shown to indicate the nature of the input impedance for various lengths of short-circuited line.

Just as the maximum magnitude of E or I is the sum of the magnitudes of incident and reflected waves, so the minimum magnitude of E or I is equal to the difference of the magnitudes of incident and reflected waves.

$$|E_{min}| = |E^+| - |E^-| \tag{22.8}$$
$$|I_{min}| = |I^+| - |I^-| \tag{22.9}$$

Also

$$|E_{min}| = R_c|I_{min}| \tag{22.10}$$

In the short-circuited line, $|E_{min}|$ is zero because the incident and reflected waves have equal magnitudes.

At the point on the line where E has a minimum value, I has a maximum value, and vice versa. The points where E has minimum values are $\lambda/4$ distant from the nearest points where I has minimum values. The maximum and minimum of current and voltage recur at half-wavelength intervals.

If the line is opened at any point distant d from the short circuit, the input impedance looking toward the short circuit is the ratio of E to I at that point. For d less than $\lambda/4$, Figs. 22.1, 22.3, and 22.5 show that E leads I by $\pi/2$ radians or 90°; the input impedance is inductive. Therefore, a short-circuited line less than $\lambda/4$ in length has an input impedance that is inductive.

If the line is opened at a distance d greater than $\lambda/4$ but less than $\lambda/2$, the current at the input terminals leads the voltage by $\pi/2$ radians or 90°, as may be seen from the same diagrams. Therefore, the input impedance of a short-circuited line whose length is between $\lambda/4$ and $\lambda/2$ is capacitive. As the line length increases, the input impedance is alternately inductive and capacitive as the input terminals move farther and farther from the terminal short circuit. This is indicated by the inductances and capacitances sketched in Fig. 22.5.

This information may be obtained also from the general equations of Sec. 18. With $E_R = 0$, (18.7) becomes

$$E = jI_R R_c \sin \beta d \tag{22.11}$$

which is the same as (22.2), since $\epsilon^{\pm j\beta d} = \cos \beta d \pm j \sin \beta d$. Similarly, from (18.8),

$$I = I_R \cos \beta d \tag{22.12}$$

Also, (18.18) becomes

$$(Z_S)_{\text{short}} = jR_c \tan \beta s \tag{22.13}$$

The input impedance is ideally a reactance, alternating in sign as βs increases. Short sections of line are used as circuit elements at radio frequencies, particularly in oscillators and amplifiers as parts of resonant circuits.

The input impedance of a short-circuited line is usually not zero. For the short-circuited dissipationless line, the input impedance is zero only when the length of the line is a multiple of a half wavelength. When the short-circuited line is one-quarter wavelength long, *i.e.*, when the input terminals are $\lambda/4$ from the short circuit ($d = \lambda/4$ in Fig. 22.5), the voltage across the input terminals is finite, but the input current is zero. Therefore, the input impedance is infinite, *i.e.*, an open circuit. This same result is obtained from (22.13), since for a line $\lambda/4$ long, $\beta s = \pi/2$ and $\tan \pi/2$ is infinite.

For the ideal dissipationless case, then, the input impedance of a short-circuited line is infinite for a line $\lambda/4$ long, and zero for a line $\lambda/2$ long. The action is somewhat analogous to parallel and series resonance in a dissipationless LC circuit.

All lines have some dissipation. Where the ideal dissipationless case indicates zero impedance, the actual impedance is not zero but is very low and is a resistance approximately equal to $R_c\alpha s$. Where infinite impedance is indicated, the actual value is very high and is a resistance approximately equal to $R_c/\alpha s$. For example, the actual input impedance of a short-circuited quarter-wavelength section of a reasonably low-loss line is about 400,000 ohms.

23. Suppression of Even Harmonics.—An application of a short-circuited quarter-wavelength line is to suppress any unwanted even harmonics in the output of a radio transmitter. A short-circuited line, one-quarter wavelength long at the desired output frequency, may be connected across the output terminals of a transmitter or across the antenna feeder at any point without placing much load on the transmitter at the fundamental or desired output frequency, since at this frequency such a section has an impedance ideally infinite, actually about 400,000 ohms. At the second-harmonic frequency, however, the short-circuited section

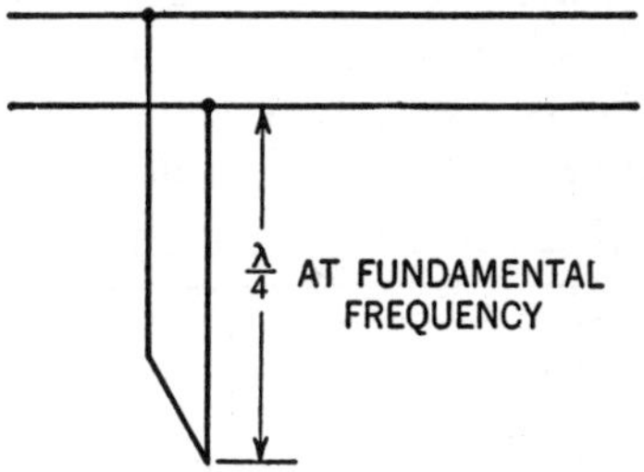

FIG. 23.1.—Mounting of quarter-wavelength harmonic-suppressing stub, open-wire construction.

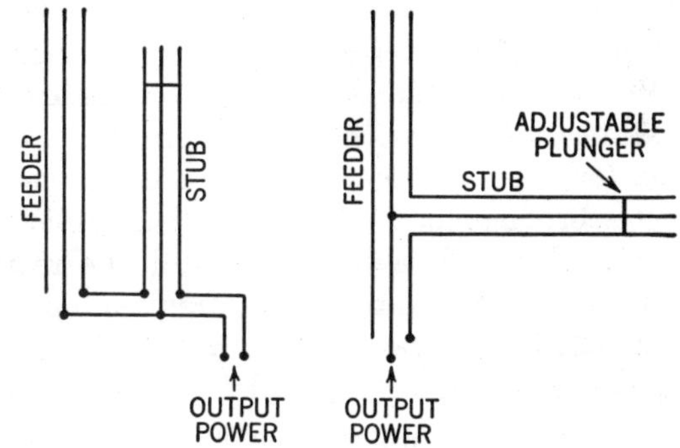

(a) (b)

FIG. 23.2.—Alternative mountings of quarter-wavelength harmonic-suppressing stub, coaxial construction.

is one-half wavelength long, and the input impedance is ideally zero, actually very low. Thus the second harmonic is short-circuited. Any even harmonic is similarly short-circuited, because the short-circuited section is a multiple of a half wavelength at any even-harmonic frequency.

The quarter-wave harmonic-suppressing stub should be so mounted that the electromagnetic field of the stub does not disturb the electromagnetic field of the line. For open-wire lines, the stub is usually mounted at right angles to the feeder, Fig. 23.1. A coaxial stub has its electromagnetic field contained within itself so that its position may be chosen with greater freedom. When used in connection with a coaxial feeder, it may be mounted parallel to or at right angles to the feeder, Fig. 23.2.

24. Insulating or Supporting Stubs.—Since the input impedance of a short-circuited quarter-wavelength section of transmission line is a very high resistance, short-circuited stubs may be used to support the line. Thus, a half wavelength of copper strap can be bent into the form of a U and used to support a two-wire line, no non-conducting insulation being required between the U support and a metal surface to which it might be attached. Also, the center conductor of a coaxial cable may be supported by coaxial stubs, Fig. 24.1. Since the input resistance supporting the center conductor of a coaxial line is $R_c/\alpha s$, where $\alpha \doteq r/2R_c$, the line may be so proportioned that the resistance of the insulating stub is a maximum. This occurs when D/a is approximately 7.9 for the two-wire line, and b/a is 9.2 for the coaxial line.

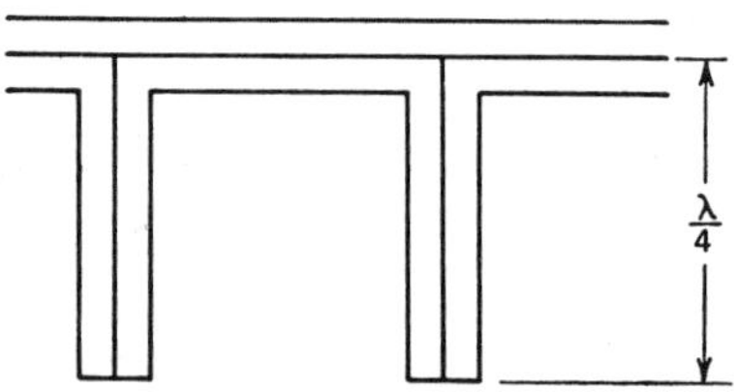

FIG. 24.1.—Insulating-stub supports for the center conductor of a coaxial line.

Another use for a short-circuited stub is in connection with a rectifying detector bridged across the line; a short-circuited quarter-wavelength stub will complete the circuit for the rectified direct current but will not load the line at radio frequencies.

25. Open-circuited Dissipationless Line.—For the open-circuited line, Z_R is infinite, and, from (21.1), $\Gamma = 1\underline{/0°}$. A voltage maximum and a current minimum occur at the open end, from (21.6). As for the short-circuited line, the reflected wave is equal in magnitude to the incident wave. The expressions for the incident and reflected components of the voltage may be obtained from the extreme right of (20.5) and (20.6), since the ratio $(Z_R \pm R_c)/Z_R$ becomes unity as Z_R increases to infinity. Then, for the open-circuited dissipationless line,

$$E = \frac{E_R}{2}\epsilon^{+j\beta d} + \frac{E_R}{2}\epsilon^{-j\beta d} \tag{25.1}$$

and from (20.8)

$$I = \frac{E_R}{2R_c}\epsilon^{+j\beta d} - \frac{E_R}{2R_c}\epsilon^{-j\beta d} \tag{25.2}$$

In Fig. 25.1 are plotted the magnitude and phase of E and I at any point on a dissipationless line $5\lambda/8$ long. Figure 25.1 is the same as Fig. 22.5, but current and voltage have been interchanged. The voltage and current, Fig. 25.1, may be obtained

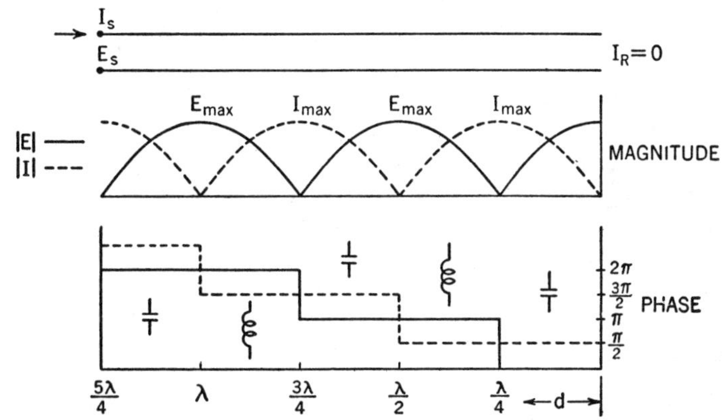

FIG. 25.1.—Voltage and current distribution for an open-circuited dissipationless line. $|E_{max}| = R_c|I_{max}|$; voltage and current magnitudes are plotted to different scales. Inductances and capacitances are shown to indicate the nature of the input impedance for various lengths of open-circuited line.

from (25.1) and (25.2), or from (18.7) and (18.8) with $I_R = 0$, and are

$$E = E_R \cos \beta d \tag{25.3}$$

$$I = j\frac{E_R}{R_c}\sin \beta d \tag{25.4}$$

The input voltage and current may be obtained by setting $d = s$ in (25.3) and (25.4). The quotient of the input voltage and current is the input impedance, or

$$(Z_S)_{\text{open}} = \frac{R_c}{j \tan \beta s} = -jR_c \operatorname{cotan} \beta s \tag{25.5}$$

The same result may be obtained from (18.18) by first dividing the numerator and denominator of the right-hand side of that equation by Z_R. As Z_R becomes infinite (open circuit), R_c/Z_R becomes zero, and the equation reduces to (25.5).

The input impedance of an open-circuited line of length less than $\lambda/4$ is capacitive. The input impedance of an open line of length greater than $\lambda/4$ but less than $\lambda/2$ is inductive. As the length increases, the impedance is alternately capacitive and inductive as the input terminals are farther and farther from the open end of the line.

Equation (25.5) and Fig. 25.1 indicate that an open-circuited quarter-wavelength dissipationless line has zero input impedance (is a short circuit) and that the input impedance of an open-circuited half-wavelength dissipationless line is infinite. Again, as in Sec. 22, there is an analogy with series and parallel resonance in dissipationless LC circuits. The effect of dissipation on the line is similar to the effect of losses in an LC circuit, in that, when the impedance is ideally zero, it is actually a very low resistance ($R_c\alpha s$ for the line); and when the impedance is ideally infinite, it is actually a very high resistance ($R_c/\alpha s$ for the line). Nevertheless, an open-circuited quarter wavelength line is practically a short circuit.

26. Suppression of the Third Harmonic.—An unwanted third harmonic in the output of a radio transmitter may be suppressed by an arrangement shown in Fig. 26.1. An open section of line $\lambda/12$ long for the fundamental frequency is $\lambda/4$ long for the third harmonic. Therefore, if such an open section is connected to the output terminals of a transmitter, the output will be short-circuited for the third harmonic. At the fundamental frequency, the open $\lambda/12$ section has a capacitive impedance, the effects of which are minimized by the inductive impedance of the $\lambda/6$ closed section placed in parallel. At the fundamental frequency the $\lambda/6$ closed section will be in parallel resonance with the $\lambda/12$ open section.

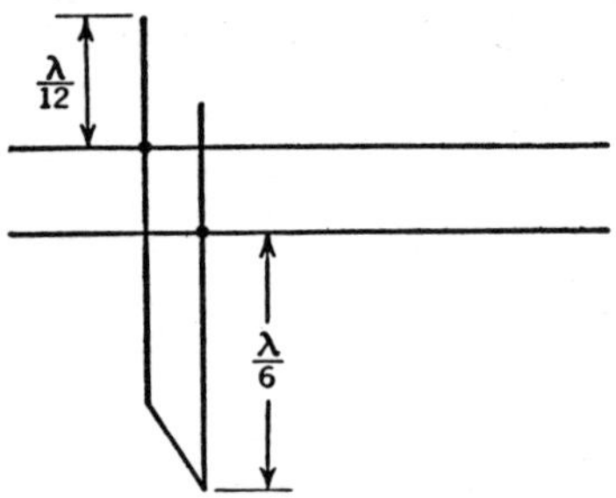

FIG. 26.1.—Stub arrangement that short-circuits the third harmonic of the frequency corresponding to λ and is antiresonant for the frequency corresponding to λ.

The antiresonance or parallel resonance occurring in the circuit of Fig. 26.1 is a special example of the fact that a short-circuited quarter-wavelength line presents to any connection made along its length the equivalent of an antiresonant circuit at the fundamental frequency. In the dissipationless case, the impedance looking toward the open part is purely capacitive; the impedance looking toward the closed part is purely inductive and of equal magnitude, as may be verified from (22.13) and (25.5). The resulting anti-resonant impedance is ideally infinite; actually it is finite and large except when the connection is made near the closed end.

27. Effect of Terminations on Effective Length.—The effective length of a line often may be shortened or lengthened by a suitable termination. Several examples now will be considered.

The physical length of a short-circuited line may be reduced by terminating the line in a pure reactance. For example, the short-circuited line of Fig. 27.1*a* may be shortened by cutting off a length d' and connecting a reactance equal to the reactance of

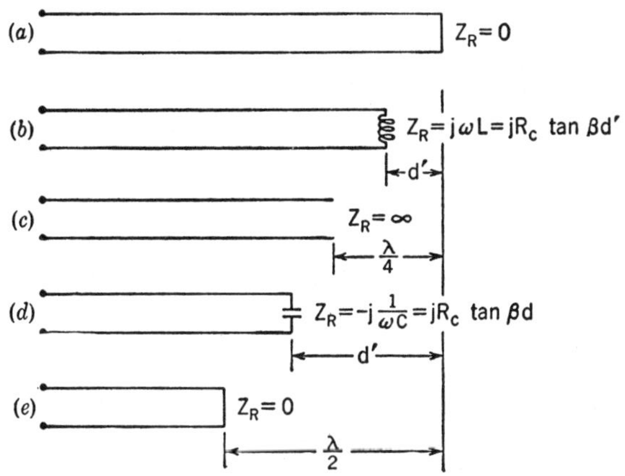

FIG. 27.1.—Five lines having equal input impedances (losses neglected).

the short-circuited section so removed. If the length cut off is less than $\lambda/4$, the reactance to be connected in place of the removed section is inductive, Fig. 27.1*b*. If the removed length is equal to $\lambda/4$, the line may be left open, Fig. 27.1*c*, since the input impedance of the removed section is ideally infinite. If the removed length is greater than $\lambda/4$ but less than $\lambda/2$, the reactance to be substituted is a capacitance, Fig. 27.1*d*. If a half wavelength is removed, the line must be short-circuited again.

The physical length of an open-circuited line may also be shortened by terminating the line in a pure reactance, Fig. 27.2. If less than a quarter wavelength is removed, a capacitance is substituted, Fig. 27.2*b*. If a quarter wavelength is removed, a short circuit is substituted, Fig. 27.2*c*, since the input impedance of the removed quarter wavelength section of open line is ideally zero. If more than a quarter wavelength but less than a half wavelength is removed, an inductance is substituted. If a half wavelength is removed, this line must be open-circuited again.

The open- and short-circuited equivalents shown in Figs. 27.1 and 27.2 are especially applicable where stubs are employed. A closed stub may be replaced by an open stub that is $\lambda/4$ longer or, where possible, is $\lambda/4$ shorter. Similarly, an open stub may be replaced by a closed stub that is $\lambda/4$ longer or shorter.

The conditions of Fig. 27.1*b* are also of interest in connection with closed stubs on open-wire lines. On a two-wire line, the short-circuiting wire, if it is the same size as the line wires, has an inductance approximately equivalent to a length of line $D/2$.

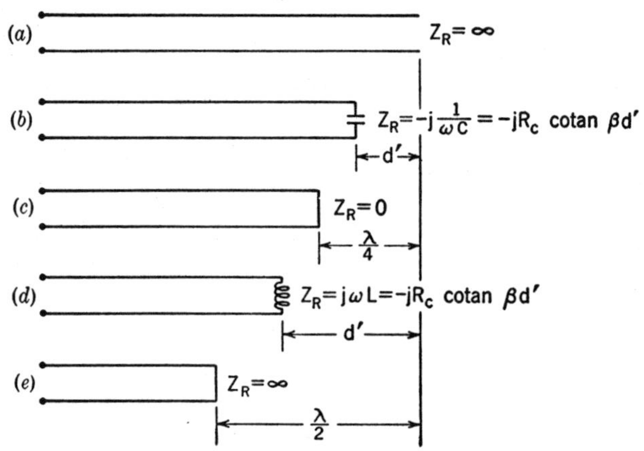

FIG. 27.2.—Five lines having equal input impedances (losses neglected).

28. Dissipationless Line with Resistive Load.—As an example, the magnitude and phase of the current and voltage for a dissipationless line terminated in a resistance equal to $3R_c$ are plotted in Fig. 28.1. The reflection coefficient (21.1) is equal to $\frac{1}{2}$ at zero angle, so that, from (21.6), a voltage maximum occurs at the load. The values may be determined by direct calculation from (18.7) and (18.8), or by a vector-addition method similar to that of Fig. 22.1, or from (21.2) and (21.4). From (20.5), the incident component of voltage is

$$E^+ = \tfrac{2}{3}E_R\epsilon^{+j\beta d} \tag{28.1}$$

and from (20.6) the reflected component is

$$E^- = \tfrac{1}{3}E_R\epsilon^{-j\beta d} \tag{28.2}$$

These, when added after the manner of Fig. 22.3, give the magnitude and phase plotted in Fig. 28.1. The current may be similarly plotted from (20.7).

In this case, the resistive load impedance is greater than R_c, and the current and voltage distribution resembles that of the open-circuited line, Fig. 25.1. In place of a minimum of zero, Fig. 25.1, there is a finite minimum in Fig. 28.1, and the phase of the voltage and current changes gradually, not abruptly as in the open-circuit condition. The higher the load resistance, the closer

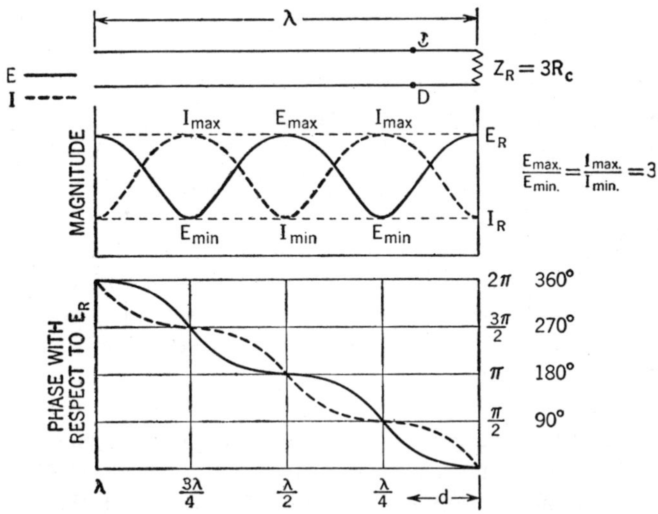

FIG. 28.1.—Voltage and current distribution for a dissipationless line terminated in $Z_R = 3R_c$. $|E_{max}| = R_c|I_{max}|$; voltage and current magnitudes are plotted to different scales. E_R is taken as reference for phase.

is the resemblance of the current and voltage distribution to that of an open-circuited line. For a resistive load greater than R_c, and also for the open-circuited line, the voltage at the load is a voltage maximum, and the current at the load is a current minimum.

If the line of Fig. 28.1 were opened at a distance less than $\lambda/4$ from the load, the input impedance of the portion connected to the load would be resistive and capacitive, since the input current would lead the voltage by less than 90°. For a distance greater than $\lambda/4$ but less than $\lambda/2$, the input impedance would be resistive and inductive. This again is similar to the open-circuited line except for the addition of resistance due to dissipation in the load. Quantitatively the input impedance would be given by (18.18).

If Z_R is equal to $R_c/3$, the standing-wave pattern is exactly the same as that of Fig. 28.1, but with current and voltage interchanged. The current and voltage distribution is similar to that for the short-circuited line, Fig. 22.5, except that the minima are finite and the phase change is gradual. The lower the resistance of the load, the closer is the resemblance to short-circuit conditions. For a resistive load less than R_c, the current in the load is a current maximum, and the voltage across the load is a voltage minimum.

29. Standing Waves.—For any type of load other than pure resistance, the current and voltage distributions resemble those of Fig. 28.1 except that maximum or minimum magnitudes of E and I do not occur at the load. For example, if the line of Fig. 28.1 were cut off at the points CD and the part to the right of CD were replaced by its input impedance, the generator at the sending end of the line would notice no difference and the distribution pattern with respect to the sending end would not be changed.

Whenever the magnitudes of voltage and current vary in the manner of Figs. 28.1, 25.1, 22.5, *standing waves* are said to exist on the line. The line is also commonly called a *resonant* line, although in some applications of lines the word "resonant" is used more strictly. A line terminated in Z_c has no standing waves and is termed a nonresonant line.

On a line with standing waves, a place where the voltage has a maximum value is called a voltage *loop*, and a place where the voltage has a minimum value is called a voltage *node*. Similar definitions hold for the current. From Fig. 28.1 and the preceding diagrams, it is seen that a voltage maximum and a current minimum always occur together at the same point, and a voltage minimum and a current maximum occur together. A voltage maximum is separated from the nearest current maximum by $\lambda/4$. The magnitude pattern on a dissipationless line repeats itself exactly every half wavelength.

30. Standing-wave Ratio.—The ratio of maximum to minimum magnitude of voltage or current is called the standing-wave ratio, denoted by ρ, or

$$\rho = \frac{|E_{\max}|}{|E_{\min}|} = \frac{|I_{\max}|}{|I_{\min}|} \tag{30.1}$$

Then

$$\rho = \frac{|E^+| + |E^-|}{|E^+| - |E^-|} = \frac{|I^+| + |I^-|}{|I^+| - |I^-|} \tag{30.2}$$

In a dissipationless line, there is no attenuation, and the magnitude of the incident component is the same along the entire line. Similarly, the magnitude of the reflected component is the same along the entire line. Therefore

$$|E^+| = |E_R^+| \tag{30.3}$$

$$|E^-| = |E_R^-| \tag{30.4}$$

where the subscript R denotes the value at the receiving, or load, end of the line. From (21.1)

$$|\Gamma| = \frac{|E_R^-|}{|E_R^+|} = \frac{|Z_R - R_c|}{|Z_R + R_c|} \tag{30.5}$$

Dividing the numerator and denominator of (30.2) by $|E^+|$

$$\rho = \frac{1 + \dfrac{|E^-|}{|E^+|}}{1 - \dfrac{|E^-|}{|E^+|}} \tag{30.6}$$

so that, from (30.3) to (30.5),

$$\rho = \frac{1 + |\Gamma|}{1 - |\Gamma|} \tag{30.7}$$

An alternative expression for ρ is

$$\rho = \frac{|Z_R + R_c| + |Z_R - R_c|}{|Z_R + R_c| - |Z_R - R_c|} \tag{30.8}$$

Special cases of this formula are useful for resistance loads.

$$\rho = \frac{R_R}{R_c} \qquad \text{when} \qquad Z_R = R_R > R_c \tag{30.9}$$

$$\rho = \frac{R_c}{R_R} \qquad \text{when} \qquad Z_R = R_R < R_c \tag{30.10}$$

31. Impedance at a Voltage or a Current Loop.—At a current or a voltage loop, the current and voltage are in phase with each other, Fig. 28.1. Therefore, if a line is opened at either a current or a voltage maximum, the impedance looking toward the load is a pure resistance. This holds true for any load.

At a voltage maximum, the resistance looking toward the load is

$$R = \frac{E_{\max}}{I_{\min}} = \frac{|E_{\max}|}{|I_{\min}|} \tag{31.1}$$

But

$$|E_{\max}| = R_c|I_{\max}| \tag{31.2}$$

so that

$$R = \frac{R_c|I_{\max}|}{|I_{\min}|} \tag{31.3}$$

and the input resistance is

$$R = \rho R_c \tag{31.4}$$

At a current maximum, the input resistance is

$$R = \frac{R_c}{\rho} \tag{31.5}$$

Thus the input impedance of a line is a resistance ρ times as great as R_c if a voltage maximum occurs at the input, or a smaller resistance $1/\rho$ times R_c if a current maximum occurs at the input. Also, a line is terminated in effect by a resistance at a voltage or current maximum.

32. Uses of Lines with and without Standing Waves.—Owing to the opposite phase shifts of the incident and reflected waves, the input impedance of lines is, in general, different from the load impedance. The input impedance of the dissipationless line is

$$Z_S = R_c \frac{Z_R + jR_c \tan \beta s}{R_c + jZ_R \tan \beta s} \tag{32.1}$$

By suitable design to obtain a proper value of R_c and selection of the length s, an impedance Z_R may be made to "look like," *i.e.*, may be "transformed" into, almost any desired impedance. In this respect, the dissipationless line is similar to the ideal transformer employed in network theory, in that it changes the ratio of voltage to current so that the input impedance is different from the load impedance. The line, however, can change the phase as well as the magnitude, so that not only the "magnitude" but the "angle" of the impedance may be "transformed." This ability of the line is extremely useful for impedance-matching purposes at radio frequencies.

Also, if Z_R is made equal to zero or infinity, short sections of line may be used as inductances and capacitances. At very high frequencies, ordinary inductors and capacitors cease to function properly, owing to distributed capacitance and inductance. Because of these effects, lead wires are made short, coil turns are widely spaced, and the circuit elements are made physically small in order to keep their maximum dimensions considerably less than a wavelength. At high frequencies, however, the wavelength is so short that ordinary construction must be abandoned. The transmission-line section is one way out of this difficulty, since, at high frequencies, sections of transmission line can be made to provide practically pure reactances without involving excessively small or excessively large dimensions.

Transmission lines may also be used as a means for measurement of power at radio frequencies, and also as a means for measuring impedances.

Where power is transmitted over distances that are large compared with the wavelength, standing waves on the transmission line are to be avoided. The arc-over of a line determines the maximum allowable voltage. For a given maximum voltage, the maximum power is transmitted over a line with no standing waves, as is shown in Sec. 33. Also, since the copper loss varies as the square of the current, the copper loss per unit length at the current loops is far greater than the loss would be if the line were terminated in R_c, and the copper loss at the nodes is less, but the total copper loss on the resonant line is greater than on the same line operated as a nonresonant line transmitting the same power.

For short lengths and for multifrequency operation, it is often simpler to operate a transmission line feeding an antenna as a "resonant" line, making all tuning adjustments at the transmitter. For greater lengths at fixed frequencies, it is common practice to employ some sort of impedance-matching device to match the antenna to the line, *i.e.*, to transform the antenna impedance into R_c, so that the antenna feeder operates as a nonresonant line. At the generator or transmitter end, similar means are employed to transform the input impedance R_c of the nonresonant line into a value that will suitably load the transmitter.

33. Measurement of Power.—It is shown in Sec. 31 that the input impedance of a transmission line at a voltage maximum is a resistance equal to ρR_c. Then the power P fed into this resistance is

$$P = \frac{|E|^2}{R} = \frac{|E_{\max}|^2}{\rho R_c} \tag{33.1}$$

or

$$P = \frac{|E_{\max}||E_{\max}|}{\rho R_c} \tag{33.2}$$

Since

$$\frac{|E_{\max}|}{\rho} = |E_{\min}| \tag{33.3}$$

the power transmitted is

$$P = \frac{|E_{max}||E_{min}|}{R_c} \tag{33.4}$$

Similarly

$$P = |I_{max}||I_{min}|R_c \tag{33.5}$$

For a given E_{max}, more power can be transmitted over a nonresonant line ($Z_R = R_c$), since $|E_{min}|$ and $|E_{max}|$ are then equal (no standing waves). For a nonresonant line

$$P = \frac{|E|^2}{R_c} = |I|^2 R_c \tag{33.6}$$

34. Measurement of Terminating Impedances.—The following is one method for measuring impedances: The unknown impedance is connected as the load impedance Z_R of a transmission line of known R_c, and the value of the impedance is determined by measuring the standing-wave ratio and the position of voltage maximum (or minimum).

When E_{max} occurs at the load, the load impedance is a resistance equal to ρR_c. When E_{min} occurs at the load (or when E_{max} occurs a distance $\lambda/4$ away from the load), the load impedance is a resistance equal to R_c/ρ. When E_{max} occurs at any other point, the ratio of Z_R to R_c is a complex number.

Let d_{max} be the distance between the load and the nearest point on the line where the voltage is a maximum. At this point, the input impedance, *i.e.*, the impedance looking toward the load, is ρR_c. Using (32.1) with Z_S equal to ρR_c and s equal to d_{max}

$$\rho R_c = R_c \frac{Z_R + jR_c \tan \beta d_{max}}{R_c + jZ_R \tan \beta d_{max}} \tag{34.1}$$

Solving for Z_R

$$Z_R = R_c \frac{\rho - j \tan \beta d_{max}}{1 - j\rho \tan \beta d_{max}} \tag{34.2}$$

where ρ is the measured ratio

$$\rho = \frac{|E_{max}|}{|E_{min}|} \tag{34.3}$$

Sometimes the point of minimum voltage may be more precisely located. If d_{min} denotes the distance from the load to the nearest point on the line where the voltage is a minimum

$$\frac{R_c}{\rho} = R_c \frac{Z_R + jR_c \tan \beta d_{min}}{R_c + jZ_R \tan \beta d_{min}} \tag{34.4}$$

and

$$Z_R = R_c \frac{1 - j\rho \tan \beta d_{\min}}{\rho - j \tan \beta d_{\min}} \tag{34.5}$$

The use of this method requires a voltage-indicating device that does not change the conditions on the line, and (34.2) and (34.5) are accurate only when the dissipation in the line is negligible compared with that in the load.

IV. IMPEDANCE MATCHING

35. Quarter-wave Transformer.—A quarter-wavelength line has the useful property of transforming an impedance into its reciprocal with respect to R_c^2. To show this, (32.1) may be solved with βs set equal to $\pi/2$. Since $\tan \pi/2$ is infinity, (32.1) would not yield directly a useful answer. If both numerator and denominator of (32.1) are divided by $\tan \beta s$

$$Z_S = R_c \frac{(Z_R/\tan \beta s) + jR_c}{(R_c/\tan \beta s) + jZ_R} \tag{35.1}$$

When $\tan \beta s$ becomes infinite

$$Z_S = \frac{R_c^2}{Z_R} \tag{35.2}$$

This equation describes the action of the quarter-wave transformer.

A common application of this principle is the matching of an antenna to a transmission-line feeder. This may be done when the impedance of the antenna is a resistance. If this resistance is denoted by R_R, the problem is to make R_R "look like" the characteristic resistance of the feeder, here denoted by R_c. The problem may be solved by inserting between the antenna and the feeder a quarter-wavelength section of specially constructed line of characteristic resistance R_c' such that

$$(R_c)_{\text{feeder}} = \frac{(R_c')^2}{(R_R)_{\text{antenna}}} \tag{35.3}$$

as in Fig. 35.1. Then

$$R_c' = \sqrt{R_c R_R} \tag{35.4}$$

If $R_R < R_c$, the characteristic resistance R_c' of the transforming section is less than R_c. If the same size of wire is used for the matching section and the feeder, the spacing of the wires of the matching section must be closer than that of the feeder, Fig. 35.1*a*, as required by (15.4). Sometimes the spacing of the wires in the matching section is the same as that of the feeder, and the radius of the wires is increased to secure the required R_c'.

If $R_R > R_c$, the characteristic resistance of the transforming section is greater than R_c. If the same size of wire is used in the matching section and the feeder, the spacing in the matching section must be greater than that of the feeder, Fig. 35.1*b*.

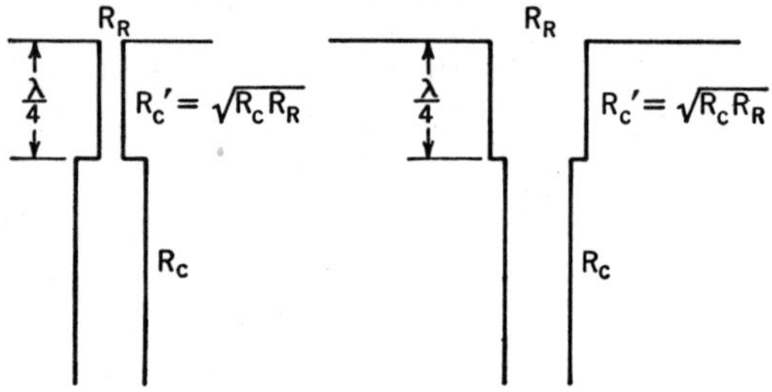

(*a*) (*b*)

FIG. 35.1.—Quarter-wave transformer for matching an antenna to a feeder, open-wire construction (not drawn to scale); (*a*) $R_R < R_c$, (*b*) $R_R > R_c$.

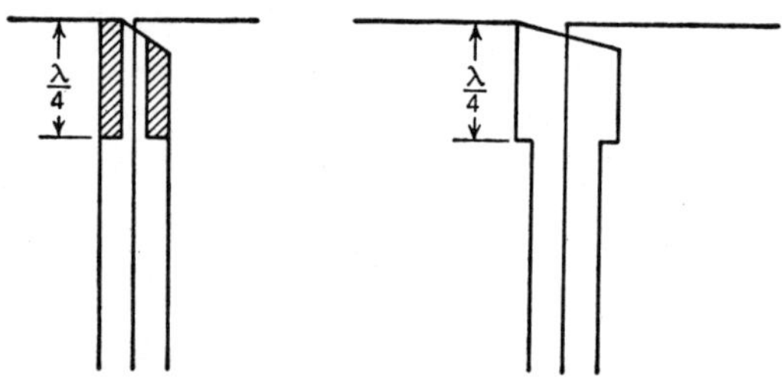

(*a*) (*b*)

FIG. 35.2.—Quarter-wave transformer for matching an antenna to a feeder, coaxial construction (not drawn to scale); (*a*) $R_R < R_c$, (*b*) $R_R > R_c$.

Figure 35.2 illustrates coaxial construction of quarter-wave matching sections. In the construction of Fig. 35.2*a*, a sleeve of appropriate outside diameter and wall thickness is shown inserted into the feeder pipe, effectively reducing the inside diameter of the outer conductor.

36. Eighth-wave Transformer.—If an eighth-wave section of dissipationless transmission line is terminated in a *pure resistance* of any value, the input impedance has a *magnitude* equal to R_c. When $s = \lambda/8$, $\beta s = \pi/4$ radians or 45°, and $\tan \beta s$ is unity. Then (32.1) and (35.1) reduce to

$$Z_S = R_c \frac{R_R + jR_c}{R_c + jR_R} \tag{36.1}$$

The numerator and denominator of (36.1) have the same magnitude; therefore

$$|Z_S| = R_c \tag{36.2}$$

Conversely, any dissipationless eighth-wave section, terminated in an impedance whose *magnitude* equals R_c, has an input impedance which is a pure resistance. For, if Z_R in (32.1) is made the conjugate of Z_S in (36.1), the input impedance with $\beta s = \pi/4$ will be the R_R which appears in (36.1). This is a consequence of the principle of conjugates.

37. Principle of Conjugates in Impedance Matching.—If a dissipationless network is inserted between a constant-voltage generator of internal impedance Z_g, and a load of impedance Z_R such that maximum power is delivered to the load, at every pair

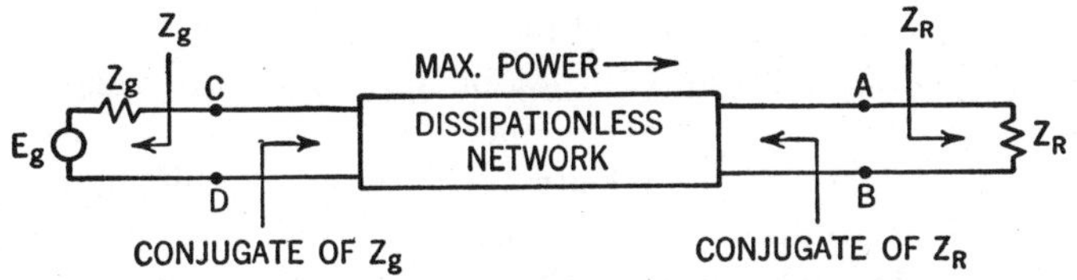

FIG. 37.1.—The principle of conjugates in impedance matching.

of terminals the impedances looking in opposite directions are conjugates of each other.

To secure maximum power output from a generator whose emf and whose internal impedance are constant, the load must have an impedance equal to the conjugate of the generator's internal impedance. In Fig. 37.1, if the dissipationless network (which could be made up of ideal lumped-constant elements or ideal transmission-line sections) is so designed that the generator is delivering maximum power into the terminals *CD*, the load impedance connected to the generator must be the conjugate of Z_g. This is expressed by saying that conjugate matching exists at *CD*. Maximum power is also fed into the terminals *AB* because no power is lost in the network. As to the terminals *AB*, the circuit to the left of *AB* is equivalent, by Thévenin's theorem, to a generator whose emf E_g' is the open-circuit voltage at the terminals *AB* and whose internal impedance Z_g' is the impedance looking to the left at *AB*. Now, Z_R must be the conjugate of Z_g'; hence Z_g' is the conjugate of Z_R. Therefore the impedance looking to the left at AB must be the conjugate of Z_R. Conjugate matching exists at AB as well as at CD.

When impedances are conjugate-matched for transmission of power in one direction, they are conjugate-matched for power transmission in the reverse direction, if no power loss occurs in the matching devices.

38. Half-wave Section.—The input impedance of a half-wave section of dissipationless line is equal to the terminal impedance. When $s = \lambda/2$, $\beta s = \pi$ radians or 180°, $\tan \beta s = 0$, and (32.1) reduces to $Z_S = Z_R$. In Fig. 28.1, the standing-wave pattern repeats itself every half wavelength, and the ratio E/I repeats itself every half wavelength. There is no impedance transformation.

However, at any point on the line the voltage and current lag by 180° the voltage and current, respectively, at a point a half wavelength nearer the generator. This property of producing a 180° phase shift in the voltage and current with no change in magnitude (for any terminating impedance) makes the half-wave low-loss section extremely useful in feeding antenna arrays.

Insulating supports should not be spaced at intervals of a half wavelength on a nonresonant line. Such supports, even when they have no leakage or dielectric loss, have the effect of placing some small capacitances across the line. If these capacitances were placed at half-wavelength intervals, their shunting effect would be cumulative, since the half-wavelength spacing would effectively place them all in parallel at one point of the line. The insulators cause the parameters of the line to be discontinuous, giving rise to reflections. These reflections add up in phase when the spacing of the insulators is one-half wavelength. If the insulators are spaced at quarter-wavelength intervals, the reflections are out of phase, and their effect is minimized. If there are many insulators per wavelength, say 10 or more, their effect can be taken as roughly equivalent to an added distributed capacitance (and conductance) averaged over the length of the line. When the line is not nonresonant, the disturbing effects of insulators may sometimes be minimized by locating the insulators at points of voltage minimum.

V. IMPEDANCE MATCHING BY MEANS OF STUBS

39. Input Admittance.—The expression for the input impedance, (18.18) or (32.1), is

$$Z_S = R_c \frac{Z_R + jR_c \tan \beta s}{R_c + jZ_R \tan \beta s} \tag{39.1}$$
$$= R_S + jX_S$$

This represents the combination of a resistance R_S in series with a reactance X_S.

In the following two sections, impedance-matching schemes are considered which employ sections of line placed in parallel with the feeder line. It is sometimes more convenient with this connection to consider the input at a given point on the feeder in terms of admittances, since the admittance of two branches in parallel can be added to yield the resultant admittance. This principle is the basis of a graphical method to be described later.

The conductance and susceptance corresponding to the impedance Z_S are obtained by taking the reciprocal of Z_S, which is

$$Y_S = \frac{1}{Z_S} = \frac{1}{R_c} \frac{R_c + jZ_R \tan \beta s}{Z_R + jR_c \tan \beta s} \tag{39.2}$$

Dividing the numerator and denominator by $R_c Z_R$

$$Y_S = G_c \frac{Y_R + jG_c \tan \beta s}{G_c + jY_R \tan \beta s} \tag{39.3}$$

or, by definition of conductance and susceptance,

$$Y_S \equiv G_S + jB_S$$

where G is the conductance and B is the susceptance. Equation (39.3) is the same in form as (39.1); impedances in (39.1) are replaced by their corresponding admittances in (39.3).

A line is "flat" or nonresonant when terminated in a load impedance R_c, or in an admittance $G_c = 1/R_c$.

40. Single-stub Impedance Matching.—It is possible to match any antenna or load to a transmission line by means of a single open or closed stub of suitable length, placed on the feeder at the proper point. The method is illustrated in Fig. 40.1 for an open stub and a two-wire feeder.

The operation of the stub may be explained as follows: When the load impedance is not equal to R_c (or when the load admittance is not equal to G_c), the magnitude and phase of E and I vary, Fig. 28.1. The admittance looking toward the load changes as the distance from the load increases. At a voltage maximum, the impedance looking toward the load is a pure resistance ρR_c; therefore the admittance is a pure conductance $1/\rho R_c$ or G_c/ρ. This conductance is less than G_c. Similarly, at a voltage minimum, the impedance is a pure resistance R_c/ρ; the admittance is a pure conductance ρG_c. This conductance is greater than G_c. At some point between a voltage maximum and the adjacent voltage minimum, the conductance is equal to G_c, and at this point the stub is located.

At this point, there is some susceptance because the admittance of a section of line is a pure conductance only at a voltage node or loop. To balance out this susceptance, the stub is added in parallel; the length and termination of the stub are adjusted so that the

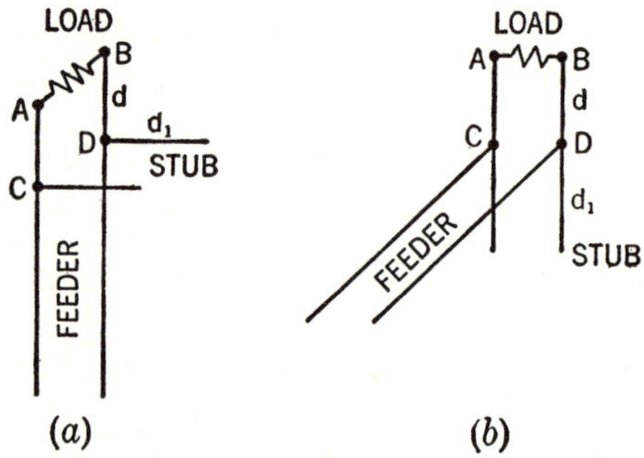

FIG. 40.1.—The use of an open stub to match a load to a transmission-line feeder, open-wire construction.

total susceptance of the parallel circuit (at the points CD in Fig. 40.1) is zero and the total admittance is a pure conductance equal to G_c. Then, at CD, the feeder is terminated by a conductance equal to G_c or a resistance equal to R_c, and standing waves on the feeder are eliminated.

The formulas for the location and lengths of the stubs are simplified by expressing the admittances in terms of the dimensionless ratio

$$\frac{Y}{G_c} = y_1 = g_1 + jb_1$$

where Y is the admittance in mhos. The quantity y_1 may be called "per unit admittance," and g_1 and b_1 may be called "per unit conductance" and "per unit susceptance." For a load impedance Z_R

$$g_{1R} + jb_{1R} = \frac{Y_R}{G_c} = \frac{R_c}{Z_R} \tag{40.1}$$

where $g_{1R} = (R_cR_R)/(R_R^2 + X_R^2)$ and $b_{1R} = -(R_cX_R)/(R_R^2 + X_R^2)$.

For an open stub, Fig. 40.1, the distance d from the load to the stub and the length d_1 of the stub are given by

$$\left.\begin{aligned} \tan \beta d_1 &= \sqrt{\frac{(g_{1R} - 1)^2 + b_{1R}^2}{g_{1R}}} \quad (40.2) \\ \tan \beta d &= \frac{g_{1R} - 1}{g_{1R} \tan \beta d_1 - b_{1R}} \quad (40.3) \end{aligned}\right] \text{for open stub}$$

If the angles βd_1 and βd are expressed in degrees

$$d_1 = \lambda \frac{(\beta d_1) \text{ in degrees}}{360} \quad (40.4)$$

$$d = \lambda \frac{(\beta d) \text{ in degrees}}{360} \quad (40.5)$$

where λ is the wavelength on the line.

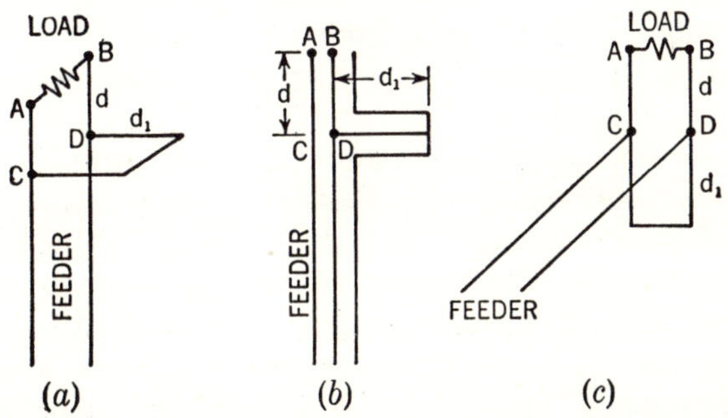

FIG. 40.2.—Closed-stub construction on two-wire and coaxial feeders.

For a closed stub, Fig. 40.2, the distance d from the load to the stub and the length d_1 of the stub are given by

$$\left.\begin{aligned} \cot \beta d_1 &= \sqrt{\frac{(1 - g_{1R})^2 + b_{1R}^2}{g_{1R}}} \quad (40.6) \\ \tan \beta d &= \frac{1 - g_{1R}}{g_{1R} \cot \beta d_1 + b_{1R}} \quad (40.7) \end{aligned}\right] \text{for closed stub}$$

If stubs less than $\lambda/4$ in length are used, a closed stub can be placed nearer the load than any other stub when $g_{1R} < 1$, and an open stub can be placed nearer the load than any other stub when $g_{1R} > 1$.

When the value of the load impedance is not known, the location and length of the stub may be determined from the standing-wave distribution on the line. If ρ is the standing-wave ratio on the feeder before the addition of the stub, and if d is measured toward the generator from the position of a voltage maximum, then, for a closed stub,

$$\left.\begin{aligned}\cot \beta d_1 &= \pm \frac{\rho - 1}{\sqrt{\rho}} \\ \tan \beta d &= \pm \sqrt{\rho}\end{aligned}\right] \quad \begin{array}{l}\text{for closed stub} \\ d \text{ measured from } E_{max}\end{array} \qquad \begin{array}{r}(40.8) \\ (40.9)\end{array}$$

These equations are obtained from (40.6) and (40.7) by setting $g_{1R} = 1/\rho$ and $b_{1R} = 0$, since at a voltage maximum the line is in effect terminated by a resistance ρR_c. If d is measured toward the generator from the position of a voltage minimum,

$$\left.\begin{aligned}\cot \beta d_1 &= \pm \frac{\rho - 1}{\sqrt{\rho}} \\ \tan \beta d &= \pm \frac{1}{\sqrt{\rho}}\end{aligned}\right] \quad \begin{array}{l}\text{for closed stub} \\ d \text{ measured from } E_{min}\end{array} \qquad \begin{array}{r}(40.10) \\ (40.11)\end{array}$$

These equations are obtained from (40.6) and (40.7) by setting $g_{1R} = \rho$ and $b_{1R} = 0$, since at a voltage minimum the line in effect is terminated by a resistance R_c/ρ.

Similar equations may be written for an open stub. Comparison of these equations would show that, when ρ is equal to 2.62, an open or closed stub of equal length may be used. When ρ is greater than 2.62, a closed stub is the shortest that may be used; and, when ρ is less than 2.62, an open stub is the shortest that may be used. However, the closed stub is preferred, because its length may be more easily adjusted and the shorting bar or piston adds mechanical support.

The "short-circuiting" wire of a two-wire closed stub has an inductance approximately equivalent to a length of line $D/2$ terminated by a zero impedance. Therefore, it is sometimes advisable, Sec. 27, to deduct $D/2$ from the lengths of the closed two-wire stubs, as given by the formulas of this section. For coaxial stubs, no such correction is necessary.

41. Double-stub Impedance Matching.—Although the single stub is extremely simple in its application, it has some disadvantages. It is not always convenient to provide an adjustment in the position of the stub, especially when coaxial cable is used. Also, an open stub may not be lengthened easily. To overcome these difficulties, two closed stubs of adjustable length but fixed spacing sometimes are used.

The load impedance involves two variables, X_R and R_R. To match this impedance to the line, two variable adjustments are required. For the single stub, the adjustments are the position and the length of the stub. For the double-stub method, the spacing of the stubs is fixed, and the two adjustments are the lengths of the two stubs. A given spacing will not accommodate all impedances, but the spacing can be chosen so that the range of impedances is sufficiently wide to allow for any slight variations in load or frequency that may be encountered with a given type of equipment.

The arrangement of the stubs is indicated in Fig. 41.1. A spacing $\lambda/2$ between the stubs is not useful, since it provides no impedance transformation. For the same reason, the spacing should not be too close. If d_2 is the length of the closed stub at

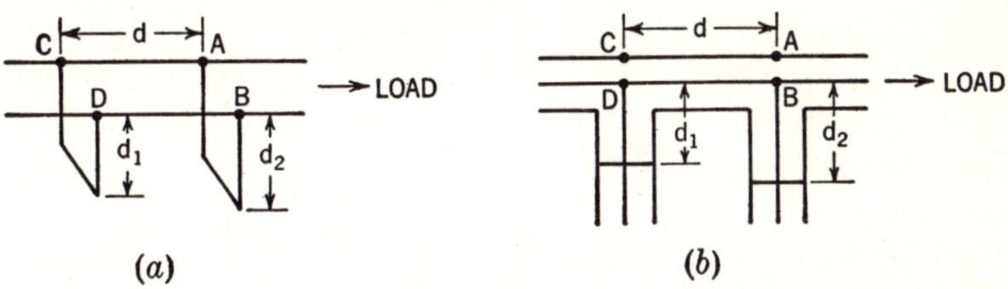

FIG. 41.1.—Double-stub impedance-matching arrangement for (*a*) two-wire line, (*b*) coaxial line.

the "load," and d_1 is the length of the closed stub at a distance d from the "load"

$$\cot \beta d_1 = \pm \frac{1}{\sqrt{g_{1R}}} \sqrt{\cot^2 \beta d - g_{1R} + 1} - \cot \beta d \tag{41.1}$$

$$\cot \beta d_2 = \pm \sqrt{g_{1R}} \sqrt{\cot^2 \beta d - g_{1R} + 1} - \cot \beta d + b_{1R} \tag{41.2}$$

The "load" which is matched to the feeder is that portion of the system which is to the right of the terminals AB, Fig. 41.1, not including the stub d_2. If there is a length of transmission line between the terminals AB, Fig. 41.1, and the antenna or load to which the power is delivered, the input impedance or admittance of this length of line is the "load" impedance or admittance to be used in (40.1) to calculate the g_{1R} and b_{1R} for (41.1) and (41.2).

The quantities under the square-root signs in (41.1) and (41.2) must be positive. Therefore the requirement that

$$\cot^2 \beta d - g_{1R} + 1 \geqslant 1$$

must be met. Hence, for a given spacing d between stubs, the greatest per unit conductance that can be matched to the feeder is

$$g_{1R}]_{\max} = 1 + \cot^2 \beta d \tag{41.3}$$

The susceptance of the "load" imposes no limitation, since it is in parallel with the susceptance of the stub d_2, which is adjustable to yield the per unit susceptance required by (41.2). Except when g_{1R} has the maximum allowable value, the choice of the + or − sign in (41.1) and (41.2) determines one of two possible combinations of d_1 and d_2 that will effect the impedance match. The same sign must be chosen in both equations. Since at a voltage maximum the feeder is terminated by a resistance ρR_c whose g_{1R} is $1/\rho$, which is less than 1, a double-stub arrangement with any useful spacing between stubs may always be used if the stub d_2 is placed at the location of a voltage maximum, and the distance d measured from this location toward the generator.

For $d = \lambda/4$, $\beta d = \pi/2$ or 90°, and (41.1) and (41.2) become

$$\left.\begin{aligned}\cot \beta d_1 &= \pm \frac{1}{\sqrt{g_{1R}}} \sqrt{1 - g_{1R}} \qquad (41.4)\\ \cot \beta d_2 &= \pm \sqrt{g_{1R}} \sqrt{1 - g_{1R}} + b_{1R} \qquad (41.5)\end{aligned}\right] \quad \text{with } d = \lambda/4$$

The per unit conductance looking toward the load at the junction of stub d_2 must be equal to or less than 1.

For $d = 3\lambda/8$, $\beta d = 3\pi/2$ or 135°, and d_1 and d_2 are given by

$$\left.\begin{aligned}\cot \beta d_1 &= \pm \frac{1}{\sqrt{g_{1R}}} \sqrt{2 - g_{1R}} + 1 \qquad (41.6)\\ \cot \beta d_2 &= \pm \sqrt{g_{1R}} \sqrt{2 - g_{1R}} + 1 + b_{1R} \qquad (41.7)\end{aligned}\right] \quad \text{with } d = 3\lambda/8$$

The per unit conductance looking toward the load at the junction of stub d_2 must be equal to or less than 2.

The location and length of the stubs may be determined from the observed standing-wave distribution on the feeder before the addition of the stubs. If, for example, the stub d_2 is located at a voltage maximum, the admittance looking toward the load is a conductance G_c/ρ, so that $g_{1R} = 1/\rho$ and $b_{1R} = 0$ at this point. Then with $d = \lambda/4$

$$\left.\begin{aligned}\cot \beta d_1 &= \pm \sqrt{\rho - 1} \qquad (41.8)\\ \cot \beta d_2 &= \pm \frac{\sqrt{\rho - 1}}{\rho} \qquad (41.9)\end{aligned}\right] \quad \begin{array}{l}\text{with}\\ d = \lambda/4\\ \text{stub } d_2 \text{ at } E_{\max}\end{array}$$

With $d = 3\lambda/8$

$$\left.\begin{aligned}\cot \beta d_1 &= \pm \sqrt{2\rho - 1} + 1 \qquad (41.10)\\ \cot \beta d_2 &= \pm \frac{\sqrt{2\rho - 1}}{\rho} + 1 \qquad (41.11)\end{aligned}\right] \quad \begin{array}{l}\text{with}\\ d = 3\lambda/8\\ \text{stub } d_2 \text{ at } E_{\max}\end{array}$$

42. Measurement of Wavelength.—The resonant properties of a short-circuited transmission line are utilized frequently for measuring the wavelength and hence the frequency. It is shown, Sec. 22, that the input impedance of a short-circuited dissipationless line is zero for lengths of line that are multiples of a half wavelength. Actually the impedance is finite but very small, being approximately equal to $R_c\alpha s$ at the half-wavelength points. If such a short-circuited line is coupled more or less loosely to an oscillator, and the short circuit is moved along the line, at certain positions the input impedance to the line will be very low; in effect the line will couple a resonant circuit to the oscillator, and there will be a marked reaction on the oscillator usually in the form of a decided decrease in the amplitude of oscillation. This reaction may be noted by observing some indicating instrument built into the

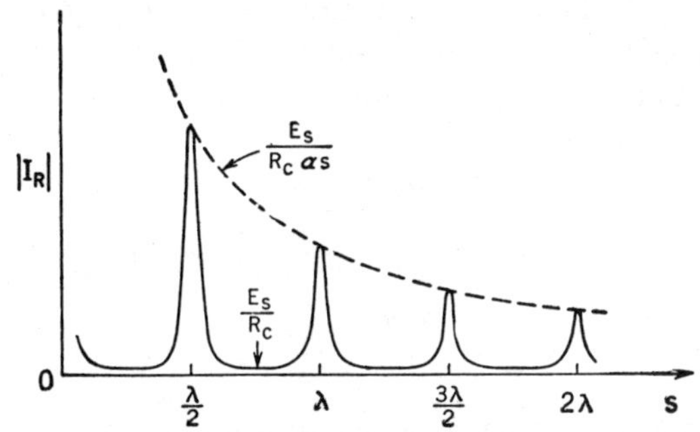

FIG. 42.1.—Current in a short-circuiting bar as a function of the distance s between generator and bar (not drawn to scale).

oscillator circuit or by noting the decrease in the oscillator output by means of a near-by receiver. The distance between two adjacent locations of the short-circuiting bar at which resonance occurs is one-half wavelength on the line.

When this measurement is performed with open-wire lines, some provision must be made for isolating the unused portion of the line. This isolation may take the form of a shield or may consist of a supplementary short-circuiting bar placed on the unused portion of the line at a distance $\lambda/4$ from the main short-circuiting bar. When a coaxial line is used, the short circuit is made by a movable plunger. The field of the short-circuited line is contained entirely within the cavity bounded by the plunger, and special means of isolation are not required.

Another method is to observe the current in a short-circuiting bar as the bar is moved along a two-wire line, provision being made to isolate the unused part of the line. Figure 42.1 shows a plot of the current in the short circuit (nowhere else on the line) as a function of the distance between the short circuit and the generator. The graph is not drawn to scale, because of the extremely great difference between the magnitudes of maxima and minima of current. The current at all minima is practically the same and equal to E_S/R_c, where E_S is the voltage applied to the input terminals of the line. The minima occur at the odd-numbered quarter-wavelength intervals. At the half-wavelength intervals, the current exhibits very sharp peaks, having a value $E_S/R_c\alpha s$, where s, at the points of current maxima, takes on values that are multiples of a half wavelength. The sharpness of the peaks may be expressed in terms of their width at the half-power point, *i.e.*, in terms of their width at those points where the current has fallen to 0.707 of the maximum. If d' denotes the distance from any one peak to any one of the near-by points at which the current has fallen to 0.707 of the value at the peak, the distance d' is given by

$$d' = \frac{\alpha}{\beta} s$$

so that the relative width of the peaks of Fig. 42.1 is

$$\frac{2d'}{s} = \frac{2\alpha}{\beta}$$

For low-loss lines in which $r \ll \omega l$ and $g \ll \omega c$,

$$\frac{2\alpha}{\beta} \doteq \frac{r}{\omega l} + \frac{g}{\omega c} = \frac{1}{Q} \tag{42.1}$$

The Q of a transmission line can be very high, of the order of 1,000, so that the position of the current peaks can be located accurately.

With proper technique, the wavelength may be measured with fairly high precision, using resonant transmission lines of coaxial or of open-wire construction. When a two-wire line is employed, this measurement is commonly named the measurement of wavelength by means of Lecher wires, since Lecher was one of the first to investigate the resonant properties of transmission lines at radio frequencies.

VI. CIRCLE DIAGRAM

The computation for numerical problems may often be simplified or eliminated by using graphs or diagrams. The accuracy depends upon the scale to which the diagram is drawn and the ease with which interpolations can be made between the lines on the diagram. The chart now to be described involves two families of circles and will be called the "circle diagram."

43. Construction of Circle Diagram.—The chart for the dissipationless line is based upon the fact that the graph of Z_S, (39.1), for any value of Z_R is a circle as s is varied. Similarly, the graph of Y_S, (39.3), is also a circle. There will be different circles for different values of Z_R or Y_R. These are the circles that surround the point 1,0 in Fig. 43.1.

All the algebraic manipulations involved in the derivation will not be given here, but the principal steps will be outlined. The reflection coefficient of (21.1) involves a magnitude and an angle, *i.e.*,

$$|\Gamma| = |\Gamma|\underline{/\psi} = |\Gamma|\epsilon^{j\psi} \tag{43.1}$$

The impedance Z_S may be expressed, from (21.2) and (21.4), with $d = s$, as

$$Z_S = R_c \frac{1 + |\Gamma|\epsilon^{j\phi}}{1 - |\Gamma|\epsilon^{j\phi}} \tag{43.2}$$

where $\phi = \psi - 2\beta s$.

To make the diagram applicable to all lines regardless of R_c, the value of Z_S is expressed in terms of R_c by

$$\frac{Z_S}{R_c} = r_1 + jx_1 \tag{43.3}$$

the subscript 1 denoting "per unit" values, with R_c as the unit. Then

$$r_1 + jx_1 = \frac{1 + |\Gamma|\epsilon^{j\phi}}{1 - |\Gamma|\epsilon^{j\phi}} \tag{43.4}$$

from which

$$|\Gamma|\epsilon^{j\phi} = \frac{r_1 - 1 + jx_1}{r_1 + 1 + jx_1} \tag{43.5}$$

The terms on each side of (43.5) are equal complex numbers and hence have equal magnitudes and equal angles. Equating the magnitudes

$$\left(r_1 - \frac{1 + |\Gamma|^2}{1 - |\Gamma|^2}\right)^2 + x_1^2 = \left(\frac{2|\Gamma|}{1 - |\Gamma|^2}\right)^2 \tag{43.6}$$

which is the equation of a circle whose center is on the horizontal axis of Fig. 43.1 at a distance $(1 + |\Gamma|^2)/(1 - |\Gamma|^2)$ from the origin, and whose radius is $2|\Gamma|/(1 - |\Gamma|^2)$. For various values of $|\Gamma|$, the circles surrounding the point 1,0 in Fig. 43.1 are obtained.

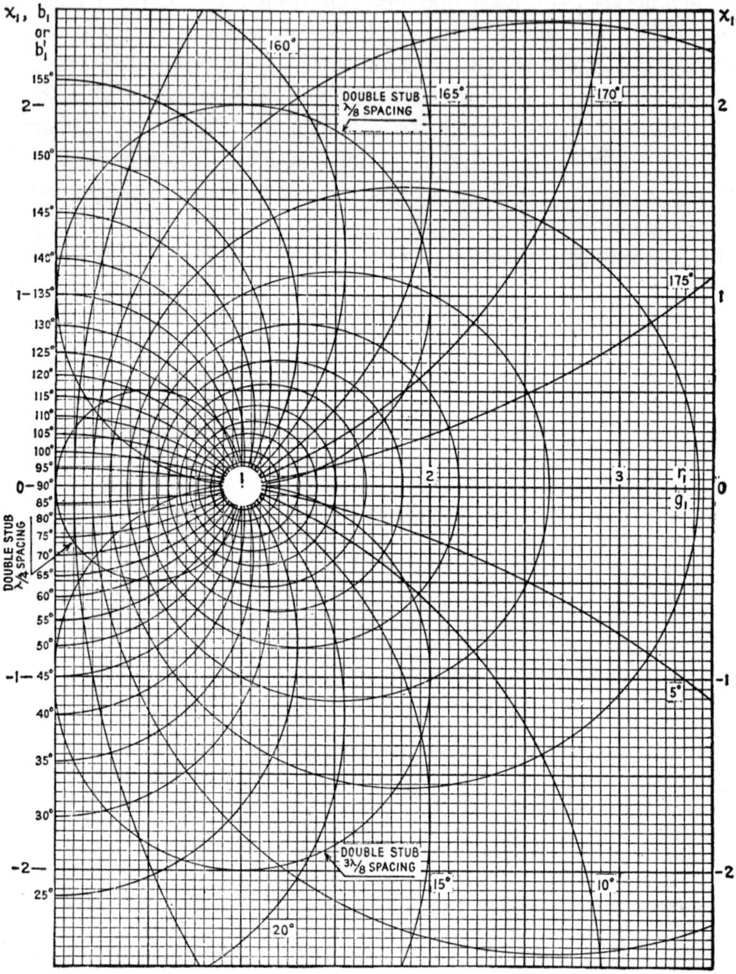

FIG. 43.1.—Circle diagram for transmission lines.

Equating the angles (actually by equating their tangents)

$$r_1^2 + \left(x_1 - \frac{1}{\tan\phi}\right)^2 = 1 + \frac{1}{\tan^2\phi} = \frac{1}{\sin^2\phi} \tag{43.7}$$

which is the equation of a circle whose center is on the vertical axis at a distance $1/\tan\phi$ below the origin, and whose radius is

$1/\sin \phi$. For various values of ϕ, the circles passing through the point 1,0 are obtained. In constructing the chart, $|\Gamma|$ is assumed real and positive, and the circles of constant ϕ are marked with the value of βs corresponding to $-\phi/2$ as determined from (43.2).

The reciprocal of Z_S is

$$Y_S = \frac{1}{Z_S} = G_S + jB_S$$

so that

$$\frac{Y_S}{G_c} = g_1 + jb_1 \tag{43.8}$$

This is also the reciprocal of (43.3) and (43.4), so that

$$\frac{Y_S}{G_c} = g_1 + jb_1 = \frac{1 - |\Gamma|\epsilon^{j\phi}}{1 + |\Gamma|\epsilon^{j\phi}} \tag{43.9}$$

Equation (43.9) is the same in form as (43.4) and leads to the same families of circles defined by (43.6) and (43.7); replacing r_1 by g_1, x_1 by $+b_1$, $+|\Gamma|$ by $-|\Gamma|$, would cause no change in the form of either (43.6) or (43.7).

Note that Y is defined as $G + jB$, where $G = R/(R^2 + X^2)$ and $B = -X/(R^2 + X^2)$. In the circle diagram in Fig. 43.1, $g_1 + jb_1$ is plotted in the same way as $r_1 + jx_1$ and the diagram may be used to determine both the normalized admittance and the normalized impedance with the same numerical values attached to the coordinate lines marked on the diagram.

44. Input Impedance.—To determine the input impedance of a line with a finite Z_R by means of the diagram, the procedure is as follows: First determine Z_R/R_c. Enter the chart at the point whose coordinates $r_1 + jx_1$ correspond to Z_R/R_c. Follow the Γ-circle passing through this point, moving clockwise in the diagram, through an angle βs as denoted by the ϕ-circles (interpolation may be necessary). Read the coordinates corresponding to this second point. Then $Z_S = R_c(r_1 + jx_1)$, where r_1 and x_1 are the coordinates of this second point.

Example.—Let $R_c = 500$ ohms, $Z_R = 1{,}000 + j750$, and $\beta s = 50°$. Now $Z_R/R_c = 2 + j1.5$. Enter the chart at this point, Fig. 44.1, which falls on the ϕ-circle marked 165°. To 165° add 50°, giving 215°, which on the diagram is equivalent to 215° − 180° = 35°. (Any multiple of 180° may be added or subtracted since the standing-wave pattern on a dissipationless line and the corresponding changes in impedance repeat every half wavelength). Following the Γ-circle that passes through the first point $2 + j1.5$, around to 35° gives the coordinates of the second point, in this case $0.77 - j1.09$. Then

$$Z_S = 500(0.77 - j1.09) = 385 - j545 \text{ ohms.}$$

The Γ-circle of this example intersects the r_1 axis at the point $3.33 + j0$. If the line were terminated in $Z_R = (3.33 + j0)R_c$ and were 165° long, Z_S would equal $(2 + j1.5)R_c$. Therefore, the load impedance of the preceding paragraph is equivalent to a length of line corresponding to $\beta s = 165°$ and terminated in a resistance $3.33R_c$. This is the significance of the 165° attached to the ϕ-circle

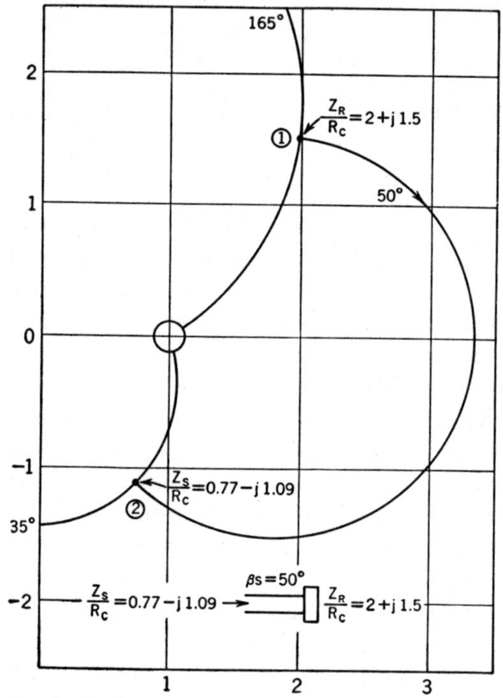

FIG. 44.1.—Solution for the input impedance of a line by means of the circle diagram.

passing through the first point in the problem. The standing-wave ratio is given by that intersection of the Γ-circle with the axis of reals, which is to the right of the point 1,0. In this example, $\rho = 3.33$.

A computation by formula (39.1) gives $Z_S = 382 - j547$. The use of the chart saves a substantial amount of time.

45. Input Admittance.—The input admittance may be obtained by a method that is practically identical with the procedure of Sec. 44, since the admittance equation (39.3) has the same form as the impedance equation (39.1).

The admittance of the load must be known or calculated from the impedance. The first step is to determine Y_R/G_c or R_c/Z_R and express the result in the form $g_1 + jb_1$. Enter the chart, Fig. 43.1, at the point whose horizontal coordinate is g_1 and whose vertical coordinate is b_1. Then follow the Γ-circle, moving clockwise in the diagram, through an angle βs as denoted by the ϕ-circles. Read the coordinates corresponding to this second point.

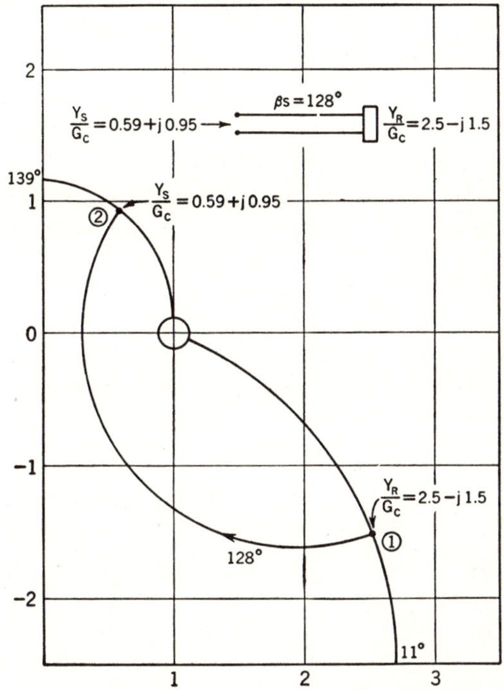

FIG. 45.1.—Solution for the input admittance of a line by means of the circle diagram.

Then $Y_S = (g_1 + jb_1)G_c$ or $(g_1 + jb_1)/R_c$ where g_1 and b_1 are the values read from the chart. Any point in the upper half of the diagram designates a capacitive susceptance, and any point in the lower half of the diagram designates an inductive susceptance.

Example.—Let $Y_R = (5 - j3) \cdot 10^{-3}$, $R_c = 500$, and $\beta s = 128°$. Then $Y_R/G_c = 2.5 - j1.5$. This point corresponds to an inductive susceptance and is to be found in the lower half of the diagram. Enter the chart at the point that is 2.5 units to the right of the origin and 1.5 units down from the origin, Fig. 45.1. The angle of the ϕ-circle passing through this point is 11° (by interpolation). To 11° add 128°, giving 139°. Follow the Γ-circle passing through the first point to the intersection with the ϕ-circle corresponding to 139°.

This second point is 0.6 unit to the right of the origin and 0.95 unit above the origin. Then $Y_S = (0.6 + j0.95)G_c = (1.2 + j1.9) \cdot 10^{-3}$ mho.

46. Open-circuit Input Impedance.—An open-circuited line is terminated in an infinite impedance, so that the Γ-circle corresponding to Z_R/R_c is infinite in radius. From the construction of the diagram, it is evident that, as the radius of the Γ-circle increases, the circles approach the left-hand margin of the diagram. In the limit, the radius becomes infinite, and the left-hand margin is the Γ-circle for an open-circuited line. Each point on the left-hand margin has associated with it an angle denoting the ϕ-circle, and a coordinate x_1. The value jx_1 equals Z_S/R_c for the open-circuited line whose βs equals the angle of the ϕ-circle passing through the point $0, x_1$ on the left-hand margin.

Example.—To find the impedance of an open-circuited line 0.1λ long whose $R_c = 72$ ohms, pick the point on the left-hand margin, Fig. 43.1, corresponding to $0.1(360°) = 36°$. The vertical coordinate of this point is -1.38. Then $Z_S = (-j1.38)R_c = -j99$ ohms.

47. Short-circuit Admittance.—A short-circuited line is terminated in an infinite conductance. The radius of the corresponding Γ-circle is infinite, as in Sec. 46. The only part of the circle appearing on the diagram is the part coinciding with the left-hand margin.

Example.—To find the input admittance of a short-circuited line whose R_c is 250 ohms and whose length is 0.42λ, enter the chart at the left-hand margin, Fig. 43.1, at the point corresponding to $0.42(360°) = 151°$. The value of $-b_1$ associated with this point is 1.80. Then $Y_S/G_c = g_1 + jb_1 = 0 + j1.80$; and $Y_S = +j1.80/R_c = +j7.2 \cdot 10^{-3}$ mho. Another way to state the solution is as follows: The vertical coordinate of the point corresponding to 151° is +1.80. Then $Y_S/G_c = +j1.80$, so that $Y_S = +j1.80/R_c = +j7.2 \cdot 10^{-3}$ mho.

48. Impedances and Admittances.—The chart may be used to compute the reciprocal of a complex number and thus may be used to determine $Y = 1/Z$ or $Z = 1/Y$. The proof of this is based upon the transforming properties of a quarter-wavelength section. For such a section

$$Z_S = \left.\frac{R_c^2}{Z_R}\right] \begin{array}{l}\text{with}\\ \beta s = 90°\end{array}$$

Dividing both sides of the equation by R_c and replacing $1/Z_R$ by Y_R and R_c by $1/G_c$

$$\frac{Z_S}{R_c} = \left.\frac{Y_R}{G_c}\right] \begin{array}{l}\text{with}\\ \beta s = 90°\end{array} \qquad (48.1)$$

For computation purposes, R_c may be assumed to have any convenient value.

The procedure to find $Y = 1/Z$ is as follows: Divide Z by any convenient real number R, so that the quotient defines a point that falls within the range of the chart. This point corresponds to the left-hand member of (48.1). Enter the chart at this point. Follow the corresponding Γ-circle through an angle 90°

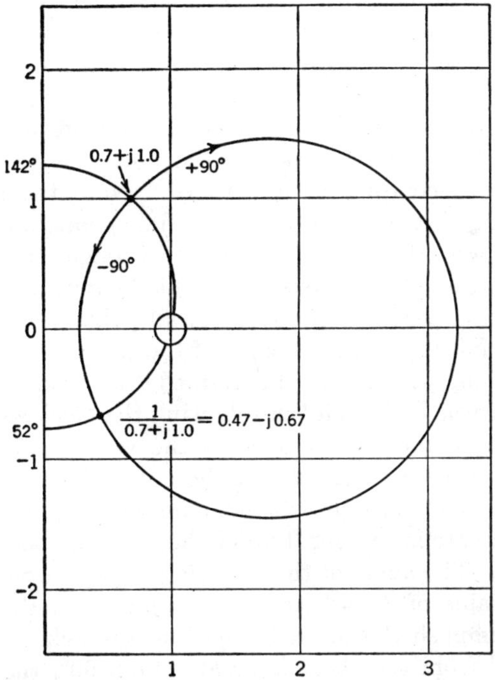

FIG. 48.1.—Graphical calculation of the reciprocal of a complex number.

(as denoted by the ϕ-circles), locating the second point. This second point denotes the complex number $g_1 + jb_1$ which is the right-hand member of (48.1). Then $Y = (g_1 + jb_1)/R_c$.

Example.—To find the admittance of the impedance $Z = 210 + j300$ ohms, divide Z by any assumed value of R_c such as 300 so that $Z/300 = 0.7 + j1.0$. Enter the chart at the point that is 0.7 unit to the right of the origin and 1.0 unit above the origin, Fig. 48.1. The corresponding ϕ is 142°. Follow the Γ-circle passing through this point to the ϕ-circle whose angle is 142° ± 90° or 52°. This second point is 0.47 unit to the right of the origin and 0.67 unit below the origin. Then $g_1 + jb_1 = 0.47 - j0.67 = Y/G_c$ from (48.1). The required Y is $(0.47 - j0.67)/300 = (1.57 - j223) \cdot 10^{-3}$ mho.

A consequence of the reciprocal relationship of (48.1) is that the Γ-circle passing through a point such as $2 + j0$ passes through its reciprocal point $\frac{1}{2} + j0$; the Γ-circle passing through the point $3 + j0$ passes through the point $\frac{1}{3} + j0$, etc. This principle is sometimes useful in locating the proper Γ-circle in the region where the lines on the diagram are necessarily crowded.

49. Single Stubs.—At the point where the stub is connected to the transmission line, the feeder is terminated by the combination of the stub in parallel with that portion of the line connected to the load or antenna. Since the admittances of parallel branches can be added, the graphical solution is carried out in terms of admittances.

Let Y_R be the complex admittance of the load or antenna, and G_c the characteristic admittance of the line. Enter the chart at the point whose coordinates are Y_R/G_c. Note the angle of the ϕ-circle passing through this point, the first point in obtaining the solution. Follow the Γ-circle passing through the first point, moving clockwise to the intersection of this Γ-circle with the vertical line on the chart denoting the locus of all points whose horizontal coordinate is unity. Note the angle of the ϕ-circle passing through this intersection, which shall be termed the second point. The total angle "turned through" in following the clockwise path from the first point to the second point is the value of βd, where d is the distance between the load and the stub. The next step is to locate a third point on the left-hand margin of the chart, whose vertical coordinate is the negative of the vertical coordinate of the second point. The angle of the ϕ-circle passing through this third point is the value of βd_1 where d_1 is the length of the closed stub that will accomplish the desired impedance match.[1] If the value of βd_1 as read from the chart is greater than 90°, the closed stub may be replaced by an open stub one-quarter wavelength shorter.

In the procedure outlined, the complex number corresponding to the second point is the value of Y/G_c looking toward the load at the location of the stub, *i.e.*, toward the terminals AB in Figs. 40.1 and 40.2, from the junction CD. The complex number corresponding to the third point is the value of Y/G_c looking into the stub at the junction CD. The sum of these two complex numbers is $1 + j0$; therefore the sum of the admittances at the junction is

[1] For two-wire lines, it may sometimes be advisable to subtract $D/2$ from the graphically determined value of d_1, to correct for the inductance of the wire closing the stub.

$G_c(1 + j0)$, and the feeder is terminated in its characteristic conductance at the junction with the stub.

Several numerical examples are worked out below. From these and from the above, it will be seen that, if stubs less than $\lambda/4$ in length are used, a closed stub less than $\lambda/4$ long can be placed closer to the load than any other stub when the real part of Y_R/G_c is less than 1. Similarly, when the real part of Y_R/G_c

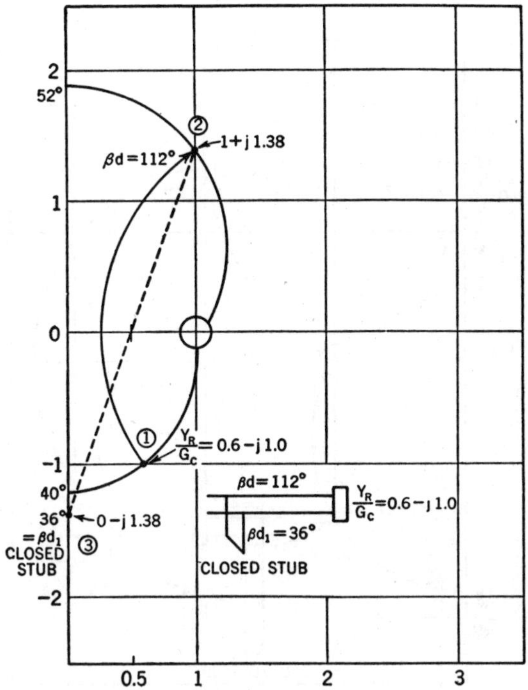

FIG. 49.1.—Graphical calculation of a closed matching stub.

is greater than 1, an open stub can be placed closer to the load than any other stub.

Example Indicating a Closed Stub.—Let $Y_R/G_c = 0.6 - j1.0$ and $\lambda = 1$ m. Enter the chart at the point 0.6 division to the right of the origin and 1.0 division down from the origin, Fig. 49.1. The angle of the ϕ-circle is 40°. Follow the Γ-circle through this first point clockwise to its intersection with the vertical line passing through the point 1,0. The intersection has the coordinates 1, +1.38; the associated ϕ-circle has the angle 152°. Then the angle βd is equal to 152° − 40° or 112°. The coordinates of the third point are 0, −1.38; the angle of the associated ϕ-circle is 36°. Then $\beta d_1 = 36°$ for a closed stub. Since $\lambda = 100$ cm,

$$d = \left(\frac{112^\circ}{360^\circ}\right) 100 = 31.1 \text{ cm}; \qquad d_1 = \left(\frac{36^\circ}{360^\circ}\right) 100 = 10 \text{ cm}.$$

Examples Indicating an Open Stub.—Let the impedance of the load be $Z_R = 160 - j120$ and the characteristic resistance of the feeder be $R_c = 500$ ohms. Then $Y_R/G_c = R_c/Z_R = 2 + j1.5$. Enter the chart at the point that is 2 divisions to the right and 1.5 divisions above the origin, Fig. 49.2. This is the first point, and the angle of the associated ϕ-circle is 165°. Follow the Γ-circle passing through the first point clockwise to the intersection with the vertical line passing through 1,0. This second point has the coordinates

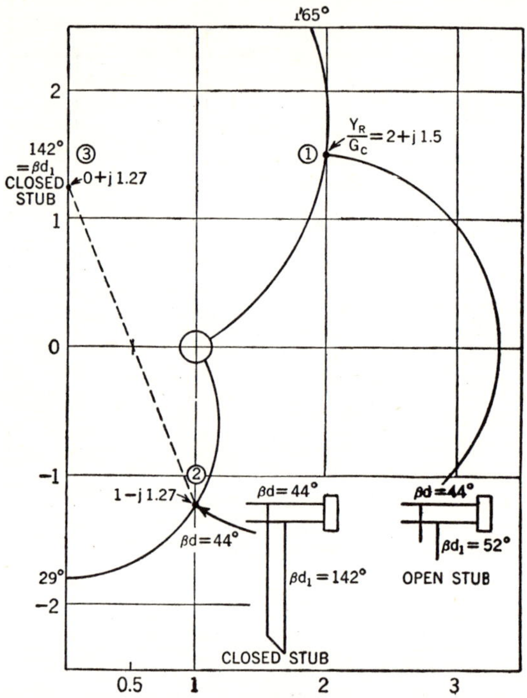

FIG. 49.2.—Graphical calculation of an open matching stub.

1, −1.27; and the associated ϕ-circle has the angle 29°. The angle "turned through" in passing from the first point to the second point is 15° + 29° (*i.e.*, from 165° to 180° and then from 0° to 29°), so that $\beta d = 44°$. The next step is to locate the third point, which is on the left-hand margin and has the vertical coordinate that is the opposite of the vertical coordinate of the second point. This third point is 0, +1.27; and the associated ϕ-circle has the angle 142°. Then $\beta d_1 = 142°$ for a closed stub, or 142° − 90° = 52° for an open stub. If the wavelength λ is 50 cm,

$$d = \left(\frac{44^\circ}{360^\circ}\right) 50 = 6.1 \text{ cm}; \qquad d_1 = \left(\frac{52^\circ}{360^\circ}\right) 50 = 7.2 \text{ cm}.$$

for the open stub.

For another example, let $R_c = 200$ ohms, $Z_R = 80 + j20$ ohms, and $\lambda = 60$ cm. Then $Y_R/G_c = R_c/Z_R = 2.35 - j0.59$. Enter the chart at the point that is 2.35 units to the right of the origin and 0.59 unit down from the origin, Fig. 49.3. The ϕ-circle corresponding to this point is 7°. Follow the Γ-circle clockwise to its intersection with the vertical line passing through 1,0. The coordinates of this second point are 1, −0.96; the associated ϕ-circle has the angle 32°. Then the angle "turned through" is 32° − 7°, or 25°, so that $\beta d = 25°$. The third point has the coordinates 0, +0.96; the associated

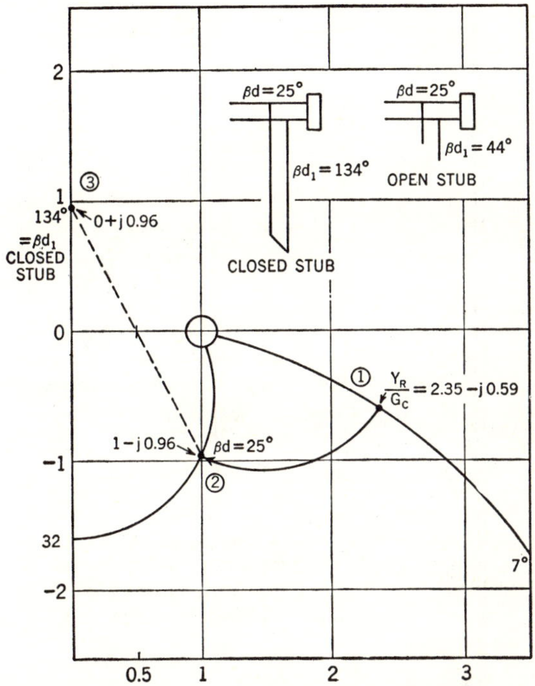

FIG. 49.3.—Graphical calculation of an open matching stub.

ϕ-circle has the angle 134°, which is βd_1 for the closed stub. Then βd_1 for an equivalent open stub is 134° − 90° = 44°. From the above,

$$d = \left(\frac{25°}{360°}\right) 60 = 4.2 \text{ cm}; \qquad d_1 = \left(\frac{44°}{360°}\right) 60 = 7.3 \text{ cm}.$$

50. Double Stubs.—In the double-stub arrangement, Fig. 41.1, the combined admittance of load and stub d_2 is transformed by the section of line d to an admittance whose per unit conductance is 1. The locus of the per unit admittances $y_1 = g_1 + jb_1$ that will be so transformed is

$$\left(g_1 - \frac{1}{2\sin^2 \beta d}\right)^2 + (\cot \beta d - b_1)^2 = \left(\frac{1}{2\sin^2 \beta d}\right)^2 \quad (50.1)$$

which is a circle of radius $1/(2 \sin^2 \beta d)$ with center at

$$g_1 = \frac{1}{2 \sin^2 \beta d} \qquad \text{and} \qquad b_1 = \cot \beta d$$

The locus circles for d equal to $\lambda/8$, $\lambda/4$, and $3\lambda/8$ are in Fig. 43.1.

Example.—Let $Y_{AB}/G_c = 1.5 - j1.0$, and $d = 3\lambda/8$. The first step is to add an admittance in the form of a closed stub of length d_2, so that the combined

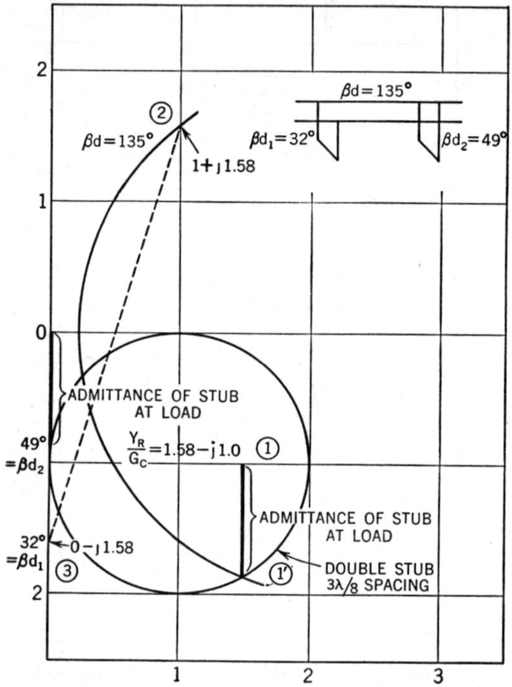

FIG. 50.1.—Graphical calculation of double matching stubs.

per unit admittance falls on the locus marked "Double Stub, $3\lambda/8$ Spacing." Let this be done by dropping vertically from point 1 at $1.5 - j1.0$ to point 1′, Fig. 50.1, at $1.5 - j1.87$. The admittance added is $-j0.87G_c$, which corresponds to a short-circuited stub 49° long, $\beta d_2 = 49°$. From this point on, the solution is the same as for the single stub with $\beta d = 135°$. Point 2 corresponds to a per unit admittance $1 + j1.58$, so that the admittance of the stub d_1 must be $-j1.58G_c$ as for the single stub, Fig. 49.1. This requires a short-circuited stub 32° long, $\beta d = 32°$.

VII. GENERAL TRANSMISSION-LINE EQUATIONS

51. Hyperbolic Form.—A more general treatment of transmission lines is one in which dissipation is taken into account, Z_c as given by (8.1) is in general not purely resistive, and the propagation constant γ as given by (7.1) has a real part that is finite. In many cases, however, the losses are so small that Z_c is almost purely resistive and the approximations of Sec. 14 hold.

When $l/r = c/g$, a condition not approached in practice except by certain specially constructed telephone lines, Z_c is purely resistive, α is constant, and β is proportional to the frequency. Such a line causes no phase or frequency distortion in the transmission of a signal and is called a distortionless line.

Dissipation on a line causes the wavelength to be shorter and the phase velocity to be less than for an ideal dissipationless line. Usually g is smaller compared to ωc than is r compared to ωl, so that the characteristic impedance is usually slightly capacitive.

The equations

$$E = A \cosh \gamma x + B \sinh \gamma x$$

and

$$I = C \cosh \gamma x + D \sinh \gamma x$$

are also solutions to (6.7) and (6.8). In terms of the rms voltage and current at the sending end

$$E = E_S \cosh \gamma x - I_S Z_c \sinh \gamma x \tag{51.1}$$

$$I = I_S \cosh \gamma x - \frac{E_S}{Z_c} \sinh \gamma x \tag{51.2}$$

and in terms of the voltage and current at the receiving end

$$E = E_R \cosh \gamma d + I_R Z_c \sinh \gamma d \tag{51.3}$$

$$I = I_R \cosh \gamma d + \frac{E_R}{Z_c} \sinh \gamma d \tag{51.4}$$

The input impedance Z_S is

$$Z_S = Z_c \frac{Z_R + Z_c \tanh \gamma s}{Z_c + Z_R \tanh \gamma s} \tag{51.5}$$

and E_S and I_S may be calculated from (18.19) and (18.20).

52. Incident and Reflected Waves.—The terms of (6.9) may still be interpreted as representing incident and reflected waves, both of which are attenuated. In terms of E_R

$$E = E_R^+ \epsilon^{+\gamma d} + E_R^- \epsilon^{-\gamma d}$$

where[1]

$$\Gamma = \frac{E_R^-}{E_R^+} = \frac{Z_R - Z_c}{Z_R + Z_c}$$

[1] Under certain conditions, *e.g.*, with a slightly capacitive Z_c and a purely inductive Z_R, it is possible to have $|\Gamma|$ greater than unity.

so that

$$E_R^+ = \frac{E_R}{2}\frac{Z_R + Z_c}{Z_R} = \frac{I_R}{2}(Z_R + Z_c)$$

$$E_R^- = \frac{E_R}{2}\frac{Z_R - Z_c}{Z_R} = \frac{I_R}{2}(Z_R - Z_c)$$

Similarly

$$I = I_R^+\epsilon^{+\gamma d} + I_R^-\epsilon^{-\gamma d}$$

where

$$\frac{I_R^-}{I_R^+} = -\Gamma$$

Also

$$I_R^+ = \frac{I_R}{2}\frac{Z_R + Z_c}{Z_c} = \frac{E_R^+}{Z_c}$$

and

$$I_R^- = -\frac{I_R}{2}\frac{Z_R - R_c}{Z_c} = -\frac{E_R^-}{Z_c}$$

The magnitude of the incident wave is attenuated in the direction toward Z_R, and the magnitude of the reflected wave is attenuated in the direction toward Z_S. Figure 52.1 shows the change in

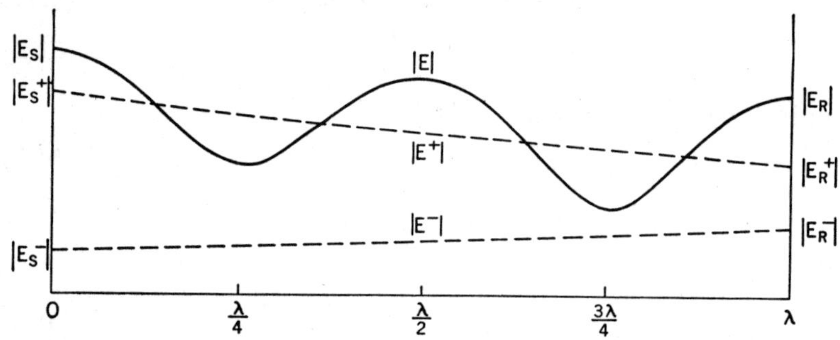

FIG. 52.1.—Voltage magnitudes on a line one wavelength long having an attenuation constant corresponding to 3.8 db per wavelength ($\alpha\lambda = 0.44$ neper) and terminated in $Z_R = 3.1\ Z_c$.

magnitude of the voltage and its incident and reflected components on a line one wavelength long having considerable dissipation.

53. Short-circuited and Open-circuited Lines.—For the short-circuited line, $E_R = 0$ and (51.3) and (51.4) become

$$E = I_R Z_c \sinh \gamma d \tag{53.1}$$

$$I = I_R \cosh \gamma d \tag{53.2}$$

and

$$Z_S = Z_c \tanh \gamma s \tag{53.3}$$

The distribution of current and voltage resembles that shown in Fig. 22.5 for magnitudes, except that the height of the maxima increases as d increases, and minima are not zero but finite, also increasing as d increases.

When the over-all attenuation αs of the line is less than 0.15 neper or 1.3 db, $\cosh \alpha d \doteq 1$ and $\sinh \alpha d \doteq \alpha d$. Then

$$\cosh \gamma d \doteq \cos \beta d + j\alpha d \sin \beta d$$
$$\sinh \gamma d \doteq \alpha d \cos \beta d + j \sin \beta d$$

If the length of the line is expressed in terms of the wavelength, $s = n\lambda$ where n is any positive number not necessarily an integer,

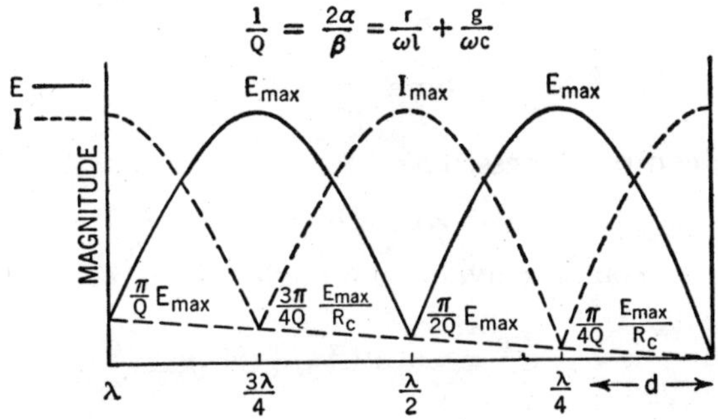

FIG. 53.1.—Current and voltage magnitudes for a short-circuited line whose over-all attenuation αs is less than 0.1 neper.

then $\alpha s = \alpha n\lambda = \alpha n(2\pi/\beta)$ and these approximations hold when

$$\alpha n \frac{2\pi}{\beta} \leqslant 0.15$$

or

$$n \leqslant \frac{0.15}{\pi} \frac{\beta}{2\alpha} \doteq \frac{Q}{20}$$

i.e., when the line is less than $Q/20$ wavelengths long. For such a line, the voltage and current vary in magnitude as in Fig. 53.1 The maxima are virtually unchanged, but the minima show a perceptible increase.

For the open-circuited line, $I_R = 0$, and

$$E = E_R \cosh \gamma d \tag{53.4}$$

$$I = \frac{E_R}{Z_c} \sinh \gamma d \tag{53.5}$$

$$Z_S = Z_c \coth \gamma s \tag{53.6}$$

Thus the curves of current and voltage are interchanged, as in the dissipationless case.

54. Line Terminated in a Finite Impedance.—When the dissipation in the load is large compared with that in the line, the current and voltage distribution resembles that for the dissipationless line, the heights of the maxima and minima increasing as the distance from the load increases. The longer the line, the smaller is the reflected component at the sending end for any fixed value of Z_R. Hence the fluctuations in the magnitude of E and I are less at the sending end than at the receiving end, as, for example, in Fig. 52.1.

The circle diagram may be used with dissipative lines also. It was shown by Kennelly that, if a complex hyperbolic angle θ is defined by

$$\tanh \theta = \frac{Z_R}{Z_c} \tag{54.1}$$

then (51.5) can be expressed as

$$Z_S = Z_R \tanh (\gamma s + \theta) \tag{54.2}$$

Similarly, if a complex hyperbolic angle $(A_R + j\Phi_R)$ is defined by

$$\frac{Z_R}{Z_c} = \coth (A_R + j\Phi_R) \tag{54.3}$$

then

$$Z_S = Z_c \coth (\gamma s + A_R + j\Phi_R) \tag{54.4}$$

or

$$\begin{aligned} r_1 + jx_1 = \frac{Z_S}{Z_c} &= \coth [\alpha s + A_R + j(\beta s + \Phi_R)] \\ &= \coth (A + j\Phi) \end{aligned} \tag{54.5}$$

where A may be defined as the combined attenuation factor and Φ the combined phase factor of the line and the load. If (54.5) is plotted on the complex r_1, x_1 plane, an orthogonal family of circles is obtained, Fig. 26.1, Chap. III. Curves of constant attenuation factor $A = A_R + \alpha s$ form the family of circles surrounding the point 1,0. Curves of constant phase factor

$$\Phi = \Phi_R + \beta s$$

form the family of circles passing through the point 1,0. The value of A in nepers and the value of Φ in degrees are marked on Fig. 26.1, Chap. III.[1]

[1] The circles of constant $|\Gamma|$ or constant standing-wave ratio in Fig. 43.1 correspond to the circles of constant A in Fig. 26.1, Chap. III. In Fig. 43.1 of Chap. I, the A's for the unmarked circles are 0.1, 0.2, 0.3, 0.4, 0.5, 0.6.

Example.—With $Z_R/Z_c = 3 + j3.25$, $s = 0.6\lambda$, $\alpha s = 0.1$ neper, the problem is to determine the input impedance. The first step is to enter the chart at Z_R/R_c. At this point $A_R = 0.15$ and $\Phi_R = 170.5°$. Since $s = 0.6\lambda$, $\beta s = 216°$. Then the point on the diagram denoting Z_S/Z_c is the point given by

$$A = 0.15 + 0.1 = 0.25; \qquad \Phi = 216° + 170.5° = 386.5° \text{ or } 26.5°$$

The r_1, x_1 coordinates of this point are $1 - j1.55$; $Z_S/Z_c = 1.0 - j1.55$.

55. Efficiency.—When the dissipative line is terminated in $Z_c = R_c + jX_c$

$$|I_R| = |I_S|\epsilon^{-\alpha s} \tag{55.1}$$

so that the efficiency η is given by

$$\eta = \frac{P_R}{P_S} = \frac{|I_R|^2R_c}{|I_S|^2R_c} = \epsilon^{-2\alpha s} \tag{55.2}$$

There is no simple and rigorous formula for the efficiency in the general case. The input and output currents can be computed from the equations in Sec. 51 by setting $x = s$ in (51.1) and (51.2), or $d = s$ in (51.3) and (51.4). In terms of (51.3) and (51.4),

$$E_S = E_R \cosh \gamma s + I_R Z_c \sinh \gamma s \tag{55.3}$$

$$I_S = I_R \cosh \gamma s + \frac{E_R}{Z_c} \sinh \gamma s \tag{55.4}$$

Then

$$\eta = \frac{P_R}{P_S} = \frac{(Re)E_R I_R^*}{(Re)E_S I_S^*} \tag{55.5}$$

where (Re) means that only the real part of the product is to be taken and I_R^* and I_S^* denote the conjugate of I_R and I_S. For a low-loss line, it is possible to show[1] that

$$\frac{P_R}{P_S} = \frac{\sinh 2A_R}{\sinh 2(\alpha s + A_R)} \tag{55.6}$$

The condition for maximum efficiency may be determined by differentiating (55.6) with respect to A_R. The condition is $A_R = \infty$ or, from (54.3), $Z_R = Z_c$. Then the efficiency is given by (55.2).

Supplementary Reading

L. F. WOODRUFF: "Principles of Electric Power Transmission," 2d ed., John Wiley & Sons, Inc., 1938.

0.7, 0.8, 0.9, 1.0, 1.2, and 1.4 in order, from left to right, of their intersection with the axis of reals between the origin and the point 1,0.

[1] R. KING, Transmission-Line Theory and Its Application, *J. Appl. Phys.*, Vol. 14, pp. 577–600, November, 1943.

J. G. BRAINERD, G. KOEHLER, H. J. REICH, and L. F. WOODRUFF: "Ultra-high-frequency Techniques," D. Van Nostrand Company, Inc., 1942.

R. I. SARBACHER and W. A. EDSON: "Hyper- and Ultra-high-frequency Engineering," John Wiley & Sons, Inc., 1943.

F. E. TERMAN: "Radio Engineers' Handbook," McGraw-Hill Book Company, Inc., 1943.

L. A. WARE and H. R. REED: "Communication Circuits," 2d ed., John Wiley & Sons, Inc., 1944.

A. H. WING and J. EISENSTEIN: Single- and Double-stub Impedance Matching, *J. Appl. Phys.*, Vol. 15, pp. 615–622, August, 1944.

R. W. P. KING: "Transmission-line Theory," McGraw-Hill Book Company, Inc., 1955; Dover Publications reprint, 1964.

CHAPTER II

ANTENNAS

Electric Circuit Theory and Electromagnetic Theory.—In order to understand the behavior of antennas and of electric circuits at ultra-high frequencies, it is essential to recognize that phenomena of a vastly more general nature are involved than are encountered in conventional electric networks. Attention is seldom called to the fact that electric-circuit theory which proceeds from Kirchhoff's laws is a highly specialized form of a more general theory. In some respects, the situation is like that in mechanics, in which the simple law of gravitation due to Newton may be looked upon as a special case of a more general law formulated in the theory of relativity. Much as Newtonian mechanics is adequate for the mechanical engineer, ordinary electric-circuit theory is accurate for the requirements of electrical power engineering and for many requirements in communications. But even as Newton's laws of motion are inadequate in dealing with atomic phenomena and some astronomical problems, so ordinary electric-circuit theory fails when applied to antennas and to most circuits that are to be used at ultra-high frequencies. The reason is that the conditions that limit the generality of Newton's laws on the one hand, or the theorems of electric-circuit theory on the other, are not satisfied. For those who have assumed that Kirchhoff's laws are perfectly general, a series of surprises is in store. They may, in fact, feel like Alice when the Red Queen was annoyed by her reluctance to believe "six impossible things before breakfast." But presently they may return through the Looking Glass and discover that they have been living in the one-dimensional Wonderland of electric-circuit theory and that Nature is as simple as this suggests only in sufficiently small spaces.

It is difficult to understand the structure of general electromagnetism without first learning the appropriate symbolism, that of mathematics. But if one is willing to accept some things on faith and to meet others with an open, perhaps even an adventurous mind, a degree of familiarity with many electromagnetic phenomena can be acquired from a qualitative discussion.[1]

The fundamental problem of all communication networks is to devise circuits such that a desired effect is produced and all interfering effects are minimized. Such an effect may consist merely of the ringing of a bell when a button is pressed, or it may be the sound of a man's voice in London when the man is speaking in Boston. Its successful accomplishment presupposes a knowledge of the laws of interaction among electric charges in motion, not only in the familiar cases involving currents in coupled coils and charges accumulating in capacitances or streaming between the electrodes of a vacuum tube, but also when a current in an antenna causes other smaller currents in countless conductors throughout the universe. It is especially in this last case that the simple laws of Coulomb and of Ampère, as well as those of Kirchhoff, no longer serve.

I. A QUALITATIVE INTRODUCTION TO GENERAL ELECTROMAGNETIC THEORY

1. Electric Charges and Currents and the Electromagnetic Field.—The equations of electromagnetism formulate in as general a manner as possible the experimentally observed fact that the electric charges contained in conductors and nonconductors always exert forces upon one another both when they are at rest and when they are in motion relative to one another. The magnitude and the direction of the forces due to one group of charges acting upon another group depend upon many factors. These include the shape and the size of the circuits in which the charges are moving or are at rest, the distances between elements of the circuits, the magnitudes and relative directions of currents, and the electrical properties of the materials that are involved.

Because of the complexity of the problem, it is convenient to express the general law of force in two parts. Thus, a mathematical structure called the electromagnetic field is first defined in terms of the distributions of charge and of current in the one circuit, and then the force on the charges in the second circuit is expressed in terms of this electromagnetic field. By repeating the process with the circuits interchanged, simultaneous equations in the distribu-

[1] An introductory mathematical formulation paralleling the qualitative one here given is in R. W. P. King, "Fundamental Electromagnetic Theory," Dover Publications, Inc., 1963.

tions of charge and of current in the two circuits can be set up. Their solution is usually very difficult but has been accomplished in a few cases. The electromagnetic field as defined in terms of the currents and charges in the first circuit consists of two vectors, an electric vector and a magnetic vector, each of which assigns a direction and a magnitude to every point in space. Except in electrostatics (when no magnetic field is required) and magnetostatics (when no electric field is needed), the electric and magnetic fields are so closely interrelated that one cannot be defined without the other. In fact, both are merely different aspects of a *single electromagnetic field.*

In the analytical description of the electromagnetic field, it is convenient to introduce two universal constants which are determined by measurement. In the rationalized mks system of units, these constants are a universal electric constant ϵ_0 (usually called the dielectric constant of space) and a universal magnetic constant μ_0 (commonly called the permeability of space). These have the numerical values

$$\epsilon_0 = 8.85 \cdot 10^{-12} \text{ farads/m} \tag{1.1}$$

$$\mu_0 = 4\pi \cdot 10^{-7} \text{ henrys/m} \tag{1.2}$$

Important combinations of these, which are also universal constants, are a characteristic velocity v_c and a characteristic resistance $\mathcal{R}_c$. These[1] are defined by

$$v_c = \frac{1}{\sqrt{\mu_0 \epsilon_0}} = 3 \cdot 10^8 \text{ m/sec} \tag{1.3}$$

$$\mathcal{R}_c = \sqrt{\frac{\mu_0}{\epsilon_0}} = 376.7 \doteq 120\pi \text{ ohms} \tag{1.4}$$

In the mks system of units, the electromagnetic field in nonferromagnetic media may be expressed in terms of the electric $\mathcal{E}$-field measured in volts/meter and the magnetic $\mathcal{H}$-field measured in amperes/meter. (When the field due to the current in a coil is calculated at low frequencies, the name ampere-turns/meter is commonly and conveniently used for the unit of $\mathcal{H}$. In the general case the added word "turns" has no significance and is therefore omitted.)

The electromagnetic field can be interpreted in one of two ways. It may be looked upon as a mathematically convenient step in the

[1] The symbol c is in common use for v_c, especially in optics. To avoid confusion with capacitance, it is not used here.

calculation of the interaction among electric charges. From this point of view, electric charges and the masses that are associated with them are treated as primary physical quantities and no particular physical significance is attached to the electromagnetic field. This interpretation will be followed in the present discussion. The second and alternative interpretation is to assume the electromagnetic field to be endowed with fundamental physical significance while electric charges are mere singularities in this field. In a similar way, masses may be regarded as singularities in a gravitational field. From the practical point of view, it is immaterial whether mathematical or physical significance is attached to the electromagnetic and gravitational fields. The present discussion is predicated throughout on the point of view that electric and magnetic fields are defined merely in order to facilitate the calculation of the distributions of charge and current and that these distributions are of primary interest in all electric circuits.

It has been stated that an important part of that fundamental problem of producing a desired effect is simultaneously preventing undesired effects. For example, in low-frequency circuits, this is accomplished by proper orientation of coils to reduce the coupling between them, or by shielding. The study of antennas is considerably simplified if first a clear understanding is obtained of just how shielding is accomplished in the light of the statement that all charges exert forces upon one another. From this point of view, it is necessary to reject the naïve belief that a metal shield protects the equipment within it from electrical disturbances in much the same way that a raincoat protects the wearer in a rainstorm. It must be assumed now that the outside currents that acted to produce undesired motions of charge in the conductors of a radio receiver before a shield was used must continue to act on those charges in the same way after the shield is put in place. The difference between the unshielded and the shielded state is not that the undesired electrical effects cannot penetrate the highly conducting shield, but that the currents that are maintained in the conducting shield itself are so distributed that their effects in producing motions of charge in the receiver are almost exactly equal and opposite to those produced by the more distant outside currents. Thus, an effective shield may be considered to provide an approximate cancellation of nearly equal and opposite effects, which are actually experienced by the charges in the wires of the receiver.

Although it may seem of little practical importance what reasoning is used to explain the shielding action of metal containers, there is much to be gained in understanding all electrical effects, in particular those related to antennas and microwave circuits, if the point of departure is always the fundamental postulate that all electrical effects are the result of moving charges acting to set other charges in motion. For example, the explanation of the readily verified fact that the currents in a coaxial line have an extremely small effect on other circuits placed near the line is not that the electrical effects cannot pass through the metallic outer conductor of the coaxial line, but that equal and opposite currents sufficiently close together always act in equal and opposite ways at outside points. This same explanation serves admirably for the two-wire line and the four-wire line, which likewise produce a very small effect on circuits placed near (but not too near) them. In open-wire lines, no metallic shields are involved.

2. The General Law of Electromagnetic Action.—A complete statement of the general law of force between moving electric charges is possible in moderately simple form only with the aid of advanced mathematical symbolism. However, the important consequences of this law can be summarized qualitatively as follows:

First, the force acting on the charges in an element of volume B (which may form part of an extended conductor belonging to a receiving antenna) due to a varying current in another small volume A (which may be part of a conductor in a transmitting antenna) decreases with increasing separation of the two elements A and B. This decrease does not obey a simple inverse-square law, as in electrostatics.

Second, the force acting on the element B at a given instant of time t_B is not determined by the distribution and motion of charge in element A at this same instant, but by the current in A at an earlier time t_A, which is given by $t_A = t_B - s_{AB}/v_c$. Here s_{AB} is the distance between the two elements, and v_c is the universal constant velocity defined in (1.3). *Electromagnetic action at a distance is not instantaneous; it is retarded.* It depends not only upon the distance between the two elements of volume that contain the charges that exert forces upon one another, but also upon the time. In effect, a certain length of time elapses before a given distribution of charge or current makes itself felt at a distant point. This suggests the convenient interpretation that electromagnetic effects due to small elements of moving charge appear to travel through space with a finite but high velocity. Since the force on the charges at B at time t_B is calculated in terms of the electromagnetic field at B due to the distribution of current and charge at A at the earlier time t_A, it is often convenient to look upon this electromagnetic field as a delivery service that rushes out from A with a package of force which it supplies to B after a time of travel

$$t_B - t_A = \frac{s_{AB}}{v_c} \tag{2.1}$$

If the electromagnetic field is given only mathematical significance, the entire picture of a propagation through space is part of the convenient but physically meaningless mathematical mechanism for calculating results that agree with experiment. If the field is assigned a physical reality, the difficulty of propagating something physically real through complete emptiness is overcome by inventing an electromagnetic ether to fill all space. The existence of such a medium has never been confirmed experimentally, and some of its properties would have to be physically absurd.

If the group of moving charges at A is separated from those at B by a homogeneous medium rather than by space, further complications are encountered. Since every medium, whether it be conducting or nonconducting, contains electric charges, the varying current at A acts upon all the charges contained in the medium, and the motions of these, in turn, exert forces upon the charges at B. If the conductors (of which the small volumes A and B are elements) are completely immersed in a homogeneous medium that extends far out beyond A and B in all directions, the net effect of the motions of charge in the medium is to reduce the apparent velocity with which electromagnetic effects are propagated from that observed in free space. If the medium is perfectly nonconducting, the apparent velocity is

$$v = \frac{v_c}{\sqrt{\mu_r \epsilon_r}} \tag{2.2}$$

where μ_r is the (dimensionless) relative permeability and ϵ_r is the (dimensionless) relative dielectric constant of the medium. For practically all nonconductors, μ_r equals 1; whereas, for common substances, ϵ_r may vary from small values such as 2.6 for polystyrene to 81 for water. If the medium is a good conductor of conductivity σ(1/ohm-m) or resistivity $\rho(=1/\sigma)$, then at frequency f

$$v(\text{m/sec}) = \sqrt{\rho f \cdot 10^7} \tag{2.3}$$

This leads to numerical values that are very small compared with v_c. Care must be taken not to confuse the apparent velocity v for electromagnetic effects transmitted entirely through an extended conducting medium with that for effects transmitted through space along (and only partly through) a good conductor.

If the attempt is made to discover the significance of a retarded rather than of an instantaneous action in terms of a periodic variation of current in London as observed in Boston, it is noted that an effect that is observed in Boston to vary in time as $\cos \omega t$, actually must be due to a current in London that varies as $\cos \omega(t - s/v_c)$. Here s is the distance between Boston and London. If numerical values are inserted, a time difference of approximately $\frac{1}{60}$ sec results between the time t when the effect on the current in a receiver is observed in Boston and the time $(t - s/v_c)$ when the current varied in the transmitter in London. From the point of view of the listener in Boston, it seems quite unimportant whether the signal received at a given instant is due to a current in London at that instant or $\frac{1}{60}$ sec earlier.

The real significance of a retarded action may be understood from the following example: Suppose two long parallel wires carrying equal currents at the same frequency in the same direction and in the same phase are separated by a distance s in air such that s/v_c is exactly one-half period. Then the force acting on a small element of conductor A due to the interaction of the current in this element with the current in a similar small element of conductor B would be one of attraction if ordinary circuit theory with quasi-instantaneous action were applicable. Actually, the force on the element of current A due to the current in element B must be calculated from the current in B at the earlier time $t - s/v_c$. If s/v_c is one-half period for the two elements under consideration, the current in B at the time $(t - \frac{1}{2}$ period) will have been actually in the opposite direction from that in A at time t. Hence the interaction will be one of repulsion rather than of attraction. Accordingly, in this particular case, the incorrect use of ordinary circuit theory would give the effect exactly opposite from that of electromagnetic theory. It is clear, therefore, that the time of retardation is of fundamental importance in the interaction of currents in circuits that are sufficiently extensive. Correct results cannot possibly be obtained in such circuits from conventional electric-circuit theory.

3. Special Case of the Near Zone.—A great simplification is possible in the general law of electromagnetic action if the following condition is satisfied:

$$\frac{\omega s_{max}}{v} = \frac{2\pi s_{max}}{\lambda} << 1 \tag{3.1}$$

where s_{max} is the maximum separation of currents or charges in the circuit or circuits that *exert a significant and uncanceled effect* on one another, v is the velocity characteristic of the medium between the currents or charges, and λ is a constant called the *wavelength* and is defined by

$$\lambda = \frac{v}{f} = \frac{2\pi v}{\omega} \tag{3.2}$$

The inequality (3.1) is called the *condition for the near or induction zone*, or for the quasi-stationary state. Whenever it is satisfied, the complicated general laws of electromagnetism reduce to the simple laws of electrostatics, magnetostatics, and direct currents or low-frequency alternating currents, *i.e.*, whenever the inequality is satisfied by a circuit or network, Coulomb's and Ampère's inverse-square laws, Kirchhoff's laws, and the network theorems that depend upon them are good approximations.

The inequality defining the conditions under which electric-circuit theory is valid may be interpreted in several ways. If it is looked upon as a condition limiting the maximum dimensions of a circuit operated at a given frequency, it specifies the near or induction zone for that circuit. If the dimensions of the circuit are considered fixed and the limiting frequency is defined by the inequality, all frequencies below this limit constitute the quasi-stationary state. It is to be noted that, if the condition for the near zone is actually satisfied by a given circuit operated at a definite frequency, the quantity $\omega s/v$ is always negligible so that it may be set equal to zero. But this is mathematically equivalent to allowing the velocity v to become infinite. Physically, v is a constant that cannot be made infinite at will, but the simplified equations resulting from the fulfilling of the condition for the near zone are actually the same as those which would obtain if v could be made infinite. Consequently, the near zone is sometimes referred to as the zone in which an infinite characteristic electromagnetic velocity may be assumed.

In comparing the nature of the restriction at different frequencies on the maximum distance between parts of circuits that are to be in the near zone with respect to each other, Table 3.1 is instructive. It gives numerical values for the extreme distance s_{max} as calculated from

$$s_{max} \gtreqless 0.01 \frac{\lambda}{2\pi} \tag{3.3}$$

The double inequality of the original condition is taken to be equivalent to requiring s_{max} to be equal to or less than 1 per cent of $\lambda/2\pi$.

TABLE 3.1

$\omega = 2\pi f$	λ			s_{max}		
	Air	Water	Copper	Air	Water	Copper
Audio 10^3	1,885 km	209 km	32.4 mm	3 km	0.33 km	0.05 mm
Radio 10^6	1.885 km	0.209 km	1.02 mm	3 m	0.33 m	$1.6 \cdot 10^{-3}$ mm
Ultra-high 10^9	1.885 m	0.209 m	0.0324 mm	3 mm	0.33 mm	$5 \cdot 10^{-5}$ mm
Super 10^{12}	1.885 mm	0.209 mm	$1.02 \cdot 10^{-3}$ mm	3 microns	0.33 micron	$1.6 \cdot 10^{-6}$ mm

Examination of the figures for air in Table 3.1 reveals that, at audio frequencies, all networks in practical use are well within the near zone, since none is commonly used that extends over distances of kilometers. At radio frequencies, most circuits except antennas and long transmission lines satisfy the near-zone condition, which for $\omega = 10^6$ requires s_{max} to be equal to or less than 3 m. At the ultra-high frequency for which $\omega = 10^9$, only extremely small circuit elements such as the interelectrode spaces in small vacuum tubes are short enough to be measured in millimeters. At still higher frequencies, practically no circuit parts of any kind fulfill the condition for the near zone. For most good dielectrics, s_{max} lies between the values for air and distilled water. The column for copper shows that, even at a frequency for which $\omega = 10^3$, s_{max} is extremely small. At audio frequencies, the near-zone condition is satisfied only for the cross sections of fine wire. At all higher frequencies, it is not fulfilled even for fine wire. It is for this reason that ordinary d-c theory of parallel circuits does not apply to parallel filaments in a conductor when used for alternating currents except at the very lowest frequencies. In determining both the distribution of current in wires and the a-c resistance of wires, electromagnetic theory must be used. It leads to the well-known formulas for skin effect.

It may not be obvious immediately why a long transmission line can be analyzed using ordinary electric-circuit theory even though its length may be many wavelengths, so that the near-zone condition is apparently not satisfied. The answer is found in the fact that elements of current that *exert a significant and uncanceled effect* on each other lie in the near zone. If the two conductors of the line carry equal and opposite currents sufficiently close together, all the forces acting on the charges in the element B in Fig. 3.1 due to the currents and charges in the remainder of the long line are negligible compared with the forces due to adjacent elements. Thus the force on charges at B due to the current at A is practically canceled by the nearly equal and opposite force at B due to the current at A'. The farther from B the two elements A and A' are, the more complete is the cancellation of their effects at B. Accordingly, if the separation b satisfies the condition for the near zone, all the currents that contribute significant and uncanceled effects at B must lie practically within the near zone from B. Thus if the condition

FIG. 3.1.—Section of a two-wire line.

$$b <<< \lambda \tag{3.4}$$

is satisfied, ordinary circuit analysis including Kirchhoff's laws may be applied to an element of the line because the current in the far-off parts of the line have a negligible effect on this element. If this inequality is not satisfied, ordinary transmission-line theory loses its validity, as explained in Chap. I.

4. Special Case of the Far Zone.—Two circuits are said to be in the far zone or radiation zone with respect to each other if the following inequalities are satisfied:

$$\frac{\omega s_{\min}}{v} = \frac{2\pi s_{\min}}{\lambda} >> 1 \tag{4.1}$$

$$s_{\min} >> 2h \tag{4.2}$$

where $s_{\min}$ is the shortest distance between any parts of the two circuits and $2h$ is the maximum extension of either circuit. The most important circuits that can be made to lie in the far zone with respect to each other are transmitting and receiving antennas. If these are many wavelengths apart, they easily satisfy the inequalities (4.1,2). The complicated general law of retarded electromagnetic action at a distance again simplifies, but in an entirely different way from that described for the near zone. In this case, there are no inverse-square laws for electric and magnetic fields but inverse-first-power laws, *i.e.*, the magnitude of the electromagnetic field diminishes as $1/s$, where s is the distance from the moving charges at A to the point B where the field is calculated.

A further simplification in the far zone is that the electromagnetic field is characterized by surfaces of constant phase which are great spheres about the center of the circuit at A. These spheres expand with a constant radial velocity which in space has the value $v_c = 3 \cdot 10^8$ m/sec. Successive surfaces that are in the same phase at any instant are all separated by radial differences which are integral multiples of λ. Thus the electromagnetic field calculated at all points in the far zone in space from the currents in a circuit of any shape at A may be described in terms of spherical surfaces about A on which the electromagnetic vectors are both in the same phase. Any given phase travels radially outward with a constant velocity v_c. *Such expanding spherical surfaces of constant phase of the electromagnetic vectors occurring at radial differences of* λ *are called electromagnetic waves.*

It is to be noted that it is the phase of the electric and magnetic vectors which is the same at all points on a given great sphere at any instant and not the amplitude. The amplitude is a function of the spherical coordinates θ, Φ, R, whereas the phase depends on R alone. However, at sufficiently great distances and over small arcs, the spherical surfaces may be considered approximately plane, and the variation in amplitude with θ and Φ may be neglected. Thus the electromagnetic field due to the current in a transmitting antenna at points near a sufficiently distant antenna which is itself short compared with this distance is characterized by practically plane surfaces of constant phase and amplitude, and these are called plane waves.

5. **General Case.**—The electromagnetic field in the near zone is characterized by an inverse-square law for amplitude and by quasi-instantaneous action. It is called the induction field. The field in the far zone is characterized by an inverse-first-power law

and by spherical waves expanding with a constant radial velocity. It is called the radiation field. It is important to bear in mind that the field in the intermediate zone, where neither the near- nor the far-zone condition is satisfied, is in general not just a superposition of an induction and a radiation field. Two mutually contradictory interpretations cannot be combined and continue to make sense. It has been shown that the induction field is equivalent to assuming an infinite velocity of propagation for electromagnetic action. The radiation zone, on the other hand, involves a finite velocity v_c. The two are mutually exclusive. The electromagnetic field in the intermediate zone cannot be partly an induction field with instantaneous action, and partly a radiation field with retarded action. It is a single field that obeys a complicated law that involves neither instantaneous action with an inverse-square law nor an inverse-first-power law with spherical surfaces of constant phase expanding with constant radial velocity. It is described in detail in Sec. 35 for a symmetrical center-driven antenna for which it assumes a simple form. In general, it does not lend itself to simple interpretation.

6. Closed and Open Circuits.—From the derivation of the word circuit, an electric circuit might be expected to consist of a closed conducting path around which a stream of electrons may flow. A circuit of this type is easily arranged in a variety of forms for use at any frequency. It consists of one or more closed loops of conductor, parts of which may be made of wires wound into coils or resistors. Or it may be a long transmission line with conducting terminations. If the circuit is sufficiently small in its extension so that all parts lie in the near zone with respect to one another, the amplitude of the current will be the same at every cross section in each conductor between branch points. If the circuit extends well beyond the near zone, as in a long transmission line, the instantaneous amplitude of current is not the same at all points along a conductor. Standing or traveling waves of electric current and charge must be expected.

If a circuit is broken by an air gap, it no longer provides a closed path for electrons. Nevertheless, alternating currents are possible, and since it is the flow of electric charges that is important, not the closed conducting path, the word "circuit" is taken to mean any conducting path along which electric charges can be set in motion, not necessarily a closed path.

If the gap in the conducting path is between two circular plates (of a capacitor, for example) separated a very short distance d

compared with their radius b, and if b in turn is very small compared with $\lambda/2\pi$, or

$$d << b << \frac{\lambda}{2\pi} \tag{6.1}$$

an alternating current maintained in the circuit charges the surfaces of the plates, one negative the other positive and vice versa as the current reverses. The charge density is uniform on the adjacent surfaces of the plates, and if the entire circuit is in the near zone there are no other points where significant concentrations of charge accumulate. Moreover, the current has the same amplitude at every cross section of the conductor around the series circuit; it vanishes in the air gap between the plates. Since the electric field at all points between the plates is defined in terms of the density of electric charge on the adjacent surfaces of the plates, the time rate of change of the electric field is proportional to the time rate of change of the electric charge, and this in turn is equal in magnitude to the current in the outer circuit. A fictitious current, called the "displacement current," is often defined in terms of the time rate of change of the electric field between the plates. It is a meaningful quantity only when the conditions (6.1) are satisfied, when it permits stating that a "current" exists completely around the circuit. In this case, the circuit is called quasi-closed, and the word "current" is not restricted to moving electric charges. This terminology is not used in this chapter; the word current is reserved for moving charges.

At radio and higher frequencies, the current is in a thin layer along the surface of the conductor (including the capacitor plates), *i.e.*, instead of flowing through the plates of the capacitor in order to charge their adjacent surfaces, the charges move radially outward along the outer surfaces, around the edges of the plates, and finally radially inward. Charge is deposited uniformly on the inner surface only if (6.1) is fulfilled. If, for example,

$$d << b \qquad \text{and} \qquad b > \frac{\lambda}{8} \tag{6.2}$$

the charge is not deposited with uniform density. Radial standing waves of electric charge are formed consisting of alternate rings of positive and negative charge a quarter period out of phase in time with the radial currents in opposite directions on the plates between them. If these plates are separated more and more so that the distance d between them becomes equal to or greater than the radius b, current can be maintained in the circuit if the frequency is raised, and electric charge is still deposited on the plates though not on their adjacent sides alone. If the size of the plates is reduced to zero and the connecting circuit is straightened into a single conductor about $\lambda/2$ in length with the generator (here assumed to be impedanceless for simplicity) connected at its center, Fig. 6.1, current persists in the generator and in the adjacent parts of the conductor. Instead of depositing electric charge on the plates of a capacitor, the current now leaves it on the *surface* of the conductor itself, principally near its ends.

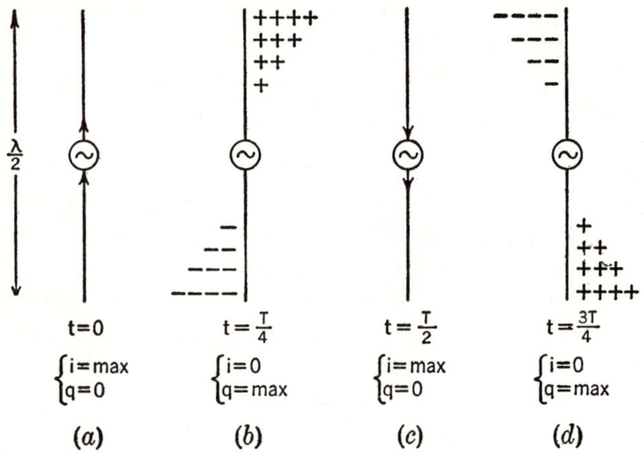

FIG. 6.1.—Current and charge in a center-driven antenna at four instants in a cycle.

Several instants in a complete cycle are illustrated in Fig. 6.1. In Fig. 6.1*a*, the current through the generator is maximum in the upward direction; no charge has accumulated anywhere. A quarter cycle later, Fig. 6.1*b*, the upward current has been reduced everywhere to zero. A maximum positive charge has been deposited along the upper end of the conductor and a maximum negative charge along the lower end. A full half cycle after (*a*) the current in the generator has increased to a maximum value downward in (*c*), while all accumulations of charge have been neutralized. A quarter cycle later in (*d*) the current is again zero everywhere; positive charge has accumulated along the lower end and negative charge along the upper end.

At first thought, it may appear quite contrary to the elementary law of force between electric charges (like charges repel, unlike charges attract one another) to have unlike charges separated as far as possible on the conductor. On capacitor plates (usually with very small separation), unlike charges are drawn as close together as possible on the opposite surfaces. Similarly, along a resonant transmission line, an accumulation of charge at a current node on one wire is always accompanied by a corresponding accumulation of charge of the opposite sign on the adjacent second wire. In both cases, unlike charges are drawn together, whereas on the antenna they appear to be driven apart. This apparent contradiction disappears if account is taken of the fact that the capacitor plates and the opposite sections of a transmission line are required to be in the near zone, whereas the ends of the antenna may *not* be in the near zone. Thus Coulomb's law with a quasi-instantaneous action applies to the charges on the plates of this capacitor or on the adjacent parts of the two-wire line.

For the antenna, on the other hand, the action between elements of charge near the opposite ends is definitely and significantly retarded. In fact, the negative charge at the extreme top in (*d*) is actually experiencing the repulsion due to the negative charge which *was* at the extreme bottom a half cycle *earlier* in (*b*), and not an attraction due to the positive charge which is at the lower end at this instant (*d*). This positive charge has no effect whatever on the negative charge that is at the top at this instant, because it takes a full half period before the effect can be experienced at any point a half wavelength distant. When the effect of the positive charge is felt at the top, the top is charged positively. Thus, without attempting to go into the details of determining the force experienced at a given instant by any particular element of charge due to all the charges in the antenna at appropriate earlier times, it is clear that the law of attraction and repulsion actually is fulfilled and is not violated if account is taken of retardation.

When the condition for the near zone is not satisfied, there may be currents in conductors that are closed or even quasi-closed with a capacitor as a gap. The amplitude of the current is not the same at different points along a conductor, because electric charge is deposited all along the surface of the conductor. Superficially it may appear that an antenna consisting of a straight conductor that is an appreciable fraction of a wavelength long and with a generator at its center may be looked upon simply as an open-end transmission line with the parallel conductors bent to lie along the same axis instead of being parallel to each other. Although there is considerable similarity between the two cases from the point of view of the approximate distribution of current, *they are nevertheless fundamentally different.* The transmission line may be analyzed to a good approximation in terms of ordinary electric-circuit theory, because equal and opposite currents are very close together. This is not true of the antenna and ordinary electric-circuit theory cannot be applied. *It is fundamentally incorrect to treat a center-driven antenna as though it were the bent-open ends of a two-wire line.*

Circuits that satisfy the condition for the near zone, either because they are sufficiently small or because they have equal and opposite currents everywhere so close together that the currents in widely separated parts of the circuit exert a negligible effect on one another, are analyzed correctly by the methods of ordinary electric-circuit theory. All other circuits must be investigated in terms of electromagnetism. This nearly always involves a study of the electromagnetic field as a useful intermediate step in determining distributions of current and charge.

II. THE DRIVEN ANTENNA AS A CIRCUIT ELEMENT

7. Properties of an Antenna.—An effective antenna is always an electric circuit that does not satisfy the conditions for the near zone. Its length is usually comparable with the wavelength associated with the driving frequency and it is designed so that equal and opposite currents are not too close together. The simplest and most common types of antenna are resonant open circuits (important exceptions are the loop and the rhombic); the distributions of current and of charge are nonuniform; current and charge have maximum values that are a quarter period out of phase and are displaced by about $\lambda/4$ along the antenna. Inductance and capacitance as used for near-zone circuits with uniform current cannot be defined, and ordinary circuit analysis does not apply. Nevertheless, a simple antenna that is driven either by a generator that is part of a conventional near-zone circuit or by a transmission line that in turn is driven by such a generator always has two input terminals. Since these are connected to a near-zone circuit, they must be very close together compared with the wavelength.

From the point of view of this driving circuit, a symmetrical antenna presents a definite impedance just like any other circuit element. This impedance may be defined as the ratio of the applied voltage at the terminals to the input current. Unsymmetrical antennas present certain difficulties considered later. If the terminals of what appears to be the antenna are not in the near zone, a part of the remaining circuit leading to the generator must be a part of the antenna, and other terminals nearer the generator that do lie within the near zone must be used in defining the impedance. The impedance of an antenna is a complicated function of its shape, of the frequency, and of its proximity to other conductors, including the earth. With symmetrical antennas, the impedance may be measured in conventional ways using bridge or substitution methods at radio frequencies and transmission-line methods at ultra-high frequencies. The calculation of the impedance of antennas is even more difficult than that of the impedance of coils. It has been accomplished approximately for some of the simpler configurations; qualitative and in some cases quantitative generalization is possible for other types. Since the determination of impedance reduces in effect to the calculation of the input current, a study of the distribution of current along an antenna is a prerequisite.

A simple and important type of antenna is a symmetrical, center-driven, straight conductor of small circular cross section. The distribution of current along an antenna of this type has been analyzed in the form of a series involving powers of the small quantity $1/\Psi$ where Ψ is in Fig. 10.2. When $h < \lambda/4$, $\Psi = \Omega - 2$ where

$$\Omega = 2 \ln \frac{2h}{a} \tag{7.1}$$

h is the half length of the antenna, and a is the radius. Fortunately the leading term in the series is very simple in form and for many purposes is entirely adequate. Actually this leading term is the hypothetical distribution of current along a perfectly conducting antenna of vanishingly small radius. Although such an antenna is physically unrealizable, it may be approximated by a copper antenna of extremely small radius compared with its length. It is simpler to consider first the distribution of current along such an antenna and then to describe the departure from this idealized distribution for an actual antenna.

8. Leading Term in the Distribution of Current and of Charge along a Center-driven, Highly Conducting Antenna of Extremely Small Radius.—If the radius of the cylindrical conductor is extremely small compared with its half length h, the instantaneous values of current and charge per unit length may be obtained to a fair approximation from the following expressions provided the antenna is no longer than a quarter wavelength in half length. Assume that the applied voltage across the terminals is either the real or the imaginary part of

$$v_0 = \hat{V}_0 \epsilon^{j\omega t} \tag{8.1}$$

and that

$$i_z = \hat{I}_z \epsilon^{j\omega t} \quad \text{and} \quad q_z = \hat{Q}_z \epsilon^{j\omega t} \tag{8.2}$$

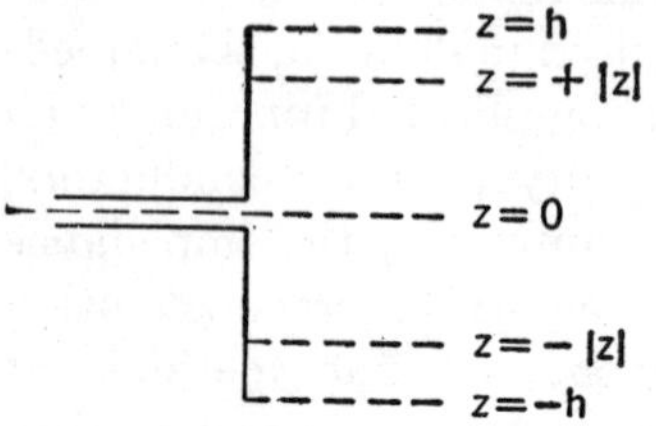

FIG. 8.1.—Location of points along the antenna in terms of the z coordinate.

where i_z is the complex instantaneous current and q_z is the complex instantaneous charge per unit length. The coordinate z is measured along the length of the antenna from the input terminals, which are very close together at the center of the antenna. Positive values of z are in the upward direction, negative values in the downward direction, Fig. 8.1. In (8.2)

$$\hat{I}_z = j \frac{2\pi \hat{V}_0}{\mathcal{R}_c \Psi} \frac{\sin \beta(h - |z|)}{\cos \beta h} \tag{8.3}$$

$$\hat{Q}_z = \pm \frac{2\pi \epsilon_0 \hat{V}_0}{\Psi} \frac{\cos \beta(h - |z|)}{\cos \beta h} \tag{8.4}$$

The following symbols are used:

$$\beta = \frac{2\pi}{\lambda} = \frac{\omega}{v_c} \tag{8.5}$$

$$\Psi = 2 \ln \frac{2h}{a} - 2 \tag{8.6}$$

Note that

$$\frac{\mathcal{R}_c}{2\pi} \doteq 60 \text{ ohms} \tag{8.7}$$

The plus sign in (8.4) applies to points in the upper half of the antenna where z is positive, the negative sign to the lower half where z is negative. The positive direction of the current is taken to be upward. Note that (8.3) is an approximation of only the reactive part of $\hat{I}_z$. The actual current is finite when $\beta h = \pi/2$.

It is often convenient to define the input value and the maximum value of the current amplitude. The input value $\hat{I}_0$ occurs at $z = 0$; $\hat{I}_{max}$ occurs at $z = \pm(h - \lambda/4)$.

$$\hat{I}_0 = j \frac{2\pi \hat{V}_0}{\mathcal{R}_c \Psi \cot \beta h} \tag{8.8}$$

$$\hat{I}_{\max} = \frac{\hat{I}_0}{\sin \beta h} \tag{8.9}$$

$\hat{I}_{\max}$ is always a fictitious current for antennas that have h less than $\lambda/4$.

The maximum amplitude of the charge per unit length occurs at the ends of the antenna where $z = \pm h$. It is

$$\hat{Q}_{\max} = \pm \frac{2\pi\epsilon_0 \hat{V}_0}{\Psi' \cos \beta h} \tag{8.10}$$

In terms of (8.8) to (8.10), (8.3) and (8.4) may be written

$$\hat{I}_z = \hat{I}_0 \frac{\sin \beta(h - |z|)}{\sin \beta h} = \hat{I}_{\max} \sin \beta(h - |z|) \tag{8.11}$$

$$\hat{Q}_z = \hat{Q}_{\max} \cos \beta(h - |z|) \tag{8.12}$$

If the imaginary part of (8.1) is chosen to represent the applied voltage with $\hat{V}_0$ real,

$$v_0 = \hat{V}_0 \sin \omega t \tag{8.13}$$

$$i_z = |\hat{I}_z| \cos \omega t = |\hat{I}_{\max}| \sin \beta(h - |z|) \cos \omega t \tag{8.14}$$

$$q_z = \hat{Q}_z \sin \omega t = \hat{Q}_{\max} \cos \beta(h - |z|) \sin \omega t \tag{8.15}$$

The distributions of current and charge as defined are indicated qualitatively in Fig. 6.1 for a so-called "half-wave dipole" with

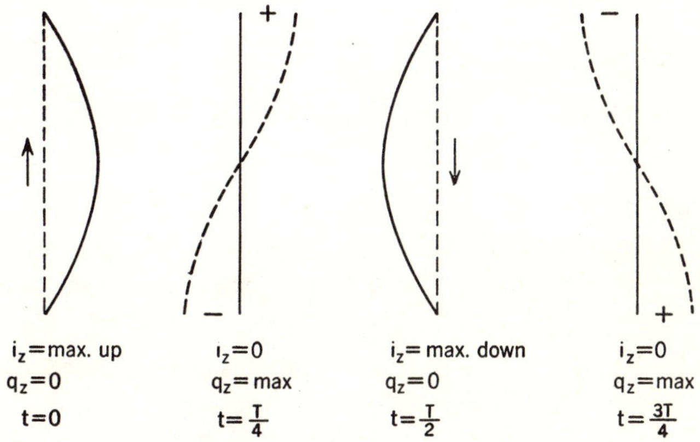

FIG. 8.2.—Sinusoidal distributions of current; cosinusoidal distributions of charge in a half-wave antenna at four instants in a cycle.

over-all length $2h = \lambda/2$. Corresponding diagrams in which the sinusoidal distributions are shown explicitly are reproduced in Fig. 8.2. In most cases, it is adequate to show only the distribution of current along a resonant antenna at the instant when the current has a maximum value in time. Diagrams for the distribution of maximum current and of maximum charge (occurring a quarter cycle later) for antennas of different lengths are shown in Fig. 8.3. For a center-driven antenna the current always vanishes at the ends where the charge per unit length always has its maximum value. The distributions are quite different along a receiving antenna, as will be described later.

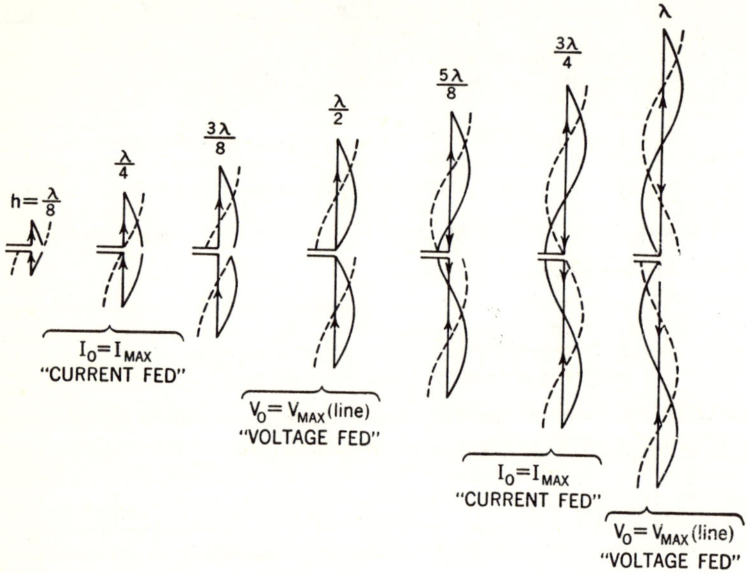

FIG. 8.3.—Sinusoidal distribution of maximum current (solid line) and cosinusoidal distribution of maximum charge (broken line) in infinitely thin center-driven antennas of different lengths $2h$.

9. Distribution of Current along a Symmetrical Antenna of Small Radius.—The distribution of current along a cylindrical antenna of small radius cannot be written in simple analytical form. The complex amplitude $\hat{I}_z$ of the current at any point z along the antenna may be expressed as

$$j\hat{I}_z' + \hat{I}_z'' = \frac{j2\pi \hat{V}_0}{\mathcal{R}_c \Psi D} [f'(z) - jf''(z)] \tag{9.1}$$

Here D is a real function of h and a but not of z; $f'(z)$ and $f''(z)$ are real functions of z. The instantaneous value of current corresponding to the voltage (8.1) is

$$i_z = \frac{2\pi \hat{V}_0}{\mathcal{R}_c \Psi D} [f'(z) \cos \omega t + f''(z) \sin \omega t] \tag{9.2}$$

the parameter Ψ is a function of the length and radius of the antenna and the frequency. It is shown in Fig. 10.3. Even when the radius is quite small, $\hat{I}_z'$ differs considerably from the sinusoidal form, and $\hat{I}_z''$ is significant. This is illustrated in Fig. 9.1 for an antenna with $h = \lambda/4$ and in Fig. 9.3 for an antenna with $h = \lambda/2$ for both a thin antenna ($\Omega = 20$) and a relatively thick antenna ($\Omega = 10$). The quantities $\hat{I}_z'/\hat{V}_0$ and $\hat{I}_z''/\hat{V}_0$ from (9.1) are plotted. Note that, in Fig. 9.1, the component $\hat{I}_z''$ is larger

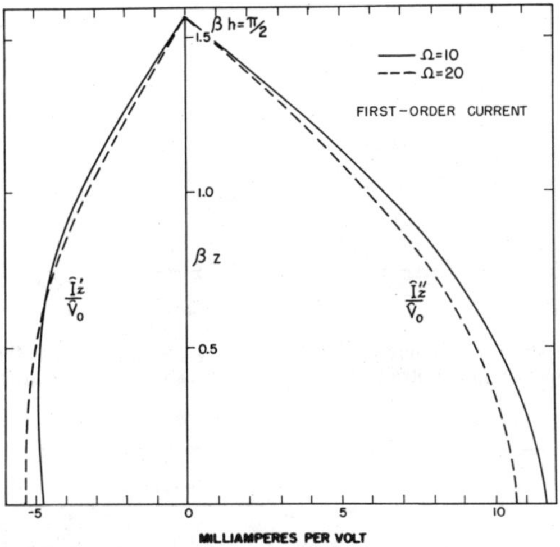

FIG. 9.1.—Components $\hat{I}_z'$ and $\hat{I}_z''$ in the complex amplitude of the current given in (9.1) for the upper half of a center-driven antenna; the normalized form $\hat{I}/\hat{V}_0$ is shown. $\Omega = 20$ corresponds to $a/h = 9.1 \times 10^{-5}$; $\Omega = 10$ corresponds to $a/h = 1.35 \times 10^{-2}$. $\beta h = \pi/2$ or $h = \lambda/4$.

than the component $\hat{I}_z'$. The amplitude of the total current at any point z along the antenna is

$$|\hat{I}_z| = \sqrt{(\hat{I}_z')^2 + (\hat{I}_z'')^2} = \frac{2\pi \hat{V}_0}{\Re_c \Psi D} \sqrt{[f'(z)]^2 + [f''(z)^2]} \tag{9.3}$$

The phase angle of $\hat{I}_z$ referred to $\hat{V}_0$ is

$$\theta = \tan^{-1}\left(\frac{\hat{I}_z'}{\hat{I}_z''}\right) = \tan^{-1}\left[\frac{f'(z)}{f''(z)}\right] \tag{9.4}$$

The function $|\hat{I}_z|/\hat{V}_0$ is shown plotted in Figs. 9.2 and 9.4 for the two cases represented in Figs. 9.1 and 9.3, respectively. The angle θ is given in Figs. 9.2 and 9.4. It is significant that, for the antenna with $h = \lambda/4$, Fig. 9.2, $|\hat{I}_z|$ is practically a cosine

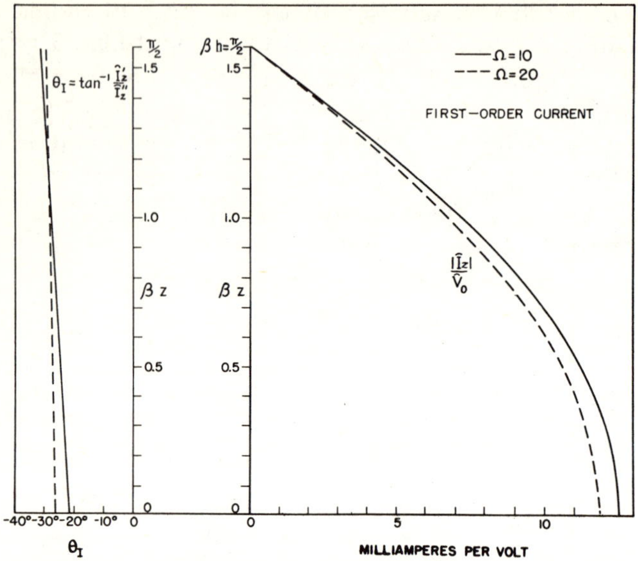

FIG. 9.2.—Phase angle θ of current $\hat{I}_z$ referred to $\hat{V}_0$ as defined in (9.4); amplitude of current per unit voltage for same antenna represented in Fig. 9.1.

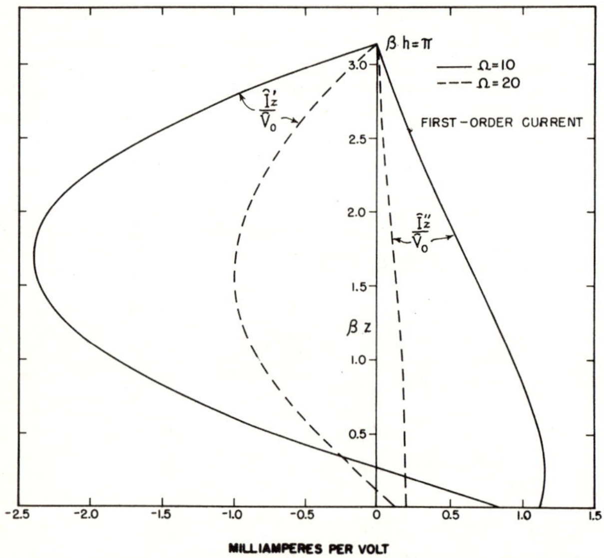

FIG. 9.3.—Like Fig. 9.1 for an antenna of greater length.

curve and θ is fairly constant at about $-30°$. Since the curve for $\Omega = 10$ in Fig. 9.1 differs only slightly from the sine curve, the

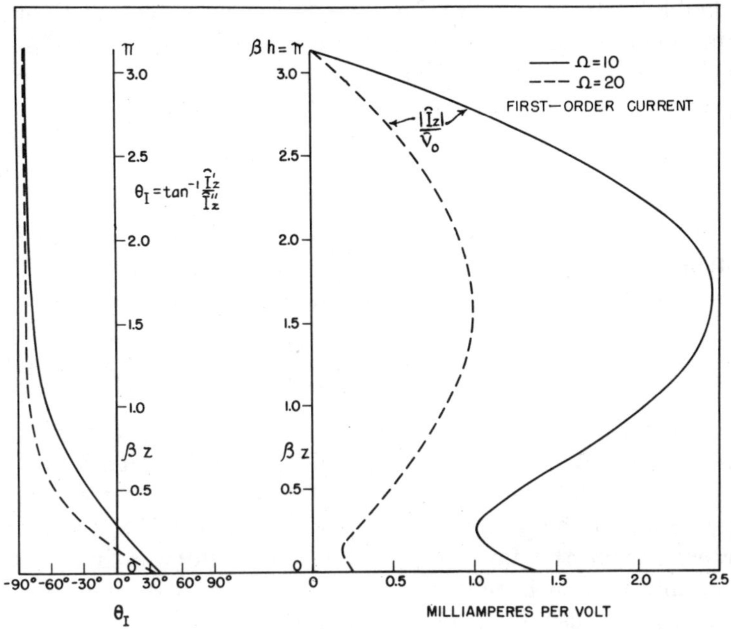

FIG. 9.4.—Like Fig. 9.2 for an antenna of greater length.

distribution of current in antennas with $h = \lambda/4$ and with a wide range of radii can be represented quite accurately by

$$\hat{I}_z = \hat{I}_0 \cos \beta z \tag{9.5}$$

For half lengths that do not greatly exceed $\lambda/4$ or are less than $\lambda/4$, the formula

$$\hat{I}_z = \hat{I}_0 \sin \beta(h - |z|) \tag{9.6}$$

is a reasonably good approximation even for moderately thick antennas if $\hat{I}_0$ is the complex input current defined by

$$\hat{I}_0 = \frac{\hat{V}_0}{Z_0} \tag{9.7}$$

where Z_0 is the input self-impedance defined in Sec. 10. For half lengths that considerably exceed $\lambda/4$, (9.6) is only a rough approximation even for relatively thin antennas.

10. Input Self-impedance of Symmetrical Center-driven Antenna of Small Radius.—The input self-impedance of a symmetrical center-driven antenna may be defined in the same way as any other impedance, provided that the terminals A and B are separated by a negligible distance compared with $\lambda/2\pi$. The potential difference maintained across the terminals by an external source is $\hat{V}_{AB}$; the complex amplitude of current entering the one and leaving the other terminal is the same if the antenna is symmetrical, or

$$\hat{I}_A = \hat{I}_B \tag{10.1}$$

Since the terminals A and B are assumed to be very close together, so that each is very nearly at the origin ($z = 0$) of a coordinate system with the z axis lying along the axis of the antenna, $\hat{I}_0$ can be written for $\hat{I}_A$ or $\hat{I}_B$, $\hat{V}_0$ for $\hat{V}_{AB}$, Z_0 for Z_{AB}.* Then

$$Z_0 = \frac{\hat{V}_0}{\hat{I}_0} \tag{10.2}$$

The function Z_{AB} or Z_0 defines the input self-impedance of the antenna if it is so far from other conductors or dielectrics that any rearrangement of these has a negligible effect on the current at every point in the antenna. This is always true if all other conductors are in the far zone of the driven antenna in question. When an antenna is center-driven from a two-wire line, its input terminals are usually separated by the spacing b of the line. Although b is assumed to be a very small fraction of a wavelength ($2\pi b/\lambda << 1$), it is not negligible. In fact the question at once arises from what point is the half length h measured? In the analysis for the self-impedance, it is assumed that the spacing of the input terminals is zero, and the half length h is one-half the distance from end to end of the antenna. This condition is approximated in practice, as in Fig. 10.1. Here an idealized line of zero length and zero spacing joins the actual line and the antenna. The combined fields of the currents in line and antenna, Fig. 10.1*b*, are the same as those of Fig. 10.1*a* because the field due to the small added element of current of length b is equal and opposite to that of the terminating length b of the line. The circuit of Fig. 10.1*b* is approximately the one analyzed. It is nearly equivalent to an antenna with a concentrated generator at its center and a line with a concentrated load at the center of a straight-wire termination. Practical arrangements may approximate it more or less closely, so that the computed results are a more or less good repre-

* In dealing with coupled antennas in Sec. 18, the symbol Z_{sm} will be used for the self-impedance of antenna m in the presence of other antennas.

sentation of an actual circuit. Note that in a thick antenna of half length h near $\lambda/2$, an appreciable concentration of charge exists on the adjacent end surfaces of the antenna halves if these are close together; this is not considered in the theory. Accordingly, the agreement between theoretical and experimental values should not be expected to be particularly good for thick, voltage-fed antennas unless an appropriate capacitance across the terminals is included. If an appreciable gap exists between the terminals as in Fig. 10.1a, the effect is roughly that of a considerable reduction in the effective capacitance of adjacent sections of the antenna as compared with the arrangement in Fig. 10.1b. Therefore

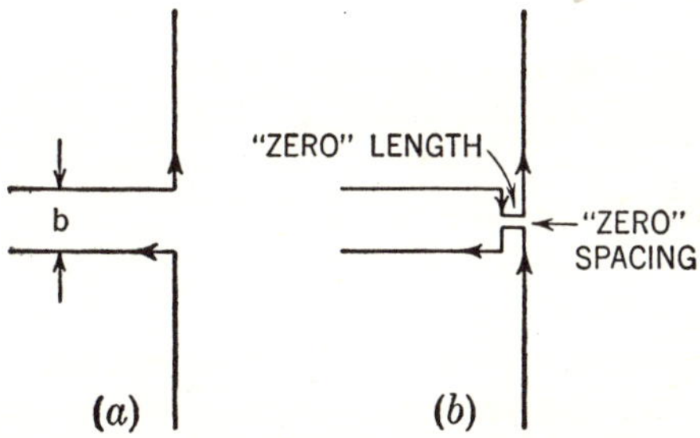

FIG. 10.1.—(*a*) Antenna driven from a two-wire line; (*b*) approximately equivalent circuit.

experimentally determined impedance obtained using the arrangement of Fig. 10.1a should agree better with the theory if a small capacitance is connected across the terminals. In any case, a lack of agreement is not necessarily due to errors in either theory or experiment but to the fact that they deal with *different problems*. Suitable terminal-zone networks of lumped elements can be designed and connected between the terminals of an ideal antenna and an ideal line to take account of end and coupling effects.

The self-impedance defined in (10.2) is by no means independent of the surrounding universe in the sense that all the power supplied at the terminals is dissipated in heating the antenna, as would be the case for an inductance coil of the ordinary low-frequency type if its self-impedance were defined in the same way. This subject is expanded later.

The input self-impedance of a cylindrical and symmetrical center-driven antenna depends upon the radius a of the antenna as well as upon the half length h, and in general is also a function of the conductivity σ of the antenna. This last fact increases the difficulty of the evaluation of the input impedance, because the terms involving σ contain the frequency under a radical. This makes it impossible to compute the input impedance of antennas in general in terms of a/λ and h/λ; each antenna must be treated individually at every frequency. Fortunately, the contributions to the impedance by all terms involving σ are small enough to be negligible for practical purposes if the conductivity

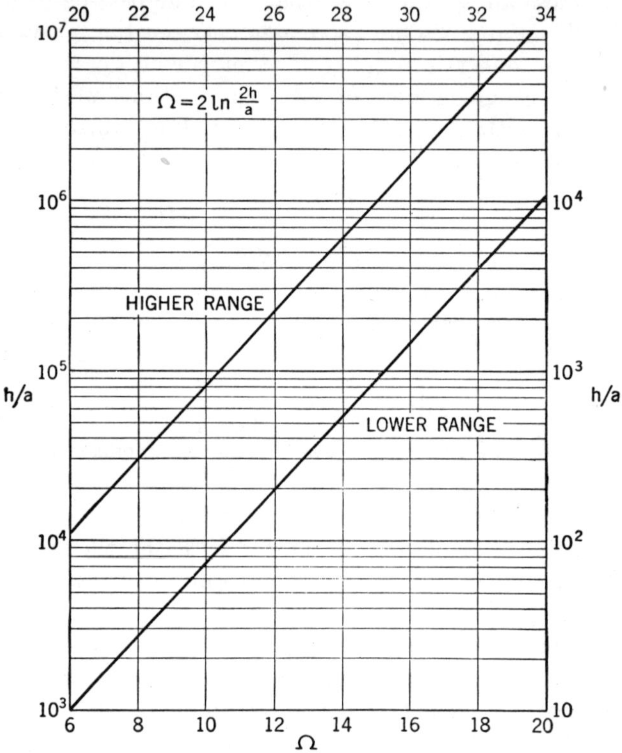

FIG. 10.2.—Parameter $\Omega = 2 \ln \frac{2h}{a}$.

is high, as for copper. Accordingly, these terms may be neglected for highly conducting antennas such as those of copper, and general curves in terms of Ω and $H = \beta h = 2\pi h/\lambda$ may be obtained.

For some purposes, the variable $\Omega = 2 \ln \dfrac{2h}{a}$ is convenient and has been used in the previous discussion. It is plotted as a function of h/a in Fig. 10.2. The related parameter Ψ is in Fig. 10.3.

A general but *approximate* formula for the input self-impedance of a cylindrical center-driven antenna that satisfies the condition

$\Omega^2 >> 1$ is

$$Z_0 = -j\,\frac{\mathcal{R}_c \Psi}{2\pi} \left[\frac{\cos H + (1/\Psi)A_1(H) + (1/\Psi)^2 A_2(H)}{\sin H + (1/\Psi)B_1(H) + (1/\Psi)^2 B_2(H)}\right] \tag{10.3}$$

where $A(H)$ and $B(H)$ are complex functions[1] of H.

General curves for the input resistance and reactance of highly conducting antennas are reproduced in the accompanying figures.

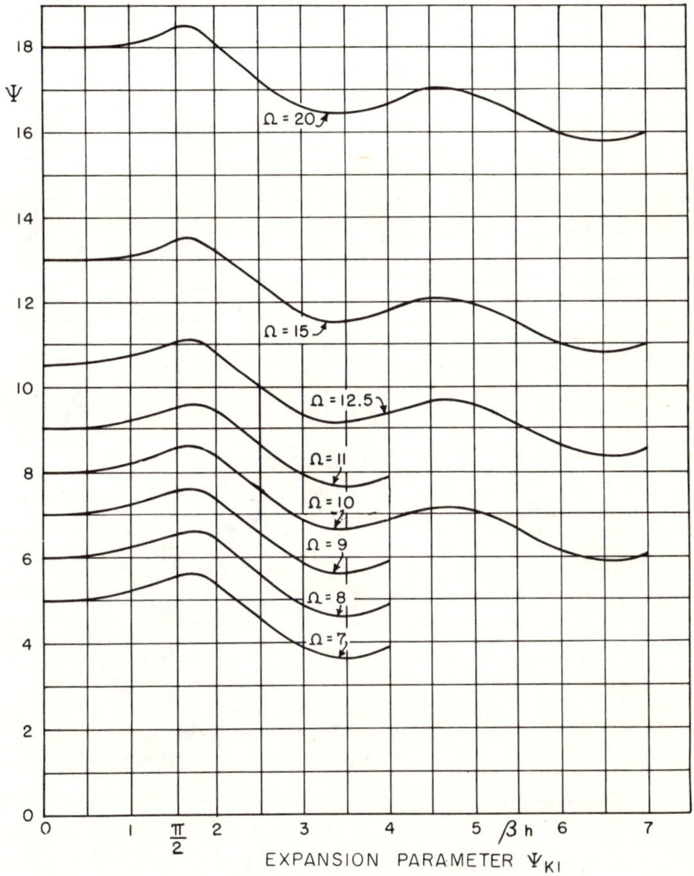

FIG. 10.3.—The parameter Ψ.

Since they assume idealized driving conditions they are accurate for antennas terminating actual transmission lines only when

[1] Formulas and tables for their evaluation are given in R. W. P. King, "Theory of Linear Antennas," Chap. II, Sec. 27, Harvard University Press, 1956.

combined with appropriate terminal-zone networks. Details are given in Sec. II.10 of the reference in the footnote on page 97. In Fig. 10.4, R_0 is plotted as a function of H using Ω as parameter. From the curves, it is to be noted that R_0 increases from very small values for small values of H to a maximum with H somewhat less

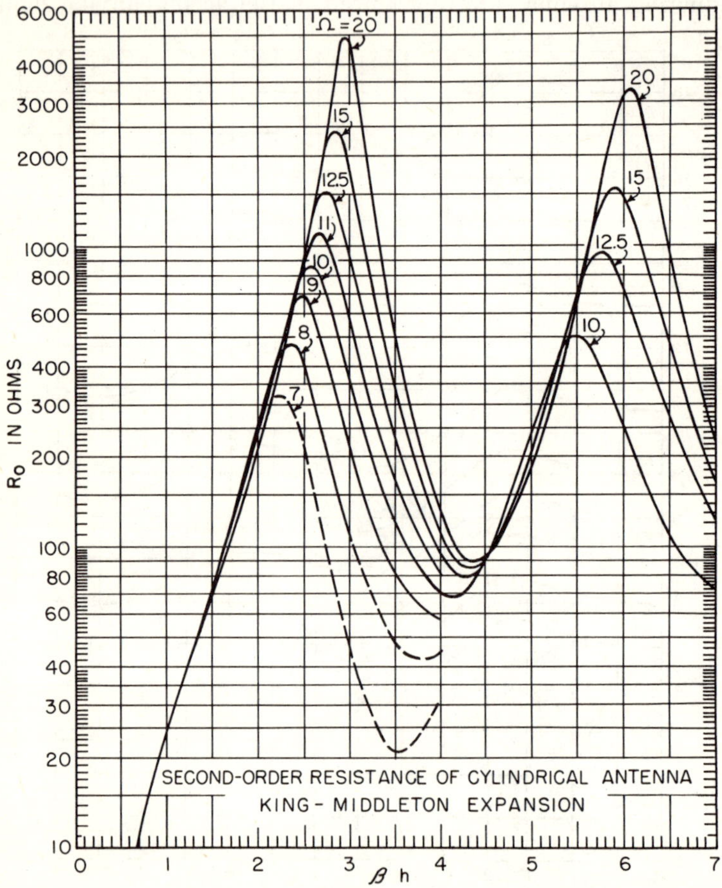

FIG. 10.4.—Resistance of antenna computed from (10.3); $H = \beta h = 2\pi h/\lambda$.

than π. For a thick antenna ($\Omega = 9$) the maximum is 700 ohms; for a thin antenna ($\Omega = 20$) it is 5,000 ohms. In practice, antennas may have maxima of R_0 ranging from 300 to 15,000 ohms. An infinitely thin antenna ($a = 0$) has a maximum $R_0 = \infty$ exactly at $H = \pi$. A minimum of between 75 and 80 ohms is reached somewhat below $H = 3\pi/4$; and a second set of maxima, slightly lower than the first, is reached at H a bit less than 2π.

In Figs. 10.5 and 10.6, X_0 is plotted as a function of H using Ω as parameter. Important points in Fig. 10.6 are vanishing values of X_0 occurring with H somewhat less than $n\pi/2$, or h some-

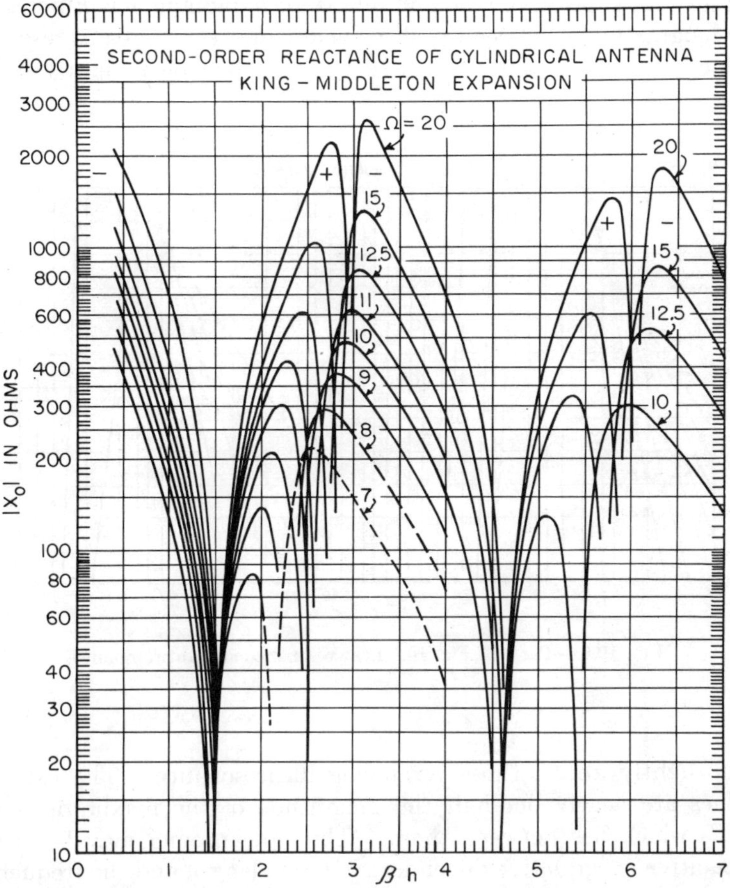

FIG. 10.5.—Reactance of antenna computed from (10.3).

what less than $n\lambda/4$, with n any integer. The condition for which X_0 vanishes with h near $n\lambda/4$ and n odd is called input resonance. It is characterized by a relatively small value of R_0 (near a minimum except for $n = 1$) and may be compared with series resonance in a conventional R, L, C circuit in the near zone. The condition for which X_0 vanishes with h near $n\lambda/2$ with n even is called input antiresonance. It is characterized by practically maximum values of R_0 and may be compared with parallel resonance in conventional low-frequency analysis. X_0 is negative (the antenna is a capacitance) for lengths below the first resonance and, as h increases, for lengths lying between each antiresonant and the next resonant value. X_0 is positive (the antenna is an inductance) as h increases, for lengths lying between each resonant and the next antiresonant value. Extreme values of X_0 occur for values of h slightly below

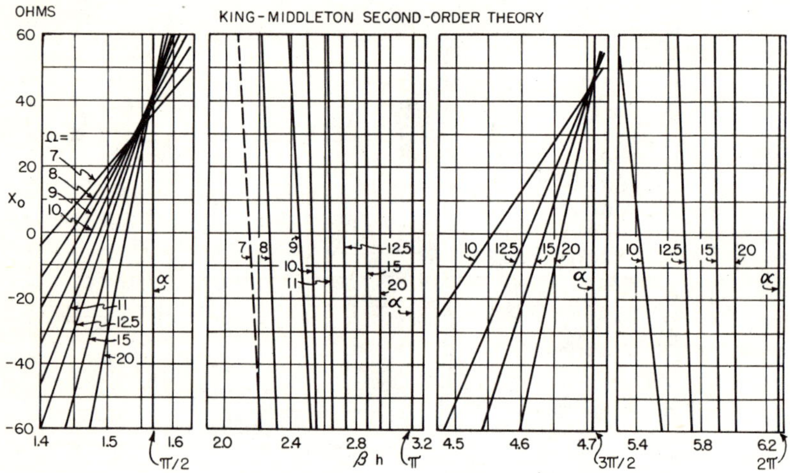

FIG. 10.6.—Parts of Fig. 10.5 near resonance and antiresonance.

and slightly above those producing antiresonance. The extreme values are nearly one-half the magnitude of the maximum value of R_0 which lies between them. Thus an antenna may be either capacitive or inductive depending upon its length and the frequency at which it is operated. A so-called "short antenna" is always capacitive. Its impedance is given by

$$Z_0 = 20H^2 - j\frac{60}{H}\left[\frac{\Omega - 2}{1 + (2ln2)/(\Omega - 2)}\right] \tag{10.4}$$

For values of H that satisfy the condition $H < 1$, (10.4) is a good approximation of Z_0 for thin antennas.

Figures 10.7 to 10.9 give the same data as Figs. 10.4 and 10.5

but with Ω used as variable and H as parameter. The curves marked H_R are plotted with H adjusted to maintain resonance or

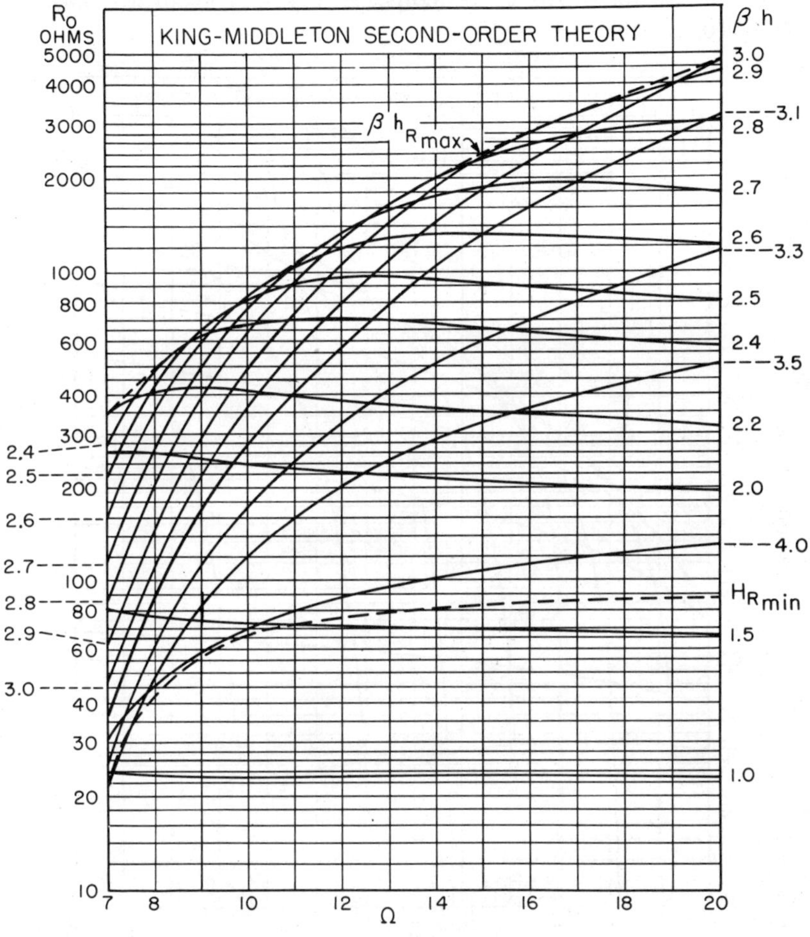

FIG. 10.7.—Resistance of antenna. The parameter numbering the curves is $H = \beta h = \frac{2\pi h}{\lambda}$, the variable is $\Omega = 2 \ln \frac{2h}{a}$.

antiresonance as Ω is varied. For many purposes, these curves are more convenient than those of Figs. 10.4 and 10.5. Because the curves are smooth and slowly varying, extrapolation to considerably larger values of Ω than 20 should give good approximations. The range of the variable Ω for practical antennas lies between 6 or 7 and 30. The analysis from which all the impedance curves were calculated assumed that $1/\Omega^2$ was negligible compared with unity. Accordingly, impedances for smaller values of Ω than $\Omega = 10$ are expected to be considerably less accurate than for $\Omega \geq 10$.

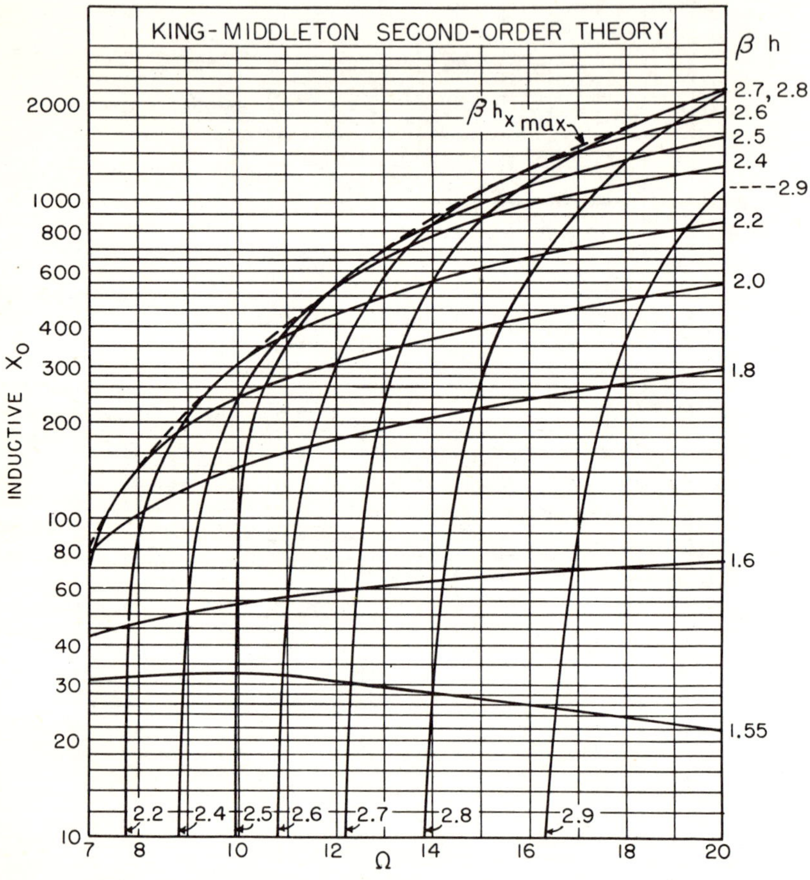

FIG. 10.8.—Inductive reactance of antenna.

In Fig. 10.10, critical values of R_0 and X_0 are shown as a function of Ω. For the range of Ω from 10 to 20, $(R_0)_{\text{res}}$ varies from 70 to 72 ohms for the first resonance ($n = 1$). It reaches the value 73.13 ohms for an infinitely thin antenna only. For the second resonance ($n = 3$), $(R_0)_{\text{res}}$ varies from 94 to 104 ohms with 105.5 the asymptotic value for the infinitely thin antenna

($a = 0$, $\Omega = \infty$). For half-wave dipoles, $H = \pi/2$, R_0 varies from about 87 ohms for $\Omega = 10$ to 78 ohms when $\Omega = 20$. Note that this resistance increases as the antenna is made thicker. In Fig. 10.10 the extreme values of both resistance and reactance are

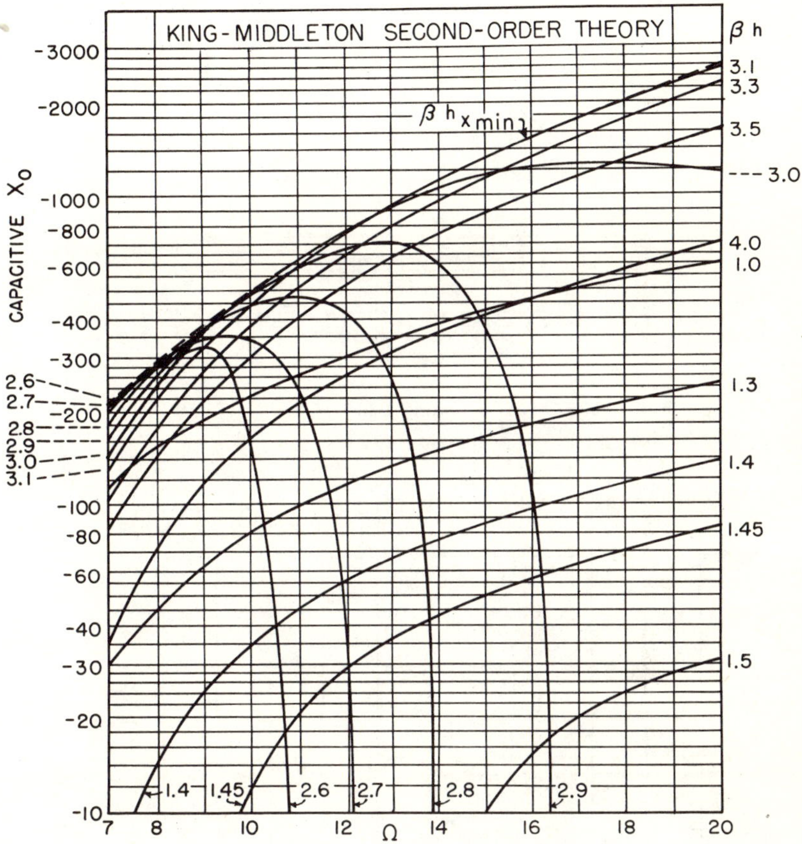

FIG. 10.9.—Capacitive reactance of antenna.

given as functions of Ω. As Ω increases from 10 to 20, $(R_0)_{max}$ varies from 855 to 4,950 ohms. It approaches infinity for $a = 0$. Resonant and antiresonant values of H_R may be determined from Fig. 10.11. $n\pi/2 - H_R$ is the amount by which the half length in radians of an actual antenna must be decreased below the value $n\pi/2$ for an infinitely thin antenna in order to maintain resonance

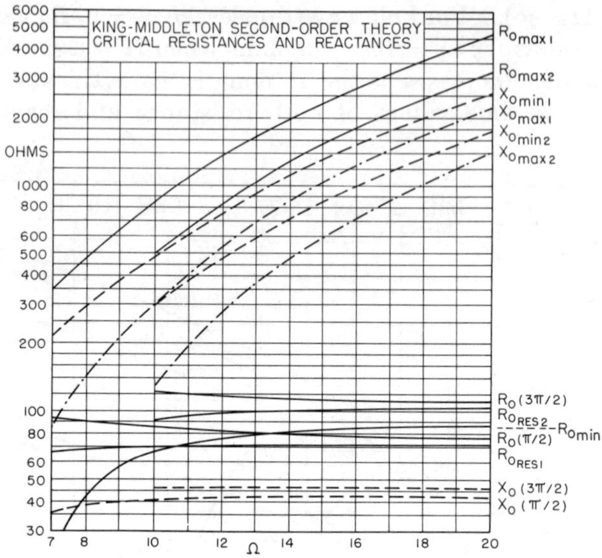

Fig. 10.10—Critical resistances and reactances of antenna.

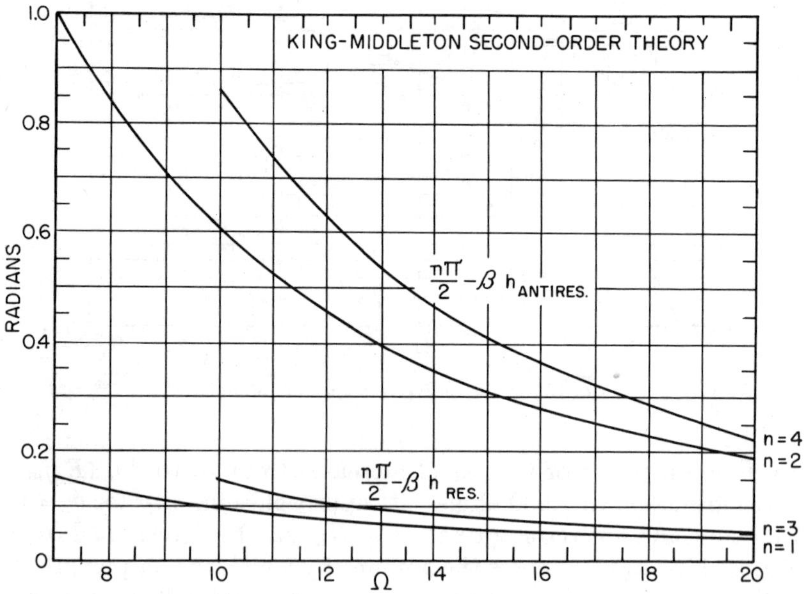

Fig. 10.11.—Resonant and antiresonant lengths of antenna.

(n odd) or antiresonance (n even). The actual lengths h_r to produce resonance or antiresonance are readily computed using

$$h_r = H_R \frac{\lambda}{2\pi} \tag{10.5}$$

Note that h_r approaches $n\lambda/4$ as the radius a vanishes and Ω becomes infinite. In Fig. 10.12, the admittance $Y_0 = 1/Z_0$

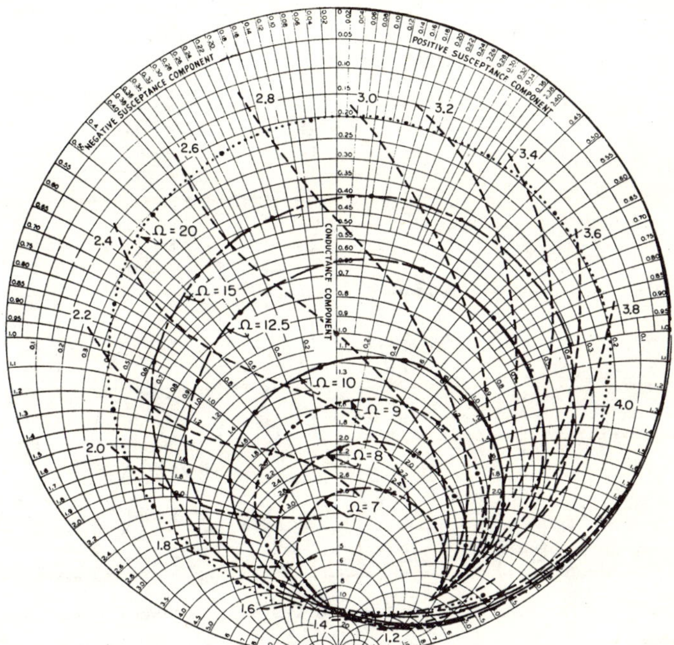

INPUT ADMITTANCE OF CENTER-DRIVEN DIPOLE
MHOS x 10^{-3} vs βh
AFTER KING-MIDDLETON

FIG. 10.12.—Admittance $Y_0 = 1/Z_0 = G_0 + jB_0$ of antenna.

$= G_0 + jB_0$ is shown on a Smith chart for a range of values of Ω. The parameter is βh.

The values of X_0 at $H = \pi/2$, $h = \lambda/4$, are readily obtained from Fig. 10.6. It is seen that X_0 varies more slowly with H as Ω is decreased. It is evident that, as Ω is increased, the curve approaches the vertical. If the half length h is fixed at $\lambda/4$ so that $H = \pi/2$, and the radius a is chosen smaller and smaller, X_0 approaches the value 42.5. On the other hand, if the value of h is determined for which X_0 vanishes with $a = 0$, then $h = \lambda/4$.

This apparent contradiction results from the fact that, as the radius a is chosen smaller and smaller and the curve of X_0 as a function of H (as in Fig. 10.6, for example) becomes more and more nearly vertical, the point where X_0 crosses the zero line moves nearer to $H = \pi/2$, while at the same time the value of X_0 at $H = \pi/2$ approaches 42.5. In the limit as a is made equal to zero, the reactance curve is vertical, and X_0 at $h = \lambda/4$ or $H = \pi/2$ therefore becomes indeterminate and may have any value, including 42.5. Actually, any antenna with a finite but small radius has a value of

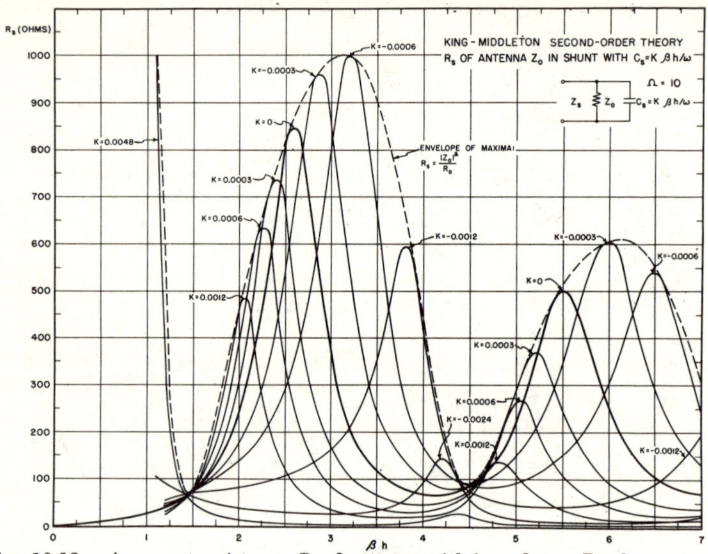

FIG. 10.13.—Apparent resistance R_s of antenna with impedance Z_0 when connected in parallel with a lumped capacitance $C_s = \dfrac{K\beta h}{\omega}$ where K is a constant.

X_0 close to 42.5 when $h = \lambda/4$ or $H = \pi/2$, and X_0 vanishes for h less than $\lambda/4$. For antennas of nonvanishing radius, R_0 is *less* than 73.13 when $X_0 = 0$ (first resonance) at a half length h less than $\lambda/4$, but R_0 is greater than 73.13 ohms when $\beta h = \pi/2$.

If a lumped capacitance is connected across the terminals of an antenna the impedance of the parallel combination differs from that of the antenna alone in the following ways: the resistance curves (Fig. 10.4) are lowered and shifted to the left; the reactance curves (Fig. 10.5) are shifted to the left and reactances, especially near antiresonance, are made more capacitive as shown in Figs. 10.13 and 10.14. The effect of lumped negative capacitances is also shown since the terminal-zone network for an antenna terminating a transmission line includes a negative capacitance in parallel with the antenna. Note that the effect of a negative capacitance is opposite to that of a positive capacitance. In particular, resistances at antiresonance may be increased greatly and the length of the antenna at which antiresonance occurs may be shifted to longer rather than shorter values.

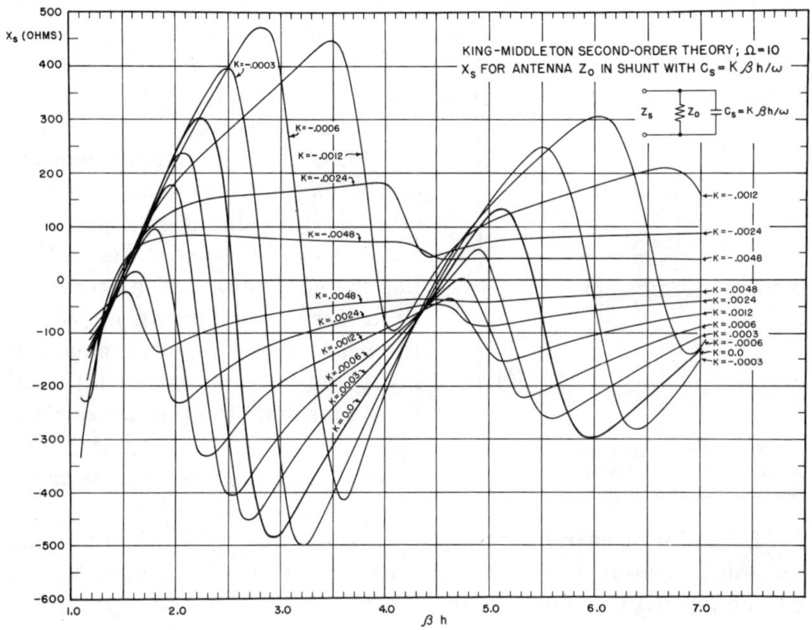

FIG. 10.14.—Apparent reactance X_s of antenna with impedance Z_0 when connected in parallel with a lumped capacitance $C_s = \frac{K\beta h}{\omega}$ where K is a constant.

11. Broad-band Antennas.—An examination of the curves for the input reactance of a cylindrical center-driven antenna (Figs. 10.5, 10.6) shows that the variation of the reactance X_0 with frequency (note that $H = 2\pi fh/v_c$ is directly proportional to frequency) is much less rapid for thick antennas than for thin ones. Accordingly, if an antenna is to be operated over a fairly wide band of frequencies, as in television, it has a smaller change in input reactance with frequency if it is thick than if it is thin. This means that a better impedance match can be obtained over a wider band of frequencies.

12. Unsymmetrical Antennas—Special Case of Antenna Erected Vertically on a Highly Conducting Half-space.—If an antenna is not symmetrical with respect to its input terminals from the point of view of the geometrical arrangement of the conductors forming its halves or with respect to the transmission line from which it is driven, a variety of complicating effects may be observed. In some instances, these are limited to an unsymmetrical distribution of current; in other instances, coupled-circuit effects arise so that their description is best postponed to a later section. One important and simple unsymmetrical antenna may be introduced conveniently at this point because of its similarity to the symmetrical center-driven type.

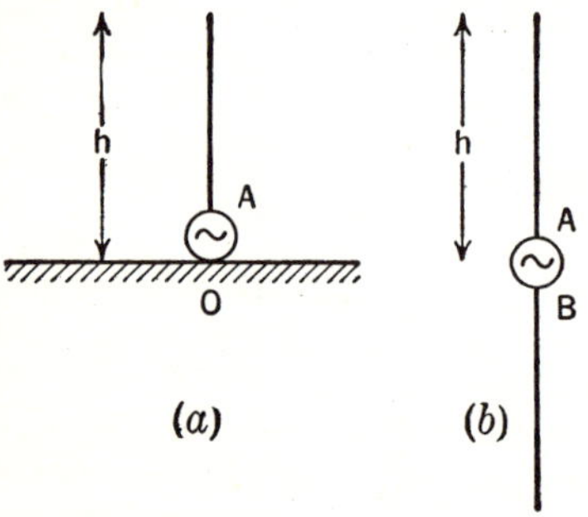

FIG. 12.1.—Antennas having the same distribution of current in the parts above the terminals A.

Consider a cylindrical conductor of length h and small radius a erected vertically over a perfectly conducting half-space, Fig. 12.1a. Let the conductor end at terminal A, a very short distance (within the near zone) above the conducting plane. Let a second terminal be connected to the conducting plane at O directly below A. The driving potential difference is maintained between A and O.

If the input current I_A at A is directed upward in the antenna, an equal current I_0 is also upward from O in the generator. It can be proved that the distribution of current along the cylindrical conductor is exactly the same as though it were the upper half of the symmetrical antenna of Fig. 12.1b. Simultaneously, the current entering at O is radially inward along the surface of the conducting plane. As the current in the vertical wire goes through its usual cycle of reversal every half period distributing alternately positive and negative charge along the wire, alternate rings of negative and positive charge travel radially outward along the conducting plane with a radial phase velocity that approaches asymptotically the characteristic velocity $v_c = 3 \cdot 10^8$ m/sec as the radius of the ring becomes sufficiently large (Sec. 36). For present purposes, it is sufficient to note that the forces acting on an individual conductor, such as a receiving antenna, placed anywhere in the upper half-space are due not only to the charges moving in the cylindrical conductor but also to those moving everywhere in the conducting plane on which this conductor is erected. The antenna consists of the vertical conductor *and the entire conducting plane* on which it is erected.

The input impedance of the base-driven antenna of *full* length h and radius a over a perfectly conducting half-space is one-half that of the symmetrical center-driven antenna of *half* length h and radius a. Accordingly, all curves computed for the center-driven antenna apply directly to the base-driven antenna if all values of resistance and reactance are divided by two.

Just as with the symmetrical antenna, it is not practically possible to drive an antenna by a concentrated generator, in this case placed at its base. At low frequencies and with thin

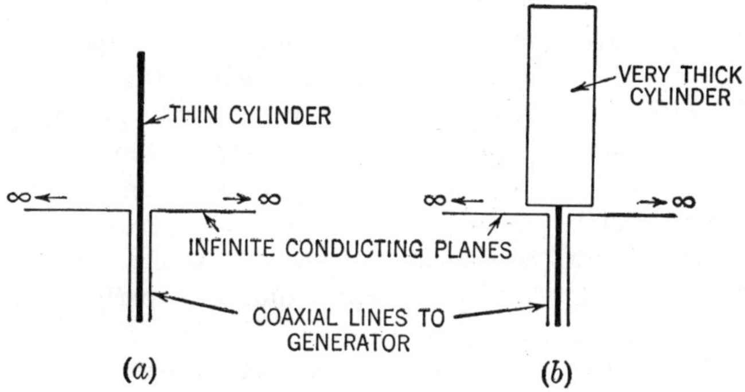

FIG. 12.2.—Methods of base-driving antennas erected vertically over a conducting half-space.

antennas, practical arrangements using a transmission line are easily constructed to approximate the idealized case assumed in the theory. At high frequencies, this becomes increasingly difficult. For example, in Fig. 12.2*a*, the driving forces in a voltage-fed antenna are distributed along the vertical wire over distances comparable with the diameter of the coaxial line, rather than being concentrated. If this diameter is not extremely small, the theoretical case is not well approximated, and a terminal-zone network with a negative capacitance is required. In the case shown in exaggerated form in Fig. 12.2*b*, the driving forces correspond more closely to the theoretically assumed case, insofar as the cylindrical surface is concerned. On the other hand, the impedance terminating the end of the line is not the same as the impedance at the driving edge of the cylindrical surface. The latter can be measured, the former computed. Complete agreement should not be expected, particularly near antiresonance, where large concentrations of charge are maintained near the base. (The theoretical value is not merely the difference between the terminating impedance measured on the coaxial line first with the antenna in place and second with it cut down to leave only a flat disk.) It follows that the theoretical curves, especially near antiresonance, do not represent accurately the physically different arrangements available in practice. They may be depended upon to give correct values for thick antennas and coaxial lines with large openings only if suitably designed lumped networks are included.

A perfectly conducting plane of infinite extent is not realizable physically. On the other hand, at frequencies in the broadcast band, the earth (if not too dry near the antenna) behaves sufficiently like a perfectly conducting plane that satisfactory approximations often are obtained by assuming it to be perfectly conducting. The input impedance so obtained is then corrected by combining it with an effective ground resistance, which is assumed to be in series with the input impedance of the antenna. In practice, it is always desirable and therefore customary to provide an extensive radial system of wires buried in the ground in order to reduce the ground losses near the antenna where the density of current is greatest. Salt water approximates a perfect conductor even at frequencies considerably above the broadcast band. At ultra-high frequencies, however, no part of the earth's surface can approximate even roughly a perfect conductor. On the contrary, it behaves in many respects like an imperfect dielectric with losses that may be very large in moist earth or in salt water.

13. Tower Antennas over a Good Conductor.—If the base-driven antenna is not cylindrical and of relatively small radius, but consists of a self-supporting steel tower of moderately large and nonuniform cross section, the input impedance cannot be obtained from the curves for the cylindrical antenna. In its dependence upon length, a steel-tower antenna behaves in general much like a rather thick cylindrical antenna. The simple distribution of current that obtains in an infinitely thin antenna is not a satisfactory approximation. The effective longitudinal distribution of current resembles that of a thick cylindrical antenna in that it consists of two components that differ in time phase by a quarter period.

14. Top-loaded Antennas.—At the lower radio frequencies, it is often desirable or necessary to reduce the physical length of an antenna erected vertically over the earth to less than the self-resonant value, which is slightly less than $\lambda/4$. This may be dictated by greater constructional economy in land installations or by reasons of design aboardship. Such a shortening is *always undesirable* in that it diminishes the total length of conductor carrying the current, the forces due to which are effective in producing currents in distant receiving antennas. If the vertical part of the antenna must be reduced below a quarter wavelength, it will be least disadvantageous from the electromagnetic point of view if the part of the conductor near the base, which carries the large current, is retained while the part near the top, which carries a small current, is sacrificed. This can be done in numerous ways. Thus, a part of the antenna at the top may be folded over into a horizontal position to make an inverted L; it may be wound into a widely spaced, comparatively low-loss coil, Fig. 14.1; it may be replaced by a large sphere, by a disk, or by a cylinder of metal; one or more horizontal wires may be arranged with centers connected to the upper end of the remaining vertical length to make a T or flat-top antenna. If suitable dimensions are chosen in these several cases, the antenna may be shortened and yet kept in resonance with approximately the same current in its vertical part as when this is extended to a full quarter wavelength without top load. The distribution of current and charge along the vertical part of the top-loaded antenna is similar to that along the same lower part of the resonant straight antenna. The distribution cannot be exactly the same, because the forces acting on the charges in the lower part of the conductor due to the charges in the extended upper part of the straight antenna are replaced by other forces due to the charges in the differently oriented and dimensioned top load.

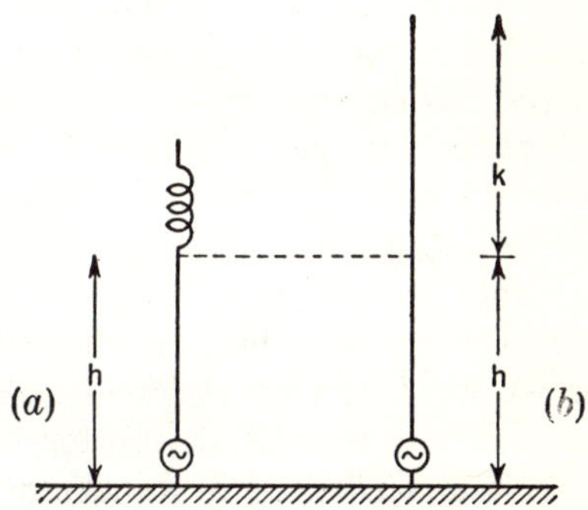

FIG. 14.1.—(*a*) Antenna with coil as top load; (*b*) antenna with straight conductor extended to be of electrical length equal to that of the coil.

Useful qualitative estimates of some of the properties of a top-loaded antenna may be obtained, by assuming a sinusoidal distribution of current along its vertical part, in the form

$$\hat{I}_z = \hat{I}_0 \frac{\sin \beta(h + k - z)}{\sin \beta(h + k)} \tag{14.1}$$

where k is the equivalent length of the top load. A coil as a top load is shown in Fig. 14.1. For an inverted L and for a very loosely spaced coil the total length of wire at resonance is roughly the same as for the straight vertical antenna. In all cases, the input resistance at resonance is lower than that of the straight vertical wire by an amount that depends upon the fraction of the antenna that is folded over, coiled, or replaced by a top load.

The discussion applies equally well to a symmetrical center-driven antenna with each end shortened by an identical end load.

15. Loading at Input Terminals.—A circuit including an antenna may always be tuned to resonance by inserting suitable series reactances in the conventional way. Since the antenna is the equivalent of a lumped impedance insofar as a voltage impressed across its closely spaced terminals is concerned, the analysis of an antenna in series with lumped impedances reduces to the ordinary near-zone circuit analysis of a series circuit.

16. Radiation.—Since all electric charges exert forces on one another according to a law of retarded action at a distance, periodically varying distributions of charge in one element of a circuit must ultimately exert forces on all charges in the universe, tending to set them in periodic motion unless currents in other elements of the same circuit or in an adjacent circuit or in an enclosing shield set up equal and opposite forces that cancel those due to currents in the element in question. In near-zone circuits, currents are entirely in closed or quasi-closed paths, all parts of which are so close together in terms of the wavelength that the resultant forces exerted by them on the charges in a distant conductor are vanishingly small. In order to exert significant forces on each other, two circuits all parts of which are in the near zone must be brought close together in terms of their own extension and hence extremely close together in terms of the wavelength. Such circuits are coupled in the conventional low-frequency sense.

If a circuit is not confined to the near zone, the forces exerted by the current in it on charges in other even quite distant conductors may be considerable unless they are canceled by equal and opposite forces produced by an adjacent current, as in a coaxial or two-wire transmission line. No such complete cancellation can occur when charges move in an antenna that is not in the near zone with respect to any other conductor. There may be a more or less complete cancellation of effects *at some angles* if the antenna is itself long enough to have current in opposite directions in different parts of its own extension, but there are then other directions in which the same component forces combine to produce a larger resultant force. Accordingly, the essential characteristic of an antenna is an extension and geometrical configuration of conductors such that periodically varying distributions of current along them can exert significant forces on charges even at great distances after an appropriate lapse of time.

Every effective antenna is by its very nature coupled to all surrounding matter in the sense that work done to maintain a periodic motion of charges in the antenna is largely work that will be done on charges that can ultimately be set in motion throughout the universe. A driven antenna may be regarded as a kind of primary circuit with an unlimited number of more or less closely coupled secondaries scattered far and wide wherever charges are set in motion. Only a small part of the energy supplied to an antenna is used to heat it. Most of the energy is transferred away from the antenna, presumably to be used eventually in heating all the innumerable imperfect conductors and imperfect dielectrics in which currents however small are maintained. *This energy which is transferred to the universe is said to be radiated.*

It is a simple matter to estimate both the smaller fraction of the energy supplied to a symmetrical center-driven antenna that is converted into heat in its own conductors, and the very much larger fraction that is transferred to other parts of the universe. It has been stated already that the curves of Figs. 10.4 to 10.9 for the input resistance and reactance of such an antenna were computed neglecting all terms involving the conductivity. This was done for mathematical reasons of simplicity in calculation and representation and is justified for all practical purposes because the large but finite conductivity of a very good conductor has a negligible effect on R_0 and X_0.

For example, in a thick copper antenna of radius 5 mm operated at 5 m wavelength, the difference in the computed values of R_0 at resonance neglecting and including ohmic resistance is less than 0.2 per cent. With a much thinner antenna of radius 1 mm operated at 100 m wavelength, the corresponding difference is nearly 3 per cent. It follows that, to a good approximation, the power $I_0^2R_0$ supplied to a highly conducting antenna (of copper), with R_0 taken from the curves of Sec. 10, is for practical purposes all radiated to the more or less closely coupled universe outside the antenna, while that used in heating the antenna itself is negligible. If the antenna is not highly conducting, the distribution of current and hence both the energy radiated and that dissipated in heat in the antenna depend significantly upon the conductivity. It is then not possible to determine the radiated energy by merely assuming an infinite conductivity.

It is instructive to compare the power supplied to a given pair of copper wires when they are arranged along the same axis to form a resonant center-driven antenna, Fig. 16.1*a*, with the power supplied when they are folded parallel to each other to form the resonant open-end section of a two-wire line, Fig. 16.1*b*. The adjustment in length for resonance differs only slightly in the two cases if a thin conductor is used. Furthermore, the distribution of current along each wire is the same to a first (sinusoidal) approximation; and the input current is also the maximum current. Therefore, the input powers given by $I_0^2 R_{in}$ are entirely comparable if I_0 is made the same in both cases. For the antenna, the input resistance at resonance is of the order of magnitude of 87 ohms (Fig. 10.7 for $\Omega = 10$ or $h/a = 74$). On the other hand, for a typical transmission line with an open end, the input resistance at resonance may be of the order of magnitude of 0.5 ohm (computed from $(R_{in})_{res} = R_c \alpha s$ with $\alpha = 2 \cdot 10^{-3}$ nepers/m, $s = \lambda/4 = 1$ meter, $R_c = 500$ ohms). The power $I_0^2 \cdot 87$ supplied to the antenna is practically all radiated; the power $I_0^2 \cdot 0.5$ supplied to the line is practically all used in heating the wires of the line. The ratio of power supplied to the line to that supplied to the antenna is a rough measure of the ratio of power dissipated in heating the antenna to that radiated, because the distributions of current in line and antenna at resonance are roughly alike. This ratio has the order of magnitude of something under 1 per cent in the numerical case considered.

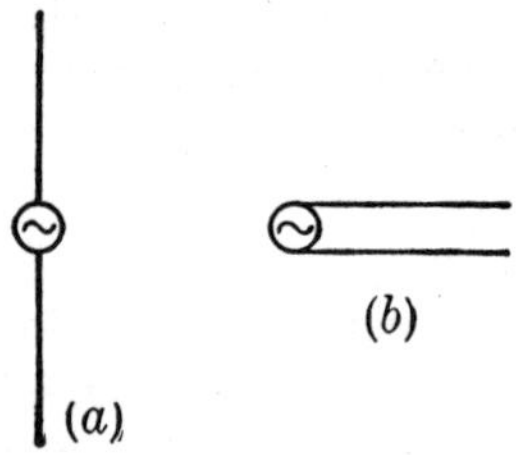

FIG. 16.1.—(*a*) Center-driven antenna; (*b*) antenna folded together to form an open-end line.

As an estimate, this is quite comparable with the values of 0.2 per cent for a thick antenna and 3 per cent for a thin antenna at different frequencies, and it illustrates in a convincing way how merely folding together the two conductors of a symmetrical antenna to form a parallel line effectively decreases the coupling to the universe and thus reduces the effective "reflected resistance" from a predominant to a negligible value. It has been assumed up to this point that none of the numberless "secondary circuits" in the universe is near the driven antenna or "primary circuit." The effect of moving one secondary, *e.g.*, another tuned antenna, near to the driven antenna (but not so close that it forms a parallel line) is considered in Sec. 19.

It has been stated that in effect an antenna is an arrangement of conductors which is electromagnetically coupled to the universe in that currents which are made to flow in the antenna exert forces at some later time on charges in matter wherever this may be. The currents that these forces produce exert, in due time, forces back on the moving charges in the antenna. But insofar as currents in the distant universe are concerned, the time within which such reactions could be felt in the antenna is far greater than any reasonable interval during which the antenna might be kept in operation. It might be supposed, therefore, that the universe outside of a relatively limited range could have no observable effect on an antenna. Actually, such a conclusion would be based on the same kind of fallacious reasoning by analogy which led to the belief that, because the velocity of a man running forward on a train must be added to the velocity of the train in order to determine his velocity with respect to the earth, the velocity of light must be similarly added to that of a moving light source. Electromagnetic phenomena seldom can be described correctly by analogy with the simpler mechanical events of daily observation. The analytically derived and physically verified fact is that, whenever a phenomenon can be described mathematically in terms of a wave equation in which there appears a *finite velocity of propagation* of a quantity which may have physical or only mathematical significance, the propagation always encounters the equivalent of a characteristic impedance, which is a pure resistance if there is no dissipation in the propagation.

Consider, for example, the circuit of Fig. 16.2a, consisting of a generator connected to an ordinary electric circuit that is link-coupled to a load Z through an artificial line of recurrent sections constituting a time-delay network. Suppose the time delay to be many days and the coupling between adjacent half coils perfect so that $M = L/2$. For simplicity, let the line be assumed dissipationless, *i.e.*, purely reactive, but its termination Z may include resistance. As far as the generator is concerned, the circuit of Fig. 16.2a is equivalent to that of Fig. 16.2b until a sufficient time has elapsed for a reflection to be received from the impedance Z.

By inserting enough sections, this can be postponed indefinitely; and, regardless of the number of sections, it will never occur if the output transformer has a one-to-one ratio and $Z = R_c = \sqrt{2L/C}$. Thus, the coupling link consisting of the dissipationless line presents a definite load at its input terminals AB long before any effect from the terminating impedance can be received. The same situation exists if the coupling link is a sufficiently long

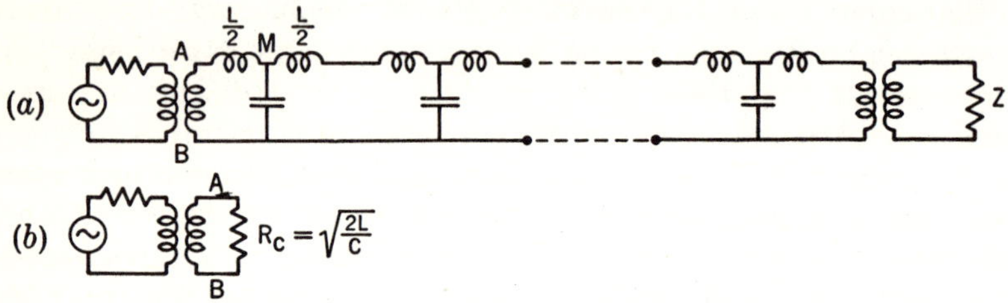

FIG. 16.2.—Dissipationless, delay network (*a*) that is equivalent to a characteristic resistance (*b*) coupled to a primary circuit.

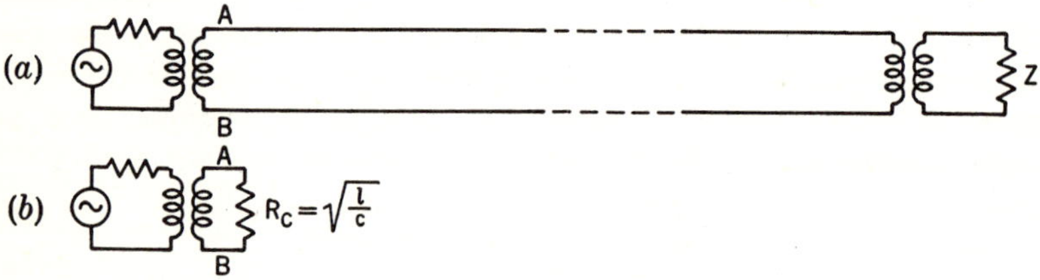

FIG. 16.3.—Dissipationless nonresonant line (*a*) that is equivalent to a characteristic resistance (*b*) coupled to the primary circuit.

FIG. 16.4.—Coupled antennas.

dissipationless transmission line, as shown in Fig. 16.3. It, too, presents a characteristic resistance $R_c = \sqrt{l/c}$ at its terminals AB until a reflection from the termination Z is received. In both cases, the impedance as seen by the generator can be varied by changing the degree and the type of coupling between generator and line.

A circuit, Fig. 16.4, which includes a generator and an antenna, necessarily coupled to the distant universe, behaves in a similar way. In this case, the driven antenna takes the place of the coupling coils at the generator; the receiving antenna, which may be any physical extension of matter in the universe, takes the place of the coupling coils at the output end; and the imperfectly conducting parts of this matter constitute the load. A coupling link in the form of a network of conductors is not required for the antenna, because the moving charges in it act directly on charges elsewhere in the universe. Note, however, that, because there is no conducting link, no one particular receiving antenna can be singled out to be the entire load. All selective or directional effects must be secured by adjusting the nature of the coupling, *i.e.*, by changing the shape and orientation of the antennas.

In spite of the fact that there is no conducting link, the generator experiences a load that is like that of a characteristic resistance coupled to the antenna. The degree of coupling to this characteristic resistance can be varied by changing the shape and size of the antenna. For example, the coupling can be reduced to a negligible value by folding the antenna into a parallel line. When this is done, the input impedance of the antenna is changed. This does not mean a change in $\mathcal{R}_c$. The existence of a characteristic resistance for electromagnetic effects is just as mysterious, but no more so than the existence of the finite velocity

$$v_c = \frac{1}{\sqrt{\mu_0 \epsilon_0}} = 3 \cdot 10^8 \text{ m/sec}$$

The simultaneous and inseparable occurrence of both is the inevitable consequence of a *mathematical representation* in terms of an electromagnetic field governed by the general wave equation. From the point of view of measurements made on antennas, space acts like a dissipationless transmission medium through which effects can be sent at a characteristic velocity, $3 \cdot 10^8$ m/sec, and which presents a characteristic resistance, $\mathcal{R}_c = 376.7$ ohms, to the propagation. This propagation in space differs from the propagation along networks and lines in that concentrations of charge do not move continuously from generator to load. Moreover, the characteristic resistance $\mathcal{R}_c$ for propagation in space is only formally similar to the characteristic resistance of a line; actually it is not a resistance in the conventional sense since it is used to describe the propagation of a field that has been assigned only mathematical significance. It is not expressible in terms of a ratio of potential difference to current. The metaphysically fascinating question as to whether the complete emptiness of space actually *is* a medium endowed with the physical properties usually associated with material media, or whether the electromagnetic field serves merely as part of a computing machine will not be considered here. Physical science seeks to provide a mathematical mechanism for predicting observable effects about nature. Whether this mechanism, if successful, is to be identified with nature is at least questionable. Such identification has been avoided consistently throughout the present discussion.

17. Radiation Resistance and Input Resistance.—A quantity called the radiation resistance of an antenna is often employed but unfortunately not always unambiguously used or even consistently defined. For this reason, great care must be exercised in making use of any data referring to radiation resistance if the data are not accompanied by a precise and complete definition of the quantity so designated. In some instances, radiation resistance is defined in one way and then curves are supplied that depend upon an entirely different definition. An attempt will be made to clarify the confusion that exists in this regard.

The total power P supplied to an antenna is used partly internally in heating the conductors of the antenna and partly externally in heating parts of the universe in which charges are set in motion. The former, P^i, may be called internally used power, the latter, P^e, externally used or radiated power. Thus,

$$P = P^i + P^e \tag{17.1}$$

If the input terminals of the antenna carry equal and opposite currents so that an input current I_0 and an input impedance $Z_0 = R_0 + jX_0$ can be defined, the input power is

$$P = |I_0|^2 R_0 \tag{17.2}$$

It is now possible to combine (17.1) with (17.2), divide both sides by I_0^2, and to define the quantities

$$R_0^i = \frac{P^i}{|I_0|^2}; \qquad R_0^e = \frac{P^e}{|I_0|^2} \tag{17.3}$$

so that

$$R_0 = R_0^i + R_0^e \tag{17.4}$$

Here R_0^i might properly be called the internal input resistance of the antenna and defined as the power dissipated in heating the antenna divided by the square of the rms magnitude of the input current. Similarly, R_0^e might well be called the external or radiation input resistance of the antenna and defined as the power transferred to media outside the antenna divided by the square of the rms magnitude of the input current. Analytically, R_0 is not easily resolved into two terms of which one may be identified with R_0^i, the other with R_0^e, except in simple special cases. This is due to the fact that both P^i and P^e depend upon distribution of current and conductivity in a complicated way. Fortunately, in highly conducting antennas, P^i is negligible compared with P^e so that approximately

$$P \doteq P^e \qquad P^i << P^e \tag{17.5}$$
$$R_0 \doteq R_0^e \qquad R_0^i << R_0^e \tag{17.6}$$

In this case, the external or radiation input resistance is to a good approximation equal to the input resistance. The curves of Sec. 10 were computed subject to (17.5), so that the input resistance R_0 given by them might be called correctly the input radiation resistance. Both expressions for R_0^e as given by (17.3) and in a more restricted form by (17.6) have been called the radiation resistance of the antenna. To avoid ambiguity, the term input radiation resistance will be used, and discussion will be limited to the simpler and more important case of (17.6).

The complex expression for the input current is of the form

$$I_0 = I_0'' + jI_0' \tag{17.7}$$

where both I_0' and I_0'' are functions of the radius and length of the antenna. Although the simple distribution of current, with I_z' a sinusoidal function of the distance z along the antenna and $I_z'' = 0$, is for some purposes a satisfactory approximation for a practical antenna, this is not true in the definition of the input impedance of such an antenna. Nevertheless, the attempt is frequently made to define in the form given in (17.3) the radiation resistance R_0^e for a center-fed antenna or a base-fed antenna over a perfect conductor, Sec. 8, under the assumption that $I_0 = I_{\max} \sin H$.

If this value of I_0 is substituted in (17.3), using (17.5),

$$R_0^e = \frac{P}{|I_{\max}|^2 \sin^2 H} \tag{17.8}$$

Since P is always finite in any antenna that has a nonvanishing radius, (17.8) requires R_0^e to become extremely large as H approaches π, 2π, etc., and to become infinite at these values. This is a physical absurdity that is the result of assuming a physically impossible distribution of current in an antenna of finite radius. This difficulty may be overcome by multiplying both sides of (17.8) by the dimensionless quantity $\sin^2 H$ and defining a new quantity R^e_m by

$$R^e_m \equiv R^e_0 \sin^2 H = \frac{P}{I_{\max}{}^2} \tag{17.9}$$

where R^e_m has the dimensions of a resistance so that it can be measured in ohms. It is *not a resistance in the ordinary sense* (except when $\sin H = 1$ and $R^e_0 = R^e_m$), because it is not the real part of an

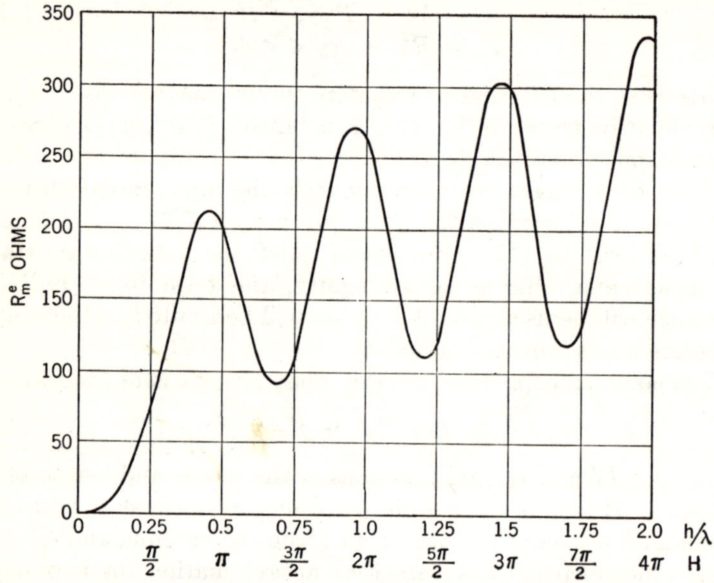

FIG. 17.1.—Radiation resistance R^e_m of center-driven antenna of zero radius referred to the maximum current I_m.

impedance that can be measured between two terminals in a circuit. It is the power supplied at a given pair of terminals divided *not* by the square of the magnitude of the current entering and leaving those terminals, but by the square of the magnitude of the current at *another point* $(h - \lambda/4)$ along the antenna (provided $h \geqq \lambda/4$) where the current is quite different, or (if $h < \lambda/4$) by an entirely fictitious current that exists nowhere. The apparent advantage of R^e_m over R^e_0 (for an infinitely thin antenna) is that, whereas R^e_0 must become infinite as H approaches π, 2π, etc., if the sinusoidal distribution of current is assumed, although incorrectly, R^e_m is always finite.

As defined in (17.9), R_m^e also has been called the radiation resistance of an antenna, and the curve of Fig. 17.1, which shows R_m^e for a center-driven antenna of half length h as a function of H and h/λ, is frequently given to show the variation of radiation resistance. For a base-driven antenna of length h over a perfectly conducting half space, the values of R_m^e in Fig. 17.1 must be divided by two. Actually, the function R_m^e applies only to an infinitely thin antenna and even for this it does not give the input resistance. Nevertheless, it is not without value in comparing the approximate power-transferring or radiating properties of antennas that are too intricate in structure to permit accurate analyses of the distribution of current so that it is necessary to be content with the rough estimate obtained by assuming arbitrarily a sinusoidal distribution. In order to distinguish R_m^e from the radiation input resistance R_0^e, which is for practical purposes the actual input resistance R_0 of highly conducting antennas, R_m^e will be called the radiation resistance referred to a maximum sinusoidal current.

III. COUPLED ANTENNAS AND TRANSMISSION LINES

18. Coupled Antennas.—The term driven, as applied to an antenna, has been used heretofore to designate an antenna that has two input terminals that are well within the near zone of each other and across which a potential difference is maintained by a generator connected either directly or through a transmission line. It was specifically required in defining the input self-impedance of such an antenna at its terminals that no other antenna or other conducting or dielectric material be nearer than the limits specified by the condition for the far zone. Subject to these conditions, the currents that are maintained in any single conductor in the far zone by forces exerted upon them by moving charges in the driven antenna may be assumed so small that their reaction upon the charges in the driven antenna can be neglected as compared with the composite effect of the distant universe as a whole, *i.e.*, each particular antenna or conductor must be sufficiently far from the driven antenna that any change in its position that does not bring it nearer produces no measurable difference in the distribution of current and in the input impedance of the driven antenna. Under these conditions, the impedance at the terminals is called the input self-impedance of the driven antenna, and each conductor in the universe is individually only loosely coupled to the antenna. This self-impedance includes the composite effect on the antenna of the universe, assuming all conductors and dielectrics to be in the far zone. This restriction now will be removed and special account taken of neighboring antennas.

Consider first a driven antenna 1 so placed that all other antennas and conductors are in the far zone. Their effects on the driven antenna are included in the composite effect of the universe as a whole, which is taken into account in the self-impedance[1] Z_0. The following equation can be written:

$$V_1 = I_{01}(Z_1 + Z_0) = I_{01}Z_{11} \tag{18.1}$$

where Z_{11} is the impedance completely around the circuit. Now let a distant conductor 2 be a driven antenna like 1. Let it be

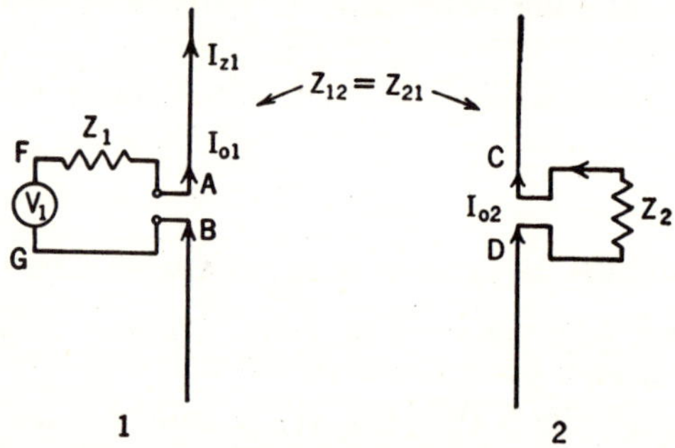

FIG. 18.1.—Circuit configuration involving two coupled antennas.

moved from the far zone with respect to antenna 1 into the intermediate or even into the near zone. Antennas 1 and 2 are now so close together that the distributions of current and the input impedances of both are significantly changed by the mutual retarded interaction of the moving charges in each. The self-impedance is altered, and a mutual impedance is required.

The following equations can be written for two driven antennas. The circuits are shown in Fig. 18.1 with the generator in circuit 2 omitted.

$$V_1 = I_{01}(Z_1 + Z_{s1}) + I_{02}Z_{12} = I_{01}Z_{11} + I_{02}Z_{12} \tag{18.2}$$
$$V_2 = I_{01}Z_{21} + I_{02}(Z_2 + Z_{s2}) = I_{01}Z_{21} + I_{02}Z_{22} \tag{18.3}$$

[1] The self-impedance of an antenna depends on the distribution of current, so that it is in general not the same when the antenna is in the presence of another antenna as it is when isolated. The symbol Z_0 has been used for the self-impedance of an *isolated* antenna. The symbol Z_{sm} will be used for the self-impedance of antenna *m in the presence of other antennas*. In general, Z_{sm} is a variable whose value depends on the configuration of neighboring antennas.

where Z_{s1} is the input self-impedance of antenna 1 in the presence[1] of antenna 2, Z_{s2} is the input self-impedance of antenna 2 in the presence of antenna 1, Z_{12} is the input mutual impedance of antenna 1 with respect to antenna 2, Z_{21} is the input mutual impedance of antenna 2 with respect to antenna 1, and Z_1 and Z_2 are lumped impedances in series with antennas 1 and 2, respectively. All self- and mutual impedances are referred to currents I_{01} and I_{02}. In all cases not involving media of variable permeability and dielectric constant, the following reciprocal relation is true (see Sec. 42):

$$Z_{12} = Z_{21} \tag{18.4}$$

Equations (18.2) and (18.3) are exactly the same in form as the equations for two coupled circuits in ordinary network theory. By use of the principle of superposition, the complex currents due to each applied voltage may be determined separately and the results combined algebraically. Accordingly, there is no loss in generality if V_2 is set equal to zero. If V_2 is zero, antenna 2 is said to be parasitic. If V_2 is not zero, its effect can be calculated by interchanging the subscripts 1 and 2 and using the principle of superposition. The equations for one driven and one parasitic antenna are

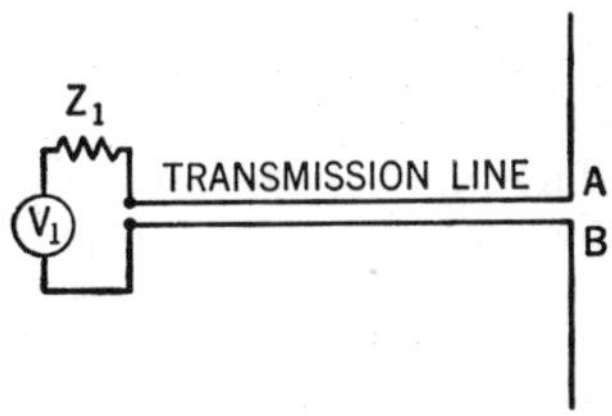

FIG. 18.2.—Antenna connected to a transmission line as load.

$$V_1 = I_{01}(Z_1 + Z_{s1}) + I_{02}Z_{12} \tag{18.5}$$
$$0 = I_{01}Z_{21} + I_{02}(Z_2 + Z_{s2}) \tag{18.6}$$

If a transmission line is connected between the input terminals AB of antenna 1 and the output terminals of the conventional (near-zone) network containing a generator, Fig. 18.2, the symbols in (18.5) must be changed as follows: Write V_{AB} (open) for V_1 where V_{AB} (open) means the open-circuit voltage across AB when the antenna is disconnected; instead of Z_1 write Z_{AB} where Z_{AB} is the impedance looking to the left at AB with the generator replaced by its internal impedance. It follows from Thévenin's theorem that (18.5) as changed is a true equation.

[1] Z_{s1} is the input impedance of antenna 1 in the presence of antenna 2 with antenna 2 open-circuited at the driving or load point where I_{02} is defined. It differs from the self-impedance Z_0 of antenna 1 when isolated, because there are currents in antenna 2 even when open-circuited at the center. The fact that I_{02} is zero does not mean that I_{z2} is everywhere zero.

By solving (18.5) and (18.6) for $Z_{FG} = V_1/I_{01}$, Fig. 18.1, the impedance Z_{FG} offered to the generator is found to be

$$Z_{FG} = (Z_1 + Z_{s1}) - \frac{Z_{12}Z_{21}}{(Z_2 + Z_{s2})} = Z_{11} - \frac{Z_{12}Z_{21}}{Z_{22}} \tag{18.7}$$

If (18.4) is used, and the following notation is introduced:

$$Z_{12} = |Z_{12}|\epsilon^{j\theta_{12}} \qquad \theta_{12} = \tan^{-1}\frac{X_{12}}{R_{12}} \tag{18.8}$$

$$Z_{22} = |Z_{22}|\epsilon^{j\theta_{22}} \qquad \theta_{22} = \tan^{-1}\frac{X_{22}}{R_{22}} \tag{18.9}$$

then

$$R_{FG} = R_{11} - \frac{|Z_{12}|^2}{|Z_{22}|}\cos(2\theta_{12} - \theta_{22}) \tag{18.10}$$

$$X_{FG} = X_{11} - \frac{|Z_{12}|^2}{|Z_{22}|}\sin(2\theta_{12} - \theta_{22}) \tag{18.11}$$

The input impedance Z_{AB} of antenna 1 in the presence of antenna 2 has the resistive and reactive parts

$$R_{AB} = R_{s1} - \frac{|Z_{12}|^2}{|Z_{22}|}\cos(2\theta_{12} - \theta_{22}) \tag{18.12}$$

$$X_{AB} = X_{s1} - \frac{|Z_{12}|^2}{|Z_{22}|}\sin(2\theta_{12} - \theta_{22}) \tag{18.13}$$

These relations reduce to those of ordinary transformer coupling if R_{s1} and X_{s1} apply to the primary coil, $Z_{12} = j\omega M$, $\theta_{12} = \pi/2$. For the transformer,

$$R_{AB}\,(\text{trans}) = R_{s1} + \frac{\omega^2 M^2}{|Z_{22}|}\cos\theta_{22} \tag{18.14}$$

$$X_{AB}\,(\text{trans}) = X_{s1} - \frac{\omega^2 M^2}{|Z_{22}|}\sin\theta_{22} \tag{18.15}$$

An important difference exists between the general case of two circuits coupled by the mutual impedance of antennas and two inductively coupled circuits in the near zone. Whereas R_{AB} (trans), (18.14), is always larger than R_{s1} by the resistance "reflected" into the primary from the coupled secondary, R_{AB} for the antenna, (18.12), may be *larger or smaller* than R_{s1}, depending on the distance between the antennas. In the transformer, the coupled secondary always represents an increased load on the primary due to the power dissipated in the secondary and supplied by the primary. With the antenna, the presence of a neighboring antenna necessarily implies an increase in the load due to the very small amount of power dissipated as heat in the coupled antenna, but it also means a change in the degree of coupling to the distant universe. This is manifested in certain directions by a complete or partial cancellation of the forces due to the currents in the two coupled antennas. A decrease in these directions in the power transferred to matter in the universe necessarily follows. This may be compensated by greater or smaller increases in other directions, so that the input resistance R_{AB} may be either larger or smaller than the self-resistance R_{s1} depending upon whether the presence of the closely coupled antenna increases or decreases the total power transferred to the distant universe. Analytically, this is determined by the algebraic sign of $\cos(2\theta_{12} - \theta_{22})$. Note that an antenna that is self-resonant, *i.e.*, adjusted to resonance when no other antenna is nearer than the far zone, is no longer resonant if another antenna is brought so near that the mutual impedance between the two is not negligibly small. The same conclusions are also true for more conventional coupled circuits.

19. Mutual Impedance of Antennas.—The mutual impedance (referred to input current) of one antenna in the presence of another depends upon the size, shape, and orientation of the conductors forming the two antennas as well as upon their conductivities. The analytical problem of deriving formulas for the mutual impedance and for the modified self-impedance of each of an unrestricted number of arbitrarily oriented antennas has not been solved either rigorously or to a reasonable degree of approximation except in the special case of two parallel antennas of the same length and radius. Most of the available data for mutual impedance have been calculated under the following assumptions:

1. The distribution of current in a driven antenna or in a parasitic antenna with arbitrary load or tuning reactance is the same when coupled to other antennas as when isolated.

2. The distribution of current for all antennas whether driven or parasitic and irrespective of load or tuning reactance is sinusoidal of the form $I_z = I_m \sin \beta(h - |z|)$. (Some writers even claim that "any convenient distribution of current" may be assumed.)

3. The self-impedance of an antenna is a constant independent of the presence of other antennas.

These three assumptions are reasonable approximations only in the special case when all of the antennas are extremely thin, and of half length $h = \lambda/4$. They are rough approximations for very thin antennas of half length h near $\lambda/4$. Curves for R_{12}, X_{12}, $|Z_{12}|$, θ_{12}, for the special case of infinitely thin symmetrical antennas of half length $\lambda/4$ are shown in Fig. 19.1 for values of b/λ (spacing over wavelength) up to 1.3. More extensive sets of curves are available in the literature.

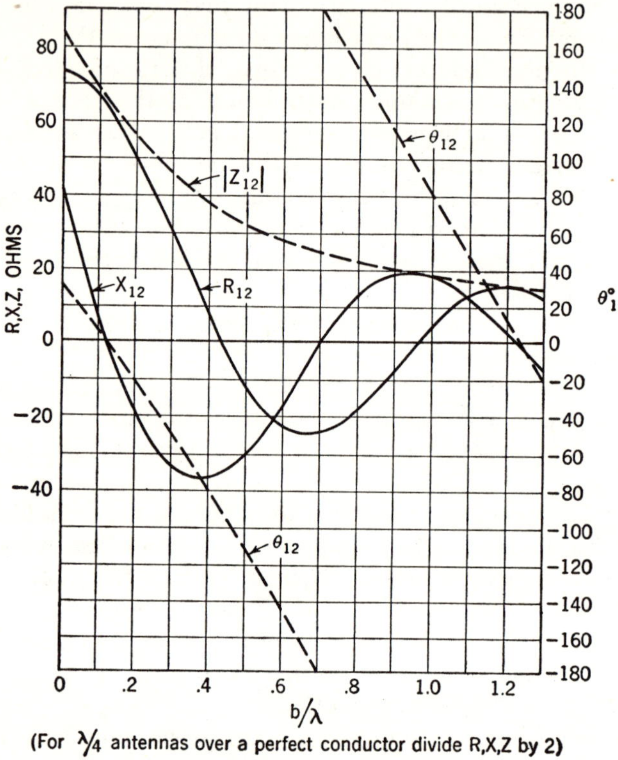

FIG. 19.1.—Mutual impedance $Z_{12} = R_{12} + jX_{12} = |Z_{12}|e^{j\theta_{12}}$ for parallel, infinitely thin antennas of half length $h = \lambda/4$.

A more accurate analysis of the problem of two coupled antennas of the same length and radius when center-driven or loaded in any way has been made proceeding from electromagnetic theory. Instead of assuming a distribution of current, the current in the two antennas is determined approximately, subject to boundary conditions. The self-impedances Z_{s1}, Z_{s2} and the mutual impedances Z_{12}, Z_{21} are defined to be the coefficients of I_{01} and I_{02} as written in (18.2) and (18.3) with $Z_1 = 0$, $Z_2 = 0$. For antennas of the same length and radius $Z_{s1} = Z_{s2}$, and by (18.4), $Z_{12} = Z_{21}$. Values of R_{s1}, X_{s1}, R_{12}, X_{12} computed in this way are given in Fig. 19.2 for antennas of half length $h = \lambda/4$ for four values of h/a corresponding to $\Omega = 2 \ln \frac{2h}{a} = 10, 15, 20, \infty$. The curves for

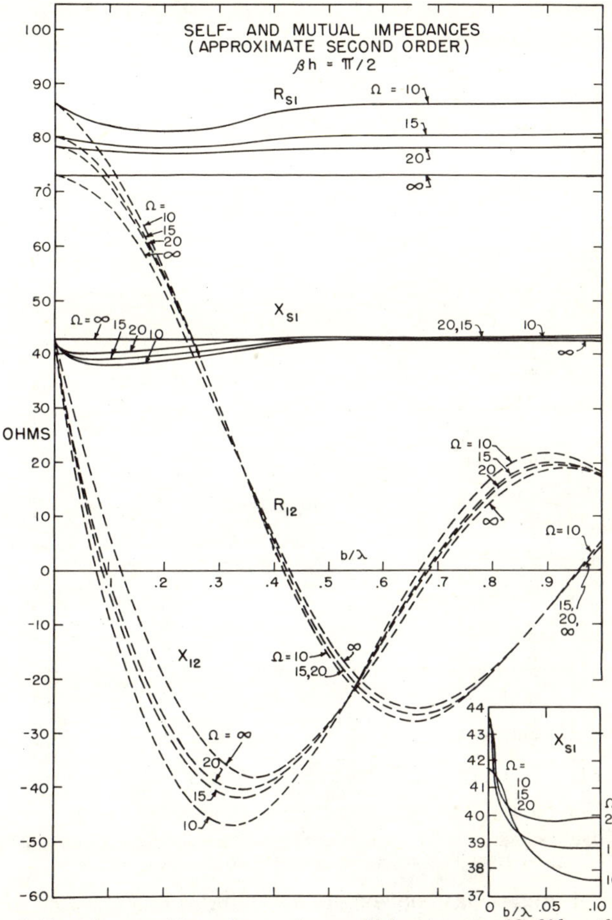

FIG. 19.2.—Self- and mutual impedances for parallel antennas of half length $h = \lambda/4$ referred to currents at the centers. The antennas are separated by a distance b.

$\Omega = \infty$ coincide with those of Fig. 19.1 for mutual impedance and with the self-impedance $Z_0 = 73.1 + j42.5$ for an indefinitely thin antenna. Values of R_0 and X_0 for an isolated antenna of the same radius may be obtained from Figs. 10.4 and 10.5. Curves for R_{s1} and X_{s1} for two parallel antennas of various lengths $2h$ are shown in Fig. 19.3. Values of R_{12} and X_{12} for the same range are shown in Figs. 19.4 and 19.5. The independent variable is $\beta b = 2\pi b/\lambda$. Note that for Figs. 19.2 to 19.5, driving conditions, tuning or load conditions do not have to be the same in the two antennas; only the lengths and radii must be the same.

Although the curves for R_{s1}, X_{s1}, R_{12}, X_{12} are correct only for two parallel antennas, a reasonable estimate for more than two

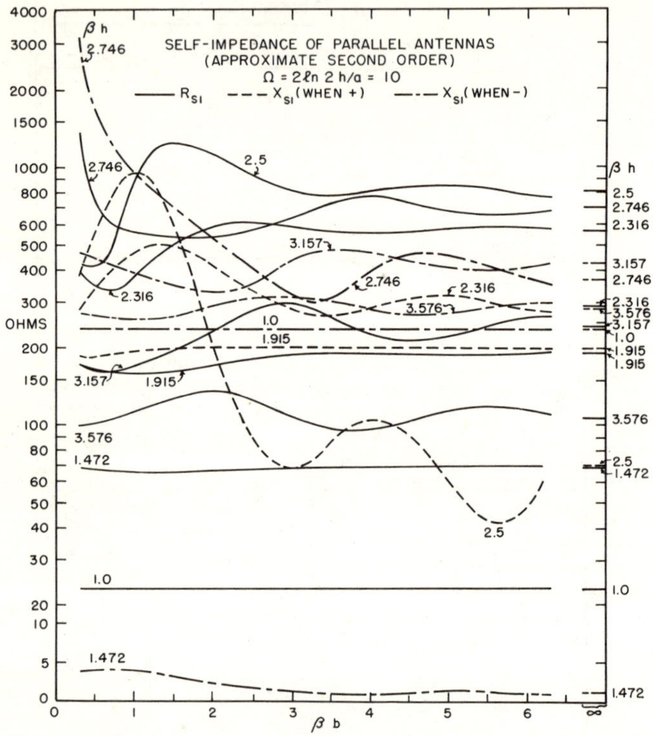

FIG. 19.3.—Self-impedances for parallel antennas with a range of half lengths h when $\Omega = 10$. The antennas are separated by a distance b.

coupled antennas may be expected using the curves of Fig. 19.2 for antennas of half length $h = \lambda/4$. This is actually not true of Figs. 19.3 to 19.5 for antennas with other half lengths.

20. Coefficient of Coupling between Antennas.—The input impedance of an antenna 1 in the presence of a single antenna 2 closer than the far zone is

$$Z_{AB} = Z_{s1} - \frac{Z_{12}Z_{21}}{Z_{22}} \tag{20.1}$$

Let a complex coefficient of coupling between antennas 1 and 2

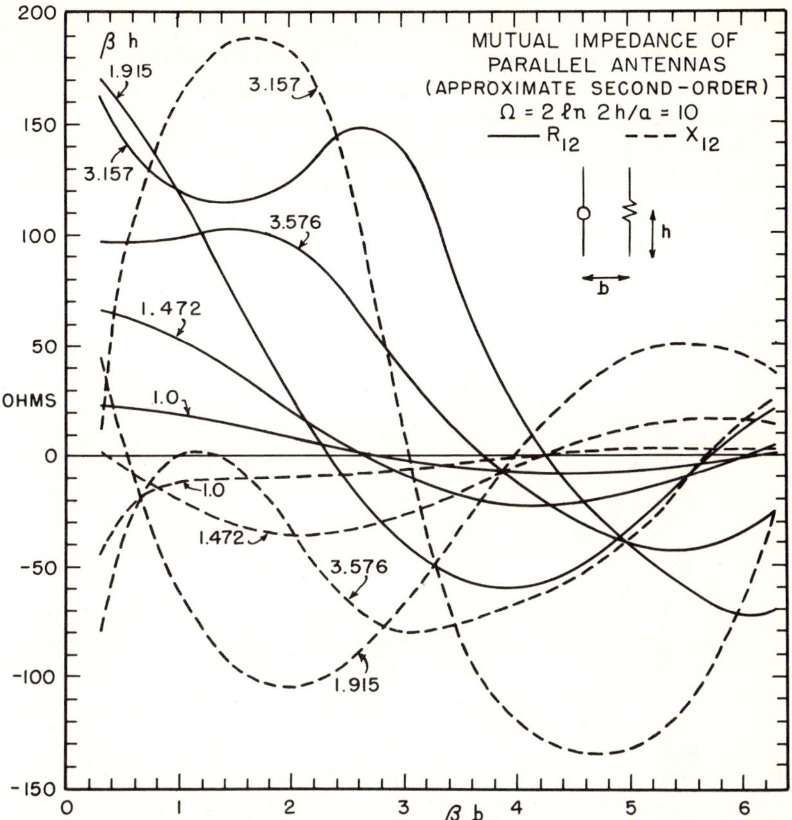

FIG. 19.4.—Mutual impedances for the antennas with self-impedances given in Fig. 19.3.

be defined by

$$k^2 = \frac{Z_{12}Z_{21}}{Z_{s1}Z_{s2}} \tag{20.2}$$

Then

$$Z_{AB} = Z_{s1}\left(1 - k^2\,\frac{Z_{s2}}{Z_{22}}\right) \tag{20.3}$$

Loose coupling may be defined by

$$|k|^2 << 1 \tag{20.4}$$

Condition (20.4) always is satisfied by antennas that are mutually in the far zone. Receiving antennas usually are loosely coupled to the transmitting antenna. The coefficient of coupling approaches but does not quite reach unity if the two antennas are self-resonant and are very close together. The resulting circuit closely resembles the two-wire line. All degrees of coupling are possible with antennas, and effects may be expected similar to those for coupled circuits of more conventional near-zone types. In particular, double resonance peaks for greater than critical coupling may be observed. Two important differences obtain: first, that

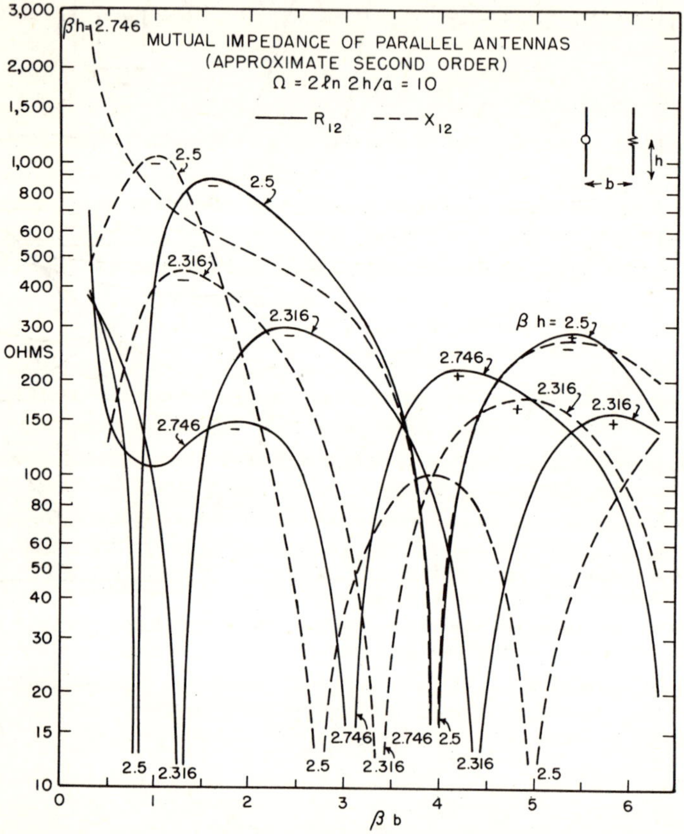

FIG. 19.5.—Like Fig. 19.4 with different values of the parameter βh.

mutual resistances may be negative as well as positive, and second, that unless the antennas are very close together resonance peaks are always blunt owing to radiation. Extensive quantitative results of coupled-circuit effects in antennas are not at present available from either theoretical or experimental investigations.

21. Coupling of Antennas and Transmission Lines.—In carrying out the mathematical analysis of two coupled antennas, no assumption was made regarding their orientation. The diagram of Fig. 18.1 suggest that they may be parallel, as is often the case in so-called "parallel arrays." Other orientations, such as antennas placed end to end, are also useful. Instead of coupling antennas only through their mutual impedance, additional coupling using coils, capacitors, or sections of transmission line (so-called "phase-reversing stubs") is often provided for special purposes. Furthermore, antennas are not always driven by potential differences maintained between two symmetrically placed terminals. Frequently they are asymmetrically driven by being coupled to each other more or less closely at one end or to a resonant or antiresonant section of a transmission line. The mathematical analysis of such cases depends on the intricate problem of providing data on mutual impedances. Consequently, calculation of the input impedance at the terminals of an antenna or of a transmission line that is coupled in one way or another to one or more antennas is not possible at present. In some cases, the radiation resistance R_m^e referred to a maximum sinusoidally distributed current has been computed. For available antennas, the distribution of current is often far from sinusoidal so that large errors in R_m^e are to be expected.

In attempting to understand in a qualitative way the operation and the distribution of current in resonant systems consisting of several coupled components of which at least one is an antenna, it is well to bear in mind that usually only highly conducting antennas and lines are involved and that in these the charges are free to redistribute themselves continuously in such a way that, at every instant and at every point along each conducting surface, the interacting tangential forces practically cancel. Accordingly, any redistribution of charge that follows the coupling of two circuits must be of such a kind as to lead to a virtual cancellation of tangential forces along all the conductors. In circuits consisting of several more or less closely coupled components, there may be more than one possible mode of oscillation. The mode that is actually excited depends upon the natural periods of the modes, upon the relative directions and magnitudes of the tangential forces due to the several interacting parts of the circuit, and upon the degree of coupling of these parts to the distant universe.

Before studying important coupled circuits involving both antennas and sections of transmission line, it is important to note that conventional transmission-line theory assumes that the currents maintained in the two conductors are at all times and at all opposite points equal in magnitude and in time-phase opposition, *i.e.*, $I_{1L} = -I_{2L}$. Whether these assumed conditions exist or not depends largely on the method used to drive the line. Thus, if the line is driven only by a generator that is symmetrically connected across the terminals at one end or symmetrically coupled at any point along the line, and if the line is symmetrically loaded, currents in the two conductors are equal and opposite, and radiation is extremely small with $b <<< \lambda$, as explained in Sec. 3. Such equal and opposite currents are called properly transmission-line currents and are designated by subscript L. On the other hand, if the line is not symmetrically driven or is not symmetrically loaded with respect to the two conductors, entirely different conditions may exist. If, for example, a two-wire line is parallel to an antenna, currents are excited in the *same* direction in the two conductors of the line; the line behaves like an antenna and radiates as if it were a single conductor with $I_{1A} = I_{2A}$. Cage antennas depend on this fact. Currents of this sort may be called antenna currents and indicated by subscript A. The line in Fig. 21.1 in which the lower half of the antenna is parallel to the line has both line and antenna currents. The resultant currents are

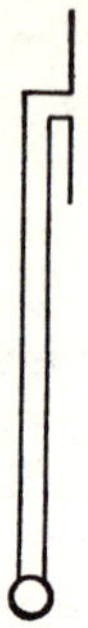

FIG. 21.1.—Antenna center-driven by line that has unbalanced currents.

$$I_1 = I_{1L} + I_{1A} \tag{21.1}$$
$$I_2 = I_{2L} + I_{2A} = -I_{1L} + I_{1A} \tag{21.2}$$

with I_1 not equal to I_2. Such a line is said to be unbalanced. The currents in every unbalanced line can be resolved into unidirectional antenna currents and equal and opposite line currents. The line currents may be analyzed by line theory; the antenna currents must be treated by antenna theory. The latter are always undesirable in a line that is used for transmission. In Fig. 21.1, the antenna currents in the line would be eliminated if the antenna were in a horizontal position with its halves symmetrical and at right angles to the line.

It may be supposed that unbalanced currents are never observed on coaxial lines, because these are enclosed in a metal shield, but this is not true. Actually, large antenna currents may be on the *outer surface of the coaxial line* while the transmission-line currents are on the inner surface of the outer conductor and the outer surface of the inner conductor. Thus, at high frequencies, a coaxial line is a *three- and not a two-conductor line.* Because of an asymmetrical structure from the point of view of driving it or loading it, antenna currents are often much more difficult to avoid on a coaxial line than on a two-wire line, as is discussed in detail in Sec. 25.

22. Collinear Array as Coupled Circuit—Phase-reversing Stubs. A number of important properties of coupled antennas may be described in terms of the collinear array. For simplicity, let this array consist of three identical self-resonant units, each something less than a half wavelength long, arranged end to end, Fig. 22.1. The central unit is driven from a resonant section of transmission

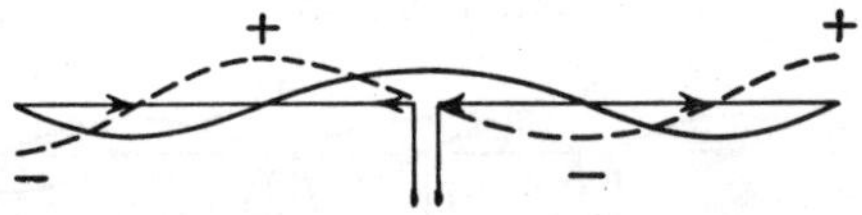

FIG. 22.1.—Center-driven collinear antenna of three elements. Solid line shows distribution of current; broken line distribution of charge. The + and − signs indicate the sign of the charge one-quarter period after maximum current shown by arrowheads.

line. Different methods of coupling the three units, Figs. 22.1 to 22.3, will be discussed in turn. Sinusoidal distributions of current and of charge per unit length are shown in solid and in dashed lines. In Fig. 22.1, the three units are connected so that they form a single resonant antenna of length near $3\lambda/2$.

In Fig. 22.2, the two outer antennas are moved away from the central one, decreasing the coupling. To a first approximation, the distributions of current and of charge per unit length along each of the three units are changed relatively little as the separations

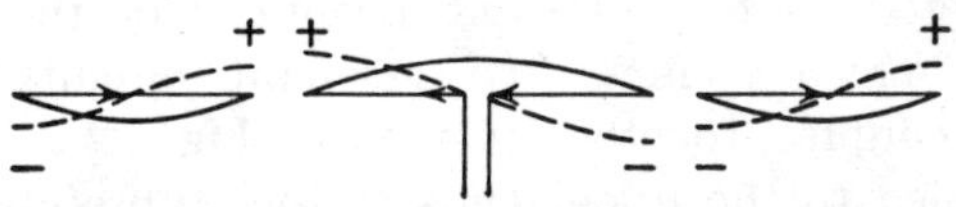

FIG. 22.2—Like Fig. 22.1 with outer $\lambda/2$ elements separated from central unit.

are increased from very small values, but the amplitudes in the outer units decrease rapidly. Adjustments in the length of the outer antennas or of the central unit produce coupled-circuit effects on the amplitude of current that resemble those in conventional coupled circuits in the near zone as the secondary or primary tuning is varied. Depending upon the degree of coupling as determined by the spacing, double or single resonance peaks may be obtained. The load on the generator driving the central unit is not correspondingly diminished when less power is transferred to the outer antennas as these are moved outward. As the degree of coupling of the central unit to the outer two is decreased, it is simultaneously increased to the distant universe, though in a smaller measure, *i.e.*, as less power is supplied to the two outer antennas, more power is radiated by the central unit. This is because the currents in the parasitic antennas are both directed opposite to the current in the central unit, so that a partial cancellation of the forces exerted by them on charges in the distant universe must result in most, but not necessarily in all, directions. The radiation resistance referred to maximum sinusoidal current for the antenna, Fig. 22.1, is 105.5 ohms, which is considerably less than three times the corresponding value, 73.1, for the central unit isolated.

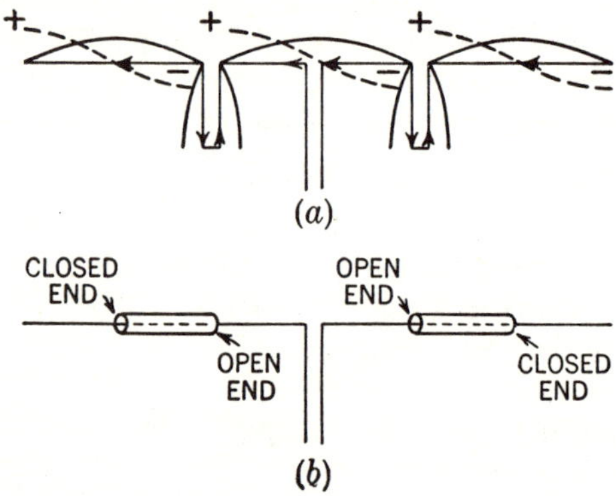

FIG. 22.3.—Like Fig. 22.2 with phase-reversing sections (*a*) of two-wire line, (*b*) of coaxial line inserted.

In Fig. 22.3*a*, a quarter-wave bridged-end section of two-wire line is connected to the adjacent terminals of the antennas of Fig. 22.2. There are now five resonant circuits. The three antennas are coupled together just as in Fig. 22.2, but they are now also coupled to the upper ends of the stub sections of transmission line. Very little power is required to drive only the stubs as coupled secondaries of the central antenna since they radiate only slightly because of the cancellation of forces arising from their almost, but not exactly, equal and opposite currents close together. When the outer antennas are joined, the situation is different. The only possible condition of resonance of all the coupled circuits in Fig. 22.3*a* requires equal and opposite currents and charges in the parallel wires of the stubs, and this reverses the currents and charges in the outer antennas. Thus the coupling forces between the stubs and the outer antennas, forces that are confined to short distances near the points of contact, are in opposition to those existing directly between the antennas that act to produce the distribution of current and charge of Fig. 22.1. But so long as the stubs are adjusted to resonance, the forces between them and the antennas are very much the stronger, and the distribution in Fig. 22.3*a* prevails.

Note that the currents in all three antennas are now in the same direction. This means that the coupling of the collinear array of three units to the distant universe is increased because cancellation of forces due to currents in different parts of the array is very much reduced for the array as a whole. Accordingly, for the same input current, much more power must be supplied to the collinear array of Fig. 22.3*a* than to the antenna of Fig. 22.1 or the coupled array of Fig. 22.2, *i.e.*, the input resistance at resonance is much greater for the collinear array than for the straight antenna of the same length. An estimate of the difference is obtained by comparing the radiation resistances referred to maximum sinusoidal current. For Fig. 22.2, R_m^e is 105.5 ohms; for Fig. 22.3*a*, R_m^e is 316.5 ohms. As a coupling device for transferring power to the distant universe, the collinear array is three times as effective as the linear radiator of the same length, $3\lambda/2$, and over four times as effective as one unit alone, assuming equal currents. An arrangement similar to that of Fig. 22.3*a* is shown in Fig. 22.3*b*. Coaxial sleeves have been substituted for two-wire stubs. Their operation is, however, much less satisfactory than that with the open-wire stubs in Fig. 22.3*a*. An alternative arrangement is with the coaxial sleeves moved in to the feeder at the center of the array but with their open and closed ends interchanged so that the open ends are still $\lambda/4$ from the center. It is important to note that the inside diameter of the coaxial sleeves must be large compared with the diameter of the antenna and their length considerably less than $\lambda/4$ if a sufficient shift in phase is to be achieved. Movable sleeves that are adjustable in length are desirable.

The collinear arrangement of Fig. 22.3*a* does not behave like a single tuned circuit as does the antenna of Fig. 22.1. It consists of five more or less closely coupled resonant circuits, and its behavior is correspondingly complex. If the tuned circuit analogue (*but not equivalent*) of a single antenna is taken to be a series resonant circuit coupled to an infinite line, as discussed in Sec. 16 and as shown in Fig. 22.4*a*, then the corresponding analogue of the collinear array approximates the circuit of Fig. 22.4*b*. It is clear that complicated coupled-circuit effects must be expected. In particular, the collinear array cannot be tuned to resonance merely by adjusting the length of the two outer units as can the antenna of Fig. 22.1. Each of the five circuits must be tuned separately, and multiple resonance peaks of current due to greater than critical coupling must be expected unless the separation of the parallel conductors in the stubs is extremely small. Because of the large load due to the coupled distant universe, resonance peaks are blunt.

If one of the outer antennas is detuned, the amplitude of its current decreases to a small value. A slight readjustment in the

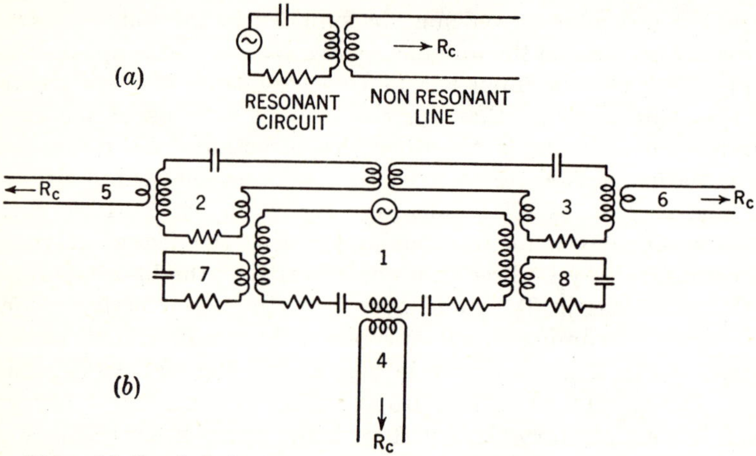

FIG. 22.4.—(*a*) Coupled circuit analogue (*not equivalent*) of antenna of Fig. 22.1. (*b*) Coupled circuit analogue (*not equivalent*) of the antenna of Fig. 22.3*a* or Fig. 22.3*b*. (1) Analogue of central antenna element of Fig. 22.3*a*. (2) Analogue of left antenna element of Fig. 22.3*a*. (3) Analogue of right antenna element of Fig. 22.3*a*. (4) Analogue of radiation load on central antenna element. (5) Analogue of radiation load on left antenna element. (6) Analogue of radiation load on right antenna element. (7) Analogue of left phase-reversing stub of Fig. 22.3*a*. (8) Analogue of right phase-reversing stub of Fig. 22.3*a*.

tuning of the rest of the circuit keeps this in resonance with no appreciable change in the distribution of current. For the same input current, the radiated power is diminished, as is the input resistance. For the same applied voltage, the amplitudes of current in all but the detuned antenna increase.

If one, or preferably both, of the coupling stubs be detuned sufficiently, the periodically varying, high concentrations of charge at their upper ends decrease in amplitude, and the current induced by tangential forces in the attached ends of the outer antennas decreases. If the current induced in the outer antennas as a result of their coupling to the attached stub is reduced below the opposing current induced as a result of their coupling to the central unit, a condition smilar to that of Fig. 22.2 is established. If the length of stubs is increased to nearly $\lambda/2$, Fig. 22.5*a*, resonance can be restored, and each stub *acts as a single conductor*, with both wires in parallel and with currents in the same direction. In Fig. 22.5*b* a

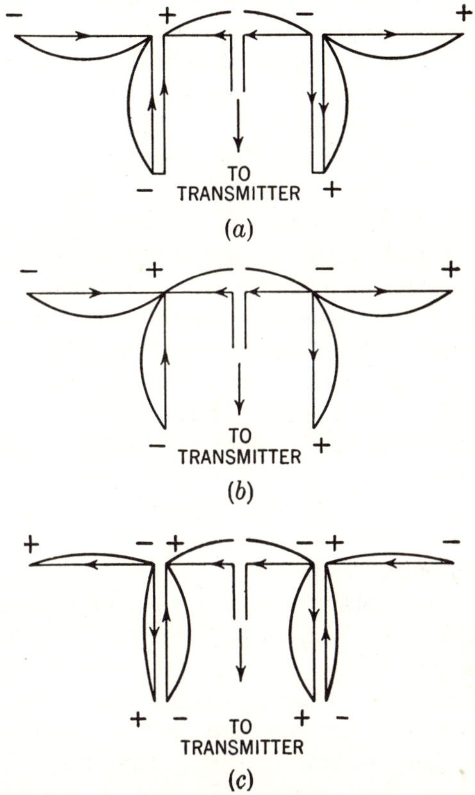

FIG. 22.5.—(*a*) Three horizontal half-wave elements coupled by half-wave short-circuited stubs which have equal currents in the same direction in both wires; (*b*) configuration equivalent to that of (*a*); like (*a*) but with the stubs open-circuited at the lower ends.

single conductor takes the place of each stub. Now there is the equivalent of five coupled antennas—three horizontal, two vertical, and a corresponding change in the coupling of the array to the distant universe. If each half-wave stub is a section of coaxial line, closed at the bottom by a metal disk connecting the inner and outer conductors, there is practically no current in the completely detuned interior. On the other hand, the outer surface is simply a thick antenna, which is self-resonant at a length somewhat less than $\lambda/2$.

If the terminating bridge (or disk) at the bottom of the stub is removed, the two-wire line (or the two-conductor interior of the coaxial line) is self-resonant in the normal way with equal and opposite currents on the two conductors. However, since the other mode of oscillation as in Fig. 22.5*a* is still possible on the two-wire stub (or on the outside of a coaxial stub) so long as the over-all length is such as to make the stub self-resonant, both modes will be excited simultaneously on the stub. In the two-wire line, the equal and opposite line currents of the one mode are superimposed on the unidirectional antenna currents of the other mode so that the distribution of Fig. 22.5*c* is obtained; in the coaxial line, the equal and opposite line currents and charges are on the inside, and the antenna currents and charges are on the outside. Since the two distributions on the stubs exert opposite forces on the charges in the outer antennas, the currents in these may be very small even though the antennas are adjusted to self-resonance. If a conducting bridge is moved up along each $\lambda/2$ stub, an adjustment may be achieved for which the opposing forces are equal so that there are *no currents on the outer antennas.* It may be concluded from this example that coupling stubs that are used for phase reversing should be chosen so that their over-all length is not near self-resonance for antenna currents.

If an open stub approximately a quarter wavelength long is connected at the center of each of the outer half-wave units, Fig. 22.6*a*, a similar problem is presented. The currents at the stub terminals due to the currents in the outer unit before the stub was inserted are equal and opposite, and transmission-line theory may be applied therefore to determine the impedance of and the distribution of current in the stub. For line currents, each stub is equivalent to a very low impedance connected at the center of the antenna. It can be adjusted to be a pure resistance of the order of magnitude of a few tenths of an ohm so that the stubs have little effect on the distribution of current in the array, *insofar as only equal and opposite transmission-line currents* are concerned. Taking these alone into consideration, the distribution of Fig. 22.6*a* is to be expected. In order to investigate the possibility of superimposed antenna currents on the stubs in Fig. 22.6*a*, it may seem reasonable to replace them by single conductors, as in Fig. 22.6*b*. However, this arrangement oscillates in the mode represented, which is quite different from the actual distribution of antenna current on the open stub shown in Fig. 22.6c. Since both the transmission-line mode of Fig. 22.6a and the antenna mode of

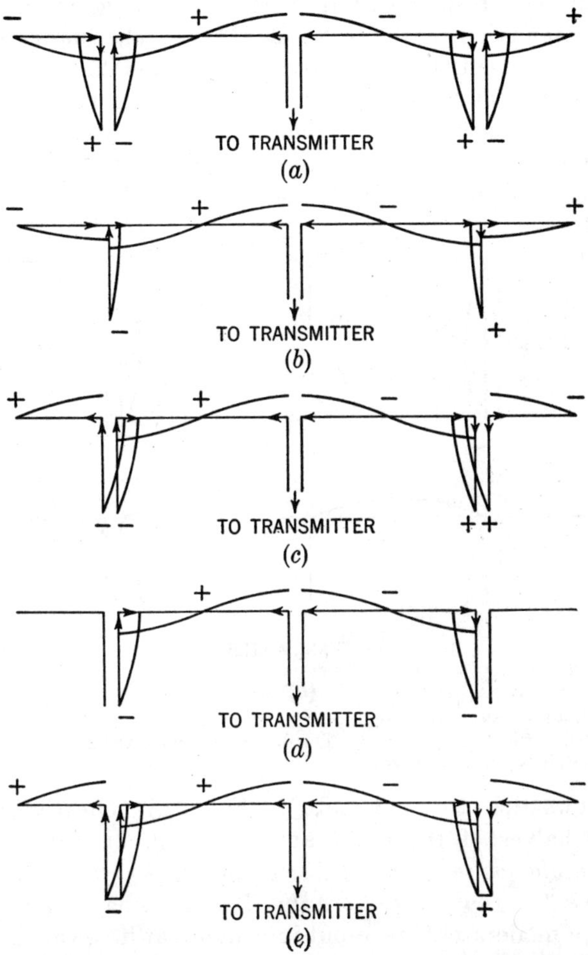

FIG. 22.6.—(a) Fictitious distribution of current assuming only equal and opposite line currents in the open-end $\lambda/4$ stubs. (b) Distribution with vertical $\lambda/4$ antennas in place of stubs. (c) Fictitious distribution assuming only antenna currents in the open-end $\lambda/4$ stubs. (d) Actual distribution obtained by superimposing (a) and (c). (e) Distribution of current with closed-end stubs; only antenna currents exist in the stubs.

Fig. 22.6c are possible, the actual distribution to be expected is a superposition of the two. This leads to a more or less complete cancellation of the currents in the outer conductor of each stub and in the outer half of the antenna attached to it, Fig. 22.6*d*. If the stubs in Fig. 22.6*a* are bridged by a conducting bar at their open ends, they present a very high impedance to transmission-line currents at their terminals, but there is no change in their ability to carry antenna currents, as in Fig. 22.6*c*. The line currents are therefore very small, and the distribution of Fig. 22.6*e* obtains.

If each stub in Fig. 22.6*a* is lengthened to a half wavelength, Fig. 22.7*a*, it presents an extremely high impedance to equal and opposite line currents at the center of each of the outer antennas. so that these currents are suppressed.

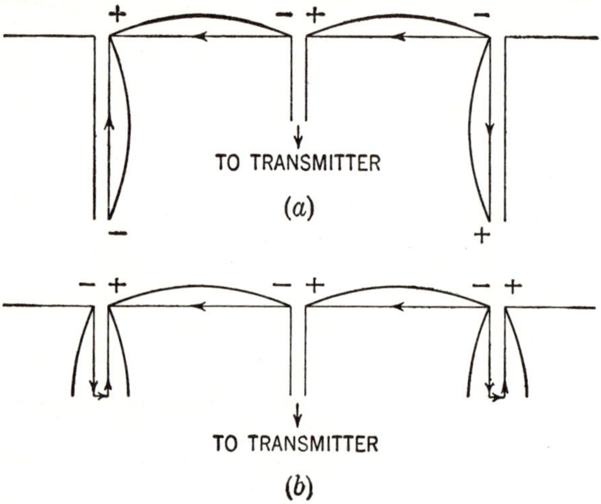

FIG. 22.7.—Possible antiresonant modes on an antenna (*a*) with $\lambda/2$ open-end stubs; (*b*) with $\lambda/4$ closed-end stubs. The outer sections of the antenna are detuned for the modes shown in both cases.

It follows that the $\lambda/2$ stubs in Fig. 22.7*a* effectively detune the outer halves of the outer antennas and, in fact, the entire antenna *insofar as the original distribution with a maximum current at the input terminals is concerned.* The antenna now has a high instead of a moderately low input impedance at its terminals. This does not mean that no current can be excited on the antenna. Actually the antiresonant distribution shown in Fig. 22.7*a* obtains. Owing to the fact that the stubs are $\lambda/2$ in length, they can oscillate with large codirectional antenna currents in each wire. Since the impedance of a $\lambda/2$ open-end stub to line currents is high, each of the $\lambda/2$ sections also can oscillate with equal and opposite line currents. A superposition of antenna currents and line currents leads to cancellation of the currents in the outer wire of each stub as shown in Fig. 22.7*a*. *This part of the circuit is detuned.* The same result is apparently also obtained with $\lambda/4$ closed-end stubs, as shown in Fig. 22.7*b*, where the natural mode of a *part* of the system is assumed to be excited. But the distribution of current shown in Fig. 22.7*b* involves only equal and opposite line currents on the stubs. Since the antenna currents of Fig. 22.6*e* also may be excited, the actual distribution is a superposition of the distributions in Figs. 22.6*e* and 22.7*b*.

If the open-wire stubs are replaced by coaxial sleeves of sufficiently large diameter somewhat shorter than $\lambda/4$, Fig. 22.8, the outer ends of the antenna are detuned as in Fig. 22.6*b*. Currents in the rest of the antenna oscillate as in Fig. 22.7*b*.

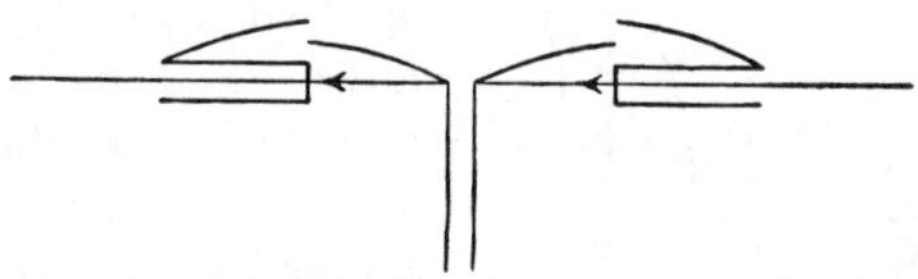

FIG. 22.8.—Coaxial sleeves to detune outer ends of antenna.

If the $\lambda/2$ stubs in Fig. 22.7*a* are each bridged by a conductor, modes are possible with distributions *along the horizontal antennas* resembling a superposition in various proportions of those of Figs. 22.6*a* and 22.7*a*. The relative amplitudes depend on the location of the conducting bridges along the stubs and on the precise location of the stubs along the antennas. With the bridges near the center of the stubs, the outer ends of the antenna may have no current, while the rest of the antenna is strongly excited, as in Fig. 22.7*a*. If the bridges are placed at the bottoms of the $\lambda/2$ stubs, both distributions may have almost equal amplitudes. The distributions along the stubs are superpositions of line and antenna currents.

The several arrangements discussed have been described in order to illustrate some of the complex effects that must be expected in antenna circuits and arrays consisting of several coupled parts with several natural modes. An important application that is considered in Sec. 27 is the problem of detuning a conductor such as the outside of a coaxial line, which is not intended to be part of an antenna but which nevertheless may have large and undesirable currents. The principle involved is clear from the cases just discussed. A high-impedance stub must be inserted in a conductor at a point where the resonant current has its maximum without the stub. Care must be exercised that another resonant mode is not made possible thereby.

23. The Coaxial Collinear Array—Phase-reversing Sleeves.—An interesting but not very effective arrangement of the collinear array makes use of the inside of a coaxial pipe as feeder, of the inside surfaces of coaxial sleeves (each about a quarter wavelength long and spaced at intervals of about a quarter wavelength) as coupling and phase-reversing stubs, and of the outermost surface, consisting partly of the coaxial pipe itself and partly of the coaxial sleeves, as the rather thick units of the antenna. A cross section of this arrangement for two collinear antennas is shown in Fig. 23.1a with the cross-sectional dimensions enlarged. Actually, both the diameter of the pipe and the separation between it and the coaxial sleeves are small compared with the wavelength. The complete array consists of the following six coupled circuits:

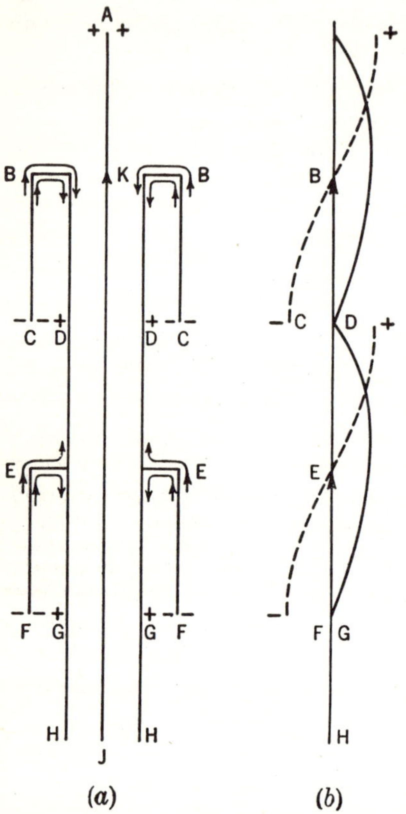

FIG. 23.1.—(a) Coaxial line feeding coaxial collinear array of two elements. (b) Distribution of current on outer surface of (a).

First, there is a coaxial feeder extending from the generator (below J in Fig. 23.1a) to K. At least a part of this line near the top must be resonant, and the entire line may be. The true line currents on the outer surface of the inner conductor and the inner surface of the outer conductor are equal and opposite and extremely close together compared with the wavelength so that the forces they exert at distant points cancel for all practical purposes.

The second coupled circuit is the self-resonant center-driven antenna ABC of which one half is the central conductor AK and the other half the outermost surface BC. (Because the section BC of the antenna is thicker than AK, it must be shortened much more below a quarter wavelength than AK if it is to be self-resonant.)

The direction of current and location of maxima at $t = 0$ are indicated by arrows; the location and sign of maximum charge a quarter cycle later also are shown.

The third coupled circuit is the self-resonant stub of length near $\lambda/4$ with open end at *CD* and closed end at *B*. It is formed by the inner surface of the sleeve and the outside of the pipe. Currents and charges are equal and opposite and very close together so that the forces they exert cancel at distant points.

The fourth coupled circuit is the antenna consisting of the outer surface *DEF*

The fifth coupled circuit is the resonant quarter-wavelength sleeve with open end at *FG* and closed end at *E*.

A sixth circuit consists of the outer surface of the pipe from *G* to its end below *H* including all attached and coupled conductors such as the generator circuit and the earth. This sixth circuit may be resonant or detuned depending upon its length and the arrangement of other attached or coupled circuits.

If power is to be transferred to conductors in the distant universe primarily by coupling to the collinear array made up of the two antennas *ABC* and *DEF*, the outside of the feeder line to *G* must be kept detuned. Methods of achieving this are considered in Sec. 27. In this case, much as in one of the outer antennas of Fig. 22.3, the antenna *DEF* is coupled both to antenna *ABC* and to the section of transmission line with open end at *CD*. The induced currents from these two are nearly opposite in direction, but the currents due to the sleeve are greater if it is adjusted to resonance, as is assumed in Fig. 23.1. If this is not true, in particular if the sleeve *BC* is removed or the open end at *CD* is closed with a disk, the retarded forces due to the moving charges in *ABC* are alone active, the current in *DEF* is reversed, and radiation is reduced.

For maximum radiation, each antenna and each sleeve must be individually tuned. (Collinear arrays with more than two coupled antennas are readily constructed by attaching units above *A* and below *H*, Fig. 23.1.) This cannot be done as accurately in the coaxial collinear array of Fig. 23.1 as in the structurally less attractive form of Fig. 22.3*a*, because the half length *BC* (outer surface) of the coaxial antenna cannot be made shorter than the half length *BC* (inner surface) of the sleeve. A section of a coaxial line that is open at one end and closed at the other always has a resonant length that differs much less from $\lambda/4$ than does the half length of an antenna of the same or greater outer diameter. Since the antenna and sleeve are closely coupled, the adjustment for maximum and as nearly as possible equal currents in the two antennas is not precisely that for self-resonance in each case. It depends on the degree of coupling and is very difficult to achieve in practice since so many variables and parameters require adjustment. In order to assure adequate coupling between the collinear antennas and the coaxial coupling section, the inner diameter of each coaxial sleeve must be much greater than the outer diameter of the conductor on which it is placed.

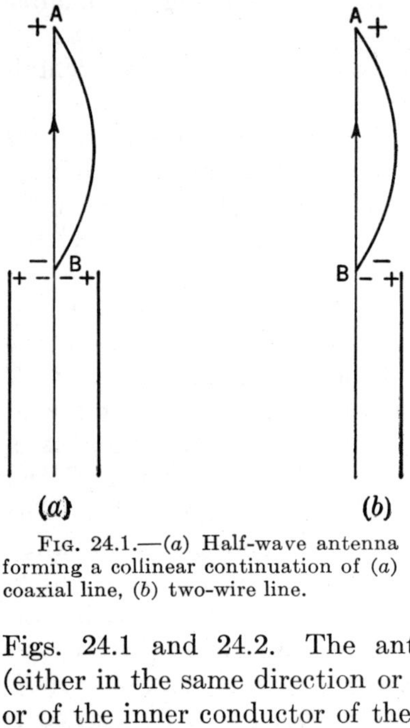

FIG. 24.1.—(a) Half-wave antenna forming a collinear continuation of (a) coaxial line, (b) two-wire line.

24. End-coupled Half-wave Antenna.—A single antenna of length near $\lambda/2$ may be center-driven from a parallel line as in Fig. 18.1 or from a coaxial line by the arrangement for the top antenna *ABC*, Fig. 23.1*a*. The sleeve *EF* is not present, and the outside of the line below *D* must be detuned by methods to be described. An antenna of this same length may be end-coupled to a resonant two-wire line or to a coaxial line or to a short impedance-transforming section of such a line using the arrangements of Figs. 24.1 and 24.2. The antenna is merely the continuation (either in the same direction or at right angles) of one of the wires or of the inner conductor of the coaxial line or pipe. Electrically, the length *AB* is a self-resonant antenna closely coupled to the resonant two-conductor transmission line inside the coaxial pipe and also to the outer surface of the pipe, or it is coupled to the two wires of the open line acting simultaneously as a transmission line and in parallel as a single conductor. If it is assumed that resonant currents on the outer surface of the coaxial line or on the two-wire line acting as a single conductor are minimized by detuning (either by adjustment in over-all length or by other methods to be described in Sec. 27), there remain only two closely coupled circuits, the antenna *AB* and the transmission line. These are coupled by the interaction of forces between charges near the junction point *B* of antenna and line. The degree of coupling depends on the spacing of the two conductors of the line; but, unless this is very small, the coupling may be so close that double resonance peaks of current are observed as the tuning of antenna or line is varied.

If the length of the line is such that the outer surface of the coaxial line or the two wires of the open line treated as a single conductor become resonant, this radiates as a coupled antenna. In the two-wire line, the resonant "antenna" current is superimposed upon the transmission-line currents so that the two-wire line acts simultaneously as transmission line and antenna. If the two conductors are enclosed in a metal shield for their entire length, the unbalanced or "antenna" current is on the outside of the shield just as on the coaxial line.

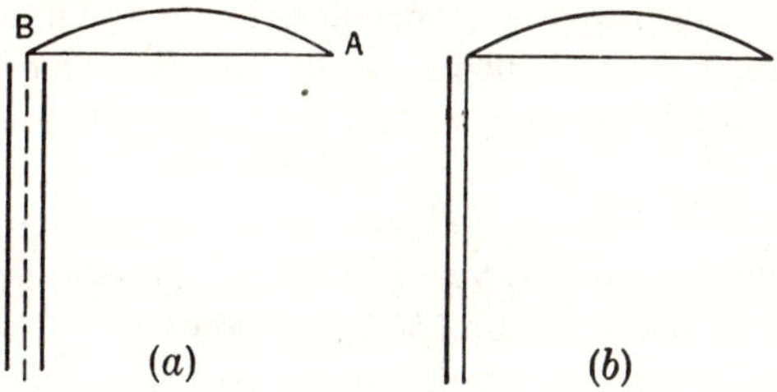

FIG. 24 2.—Arrangements similar to those of Fig. 24.1 but with the antenna at right angles to the line.

25. Unsymmetrical Antennas and Arrays.—The symmetry of an antenna or of an array is measured in terms of the geometrical arrangement of the conductors with respect to the input terminals. In order to define an input impedance for an antenna or an array in the conventional low-frequency or near-zone sense, the input terminals must be so close together that they may be connected as part of a near-zone network or, more commonly, to the end of a transmission line the conductors of which are sufficiently close together to satisfy the condition for the near zone. The currents at these terminals must be equal and opposite. In a coaxial line, an "antenna" current is on the outside of the outer conductor at sufficiently high frequencies so that separate equal and opposite transmission-line currents may be distinguished. On the other hand, in a two-wire line, an "antenna" current is in the same direction on both conductors of the line superimposed on the conventional transmission-line currents. Experimentally there is no way to separate these currents.

A long transmission line or a stub section of line does not carry a significant "antenna" current if the coefficient of coupling between the line (acting as a single conductor) and the entire array is small, and if the line with all that is attached to it is detuned. In each of Figs. 22.1 to 22.3 the antenna or array is itself symmetrical and the two-wire line is symmetrically placed, so that the retarded forces tangent to the line (which are exerted by the moving charges in the antenna on the charges in the line) cancel and the mutual impedance vanishes. The two wires of the open line thus carry equal and opposite transmission-line currents and no "antenna" current. The input impedance of the array may be defined in the conventional way and can be determined experimentally by substitution or bridge methods. The coaxial array of Fig. 23.1 and the end-fed antennas of Figs. 24.1*a* and 24.2*a* are unsymmetrical. The charges on the outer surface of the coaxial line (below G in Fig. 23.1*a*) experience uncanceled retarded forces from the moving charges in the antennas above, so that the coefficient of coupling does not vanish. The outer surface of the entire coaxial pipe (below G in Fig. 23.1*a*) must be detuned, or it will carry an antenna current and will be a part of the array. The input impedance of the entire array, which is also the terminating impedance of the coaxial line, may be defined in the conventional manner, because the inner conductors of the line carry equal and opposite currents except very near the open ends. Since a small end correction is always included with a terminating impedance of conventional type, the same may be done here. (The equations of the transmission line imply an infinitely long line. If the line is finite, the error made in assuming the constants per unit length of line to be the same near the ends as far from the ends is absorbed into the terminating impedance except when the end is open, when it appears as a small end correction in length.)

In Fig. 24.1*a* or Fig. 24.2*a*, the component of current at the junction point B of line and antenna is the component I_z'', discussed in connection with (9.1) for an antenna of nonvanishing radius and of antiresonant length near $\lambda/2$. The approximately equal and opposite current $-I_z''$ on the inner surface of the outer conductor of the line continues to the outside of the coaxial line at the upper end, even if this is detuned. For thin antennas, this current is small. A substitution or bridge method of measuring the impedance *at the terminals* of the antenna generally is not reliable, because the outside of the coaxial cable is a part of the antenna even if detuned, and theoretically, therefore, it would have to be detached as part of the antenna and leave only the inside transmission line, which is physically impossible. On the other hand, if the constants and the length of the transmission line are known accurately, the input impedance of the line at the generator may be measured and the terminating impedance calculated. A similar situation exists when the inner rod of a coaxial line projects from the metal surface of an airplane. The antenna is the *entire plane and the rod.*

In the unsymmetrical arrangements of Figs. 24.1*b* and 24.2*b*, the two conductors of the two-wire line are unbalanced if an appreciable "antenna" current is in the same direction in both conductors. It is somewhat unbalanced even if this is avoided by careful detuning of the line treated as a single antenna, because the current at the junction point B of one of the conductors of line and antenna does not vanish if the antenna has a physically realizable radius. A component of current I_z'' continues into the antenna, and this can have no equal and opposite component on the other parallel wire, because this wire ends and the current must vanish. Accordingly, the two-wire line, Figs. 24.1*b* and 24.2*b*, is slightly unbalanced at best if the antenna is thin and the line is detuned as an antenna; at worst it is entirely out of balance if the line is not detuned.

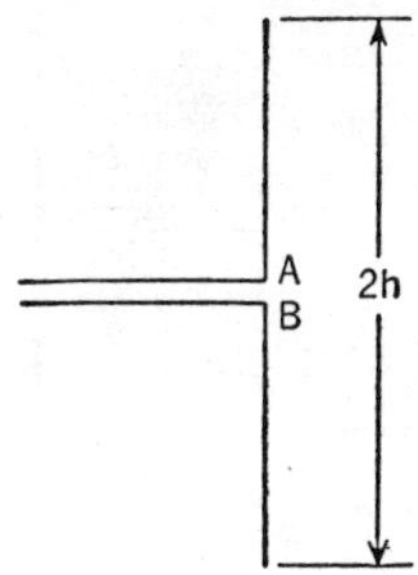

FIG. 25.1.—Antenna center-driven from two-wire line.

Up to this point, nothing has been said of the possibility of substituting a coaxial line for the two-wire line in the completely symmetrical arrangement of Fig. 25.1. It might be supposed that the circuit of Fig. 25.2*a* using a coaxial line is as symmetrical as that of Fig. 25.1 with the two-wire line. Insofar as the cancellation of *tangential* forces along the outer surface of the coaxial line is concerned, this would certainly be true everywhere except in proximity to the points A and B (where the antenna is attached) if the halves of the antenna carried the same current and charges of opposite sign distributed in the same way. The fact is, however, that the distribution of metal and hence the distribution of charge at the end of a coaxial line are unsymmetrical with respect to the halves of the antenna. In the enlarged line of Fig. 25.2*b*, the periodically charged outer conductor at B' maintains forces on the antenna between A and B'. The same forces do not act on the lower half of the antenna, and a condition of unbalance therefore exists. It can be reduced somewhat if the end of the coaxial line is cut away near the upper half of the antenna, Fig. 25.2*c*.

A further unbalance results from the fact that the axis of symmetry for electromagnetic effects due to the coaxial line is the central conductor, whereas the point of symmetry for the halves of the antenna is midway between the terminals A and B. Another way of expressing this same dissymmetry is in terms of the current into the halves of the antenna. That which is directed upward at A continues along the outer surface of the same conductor bent at a right angle, whereas that which turns down must change its rotationally symmetrical distribution along the inner surface of the outer conductor and converge to a single point B. Accordingly,

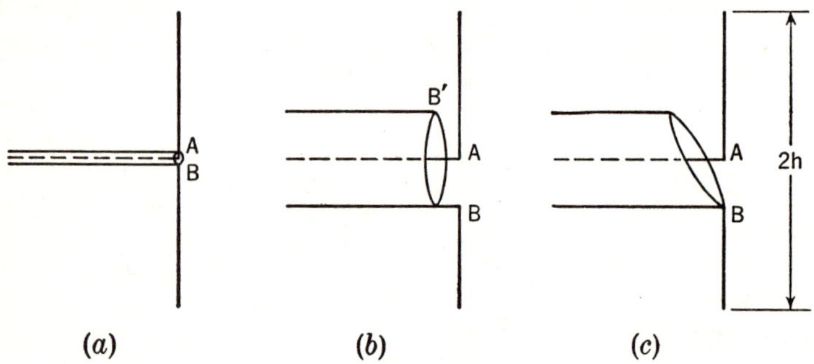

Fig. 25.2.—Antennas center-driven from coaxial lines.

the adjustment for resonance and the maintenance of similar distributions and amplitudes of current and charge in the halves of the antenna are somewhat improved if the part of the antenna that is attached to the outer conductor of the coaxial line is made shorter than the other part of the antenna, Fig. 25.2*b*,*c*.

Neither cutting away the end of the outer conductor of the coaxial line nor shortening the lower half of the antenna is sufficient to assure symmetry. In fact, the geometrical structure of Fig. 25.2*c* is so obviously asymmetrical as to make it perfectly clear that the forces acting on charges in the outer surface of the coaxial line in a direction parallel to its axis due to currents and charges in the halves of the antenna cannot be expected to cancel completely near the end of the line. If small tangential forces remain, there must be currents on the highly conducting outer surface of the coaxial line; and, if the outer surface of the line is not detuned, the currents will be large and the outer surface of the line will act as a coupled antenna.

If the radius of the coaxial line is very small (as required by the condition of the near zone) so that the distance AB in Fig. 25.2 is an extremely small fraction of a wavelength, the dissymmetry is relatively slight. Even so, currents due to resonance along the outer surface of the coaxial line may be significant. At very short wavelengths, it is not always possible to keep the radius of a coaxial line a small fraction of a wavelength, because spark-over can occur. If the distance AB is an appreciable fraction of a wavelength, as in Figs. 25.2*b* and 25.2*c*, the dissymmetry is great, and relatively large uncanceled forces may be expected to act on the charges in the outer surface of the coaxial line. This must be kept detuned therefore if resonant amplitudes are to be avoided.

26. Transmission-line Feeders.—The connecting circuit between a generator with its associated network and an antenna

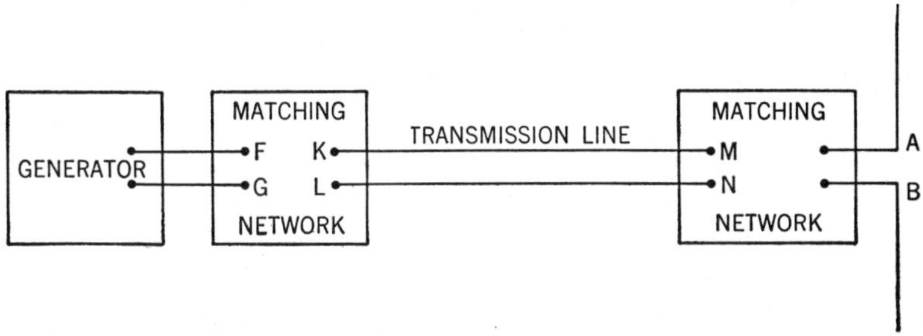

FIG. 26.1.—Transmission-line feeder with matching networks.

usually consists of a transmission line. Such a line is commonly of the two-wire, the four-wire, or the coaxial type, and it may be long or short. A typical circuit is shown schematically in Fig. 26.1. It consists of a generator with output terminals FG, a matching and tuning circuit for the generator with output terminals KL, a transmission line with output terminals MN, and a tuning and matching network for the antenna with output terminals AB which are also the input terminals of the antenna. The most effective over-all transmission of power is achieved when the impedance looking to the right at FG presents the *optimum load* for the generator, and the power losses in the line and in the matching networks are a minimum. The losses in the matching networks may be kept small by using only circuit elements with low resistance; if sections of line are used, they must be short. The losses along a transmission line are smallest when it is terminated in its characteristic impedance Z_c, and the impedance Z_{MN} looking to the right at MN should therefore be equal to Z_c. Thus, the matching network for the antenna must transform the input impedance Z_{AB} of the antenna into an impedance Z_{MN} looking to the *right* at MN and given by Z_{MN} (right) $= Z_c$. If the generator is designed to feed a load Z_{opt} for optimum performance (*e.g.* maximum efficiency, or maximum power), Z_{FG} (right) $= Z_{\text{opt}}$. Accordingly, the matching network for the generator must transform the input impedance Z_{KL} (right) of the line at KL into the desired Z_{opt} at FG. If the line is terminated in Z_c, Z_{KL} (right) $= Z_c$, and the matching network of the generator must transform Z_c looking to the right at KL into Z_{opt} looking to the right at FG. If the generator has been designed so that $Z_{\text{opt}} = Z_c$, the matching network for the generator is unnecessary. For most lines, $Z_c = R_c + jX_c$ is predominantly resistive with X_c very small and negative. Thus, for purposes of matching, $Z_c = R_c$; $X_c << R_c$.

If the condition Z_{MN} (right) $= Z_c$ is fulfilled by a suitably adjusted matching network for the antenna, the transmission line is said to be nonresonant. A nonresonant line is a line that is terminated specifically and only in its characteristic impedance. Advantages of the nonresonant line are low losses, the absence of high resonant voltages, and broader band frequency characteristics of the system as a whole. If the line is short, not over one or two wavelengths, the losses in the line are negligibly small compared with the power transferred to the antenna; and, insofar as losses are concerned, it is unimportant whether the line is or is not nonresonant. If the line is long, the losses in the line may be excessive unless it is nonresonant.

Even though terminated in its characteristic impedance, it may be impossible to make a line nonresonant if the supporting insulators or spacers are incorrectly placed. At long wavelengths, supporting insulators are usually separated a negligible fraction of a wavelength for both two-wire and coaxial lines, and no difficulty is encountered. At extremely high frequencies, the spacing of dielectric beads or other supports may be an appreciable fraction of a wavelength. If the beads occur at intervals of slightly less than $\lambda/2$, successive partial reflections occur at each bead, and the effect is cumulative. Accordingly, resonant amplitudes may be built up even though the line is correctly terminated. The contrary is true if the beads are arranged in pairs, the spacing between the two beads being approximately $\lambda/4$, and with no restriction as to the interval between successive pairs. At one frequency, a partial reflection at one bead is canceled practically by that at the next bead of the pair. An important disadvantage of the nonresonant line is that it does not lend itself readily to multiband operation, because it is not possible to provide a single matching network that will terminate the line in its characteristic impedance at more than one or at most two frequencies.

For short distances and for multiband operation, resonant transmission lines may be used. In this type of operation, the antenna-matching network is not required. The impedance Z_{KL}, Fig. 26.1, looking to the right at KL is that of the transmission line terminated in the antenna. The condition for optimum performance of the generator requires the network matching the generator to transform Z_{KL} (right) at terminals KL into Z_{FG} (right) $= Z_{\text{opt}}$ at terminals FG. The condition X_{FG} (right) $= X_{\text{opt}}$ is usually equivalent to tuning the entire circuit including the transmission line to resonance using the matching network of the generator.

27. Detuning Sleeves; Line Transformers.—In analyzing transmission lines, it is assumed that "antenna" currents are absent either in the form of an unbalanced current on open-wire lines or a current on the outer surface of a coaxial line. Such currents usually can be avoided if symmetrical center-fed systems are used in conjunction with symmetrically placed open-wire lines as feeders. They are automatically avoided where a transmission line, either open or coaxial, is very close to the surface of a highly conducting earth upon which an array of any configuration is erected. The distribution of charges and currents in the highly conducting earth is always such as to lead to a virtual cancellation of all tangential forces along its surface.

If a cancellation of tangential forces along the transmission line cannot be accomplished by a symmetrical arrangement, the currents can be minimized by detuning the line (open or coaxial) treated as a single conductor. In fixed installations, this can be accomplished sometimes by adjusting the over-all length. It often can be accomplished by following the method already mentioned in Sec. 22 in connection with Fig. 22.6*b* where part of a resonant antenna is detuned by cutting it at a point of maximum current and connecting the terminals so obtained to the open end of a high-impedance stub.

Any conductor, or group of conductors excited in parallel as a single conductor, can be detuned *with respect to a particular mode* by inserting a high-impedance stub at a point where a current maximum would be if the conductor were resonant without the stub

However, as is pointed out in conjunction with Fig. 22.6, another mode of oscillation that has a minimum of current at the open end of the inserted stub may be excited if the effective over-all length of the line from this open end of the stub is electrically equivalent to an integral number of half wavelengths. A change in this length or additional stubs suitably placed to detune this mode then are required. A number of arrangements for detuning two-wire lines and coaxial lines as single conductors are shown in Figs. 27.1 to

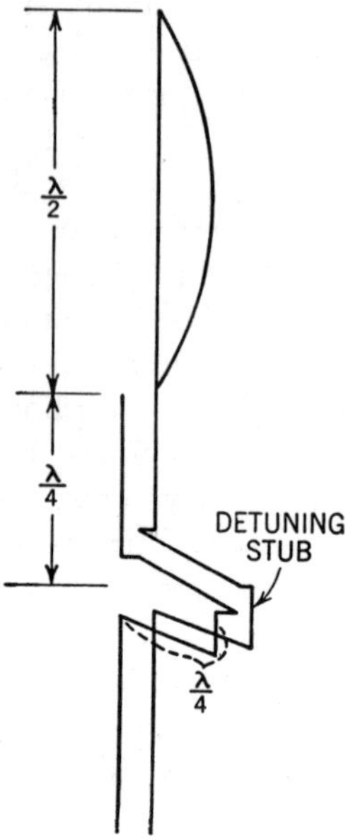

FIG. 27.1.—Arrangement for detuning a two-wire line end-driving a λ/2 antenna.

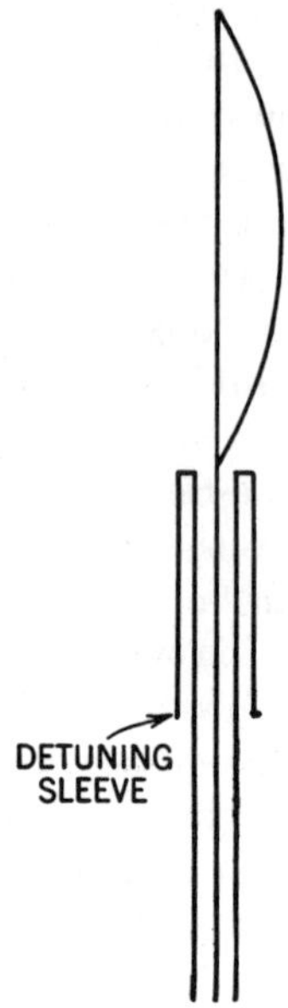

FIG. 27.2.—Arrangement for detuning coaxial line end-driving a λ/2 antenna.

27.4. In all these, the transverse dimension is exaggerated for clarity.

In Fig. 27.1, a closed-end detuning stub is connected into the two-wire line of Fig. 24.1*b*. The two sides of the stub are mutually at right angles, and there is therefore no interaction of the equal and opposite transmission-line currents; whereas, to the antenna current (which is in the same direction at corresponding points of the two parallel conductors), the two wires of the line are in parallel. A resonant antenna current would have a current node at the upper end or junction with the antenna. The first current loop, therefore, would be λ/4 from the end. If the line is cut at this point and a $\lambda/4$ closed-end stub is inserted, Fig. 27.1, the line is detuned as an antenna.

In Fig. 27.2, the same thing is accomplished for a coaxial line. A $\lambda/4$ detuning sleeve is provided with its open end a quarter wavelength from the end of the ine. The sleeve might equally well be moved down $\lambda/4$ if the upper end were open and the lower end were closed.

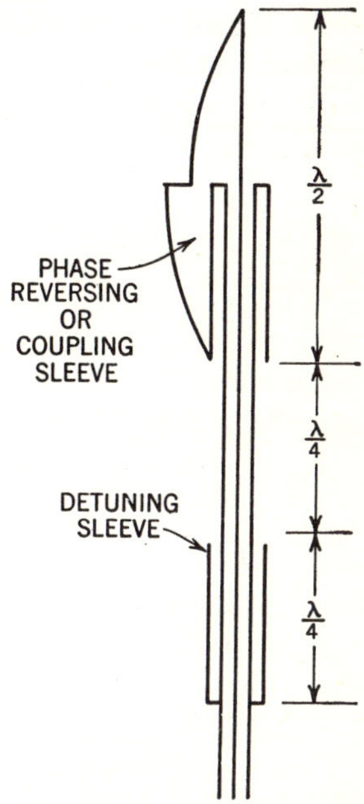

Fig. 27.3.—Detuning sleeve on a coaxial line center-driving a coaxial half-wave antenna.

In Fig. 27.3, the center-fed antenna uses the outer surface of the phase-reversing or coupling sleeve as its lower half. The inside of this sleeve is a resonant-coupled circuit, as explained in Sec. 23. If the second (lower) sleeve were closed at the top and left open at the bottom, it also would be a phase-reversing or coupling sleeve, Fig. 23.1*a*. However, with the open end at the top and the closed end at the bottom, Fig. 27.3, the lower sleeve serves to detune the outside of the coaxial line. Such a sleeve could be added in Fig. 23.1*a* to detune the feeding line.

In Fig. 27.4, the outside of the coaxial line used to center-feed an antenna is detuned by a detuning sleeve. The effects of the partly uncanceled forces acting tangentially along the outside of the line may be summarized roughly by stating that a part of the current from the lower antenna is along the outside instead of only along the inside, as it would be if no dissymmetry existed. A resonant current on the outside has its maximum directly at the end. Therefore, the open end of a coaxial $\lambda/4$ detuning sleeve must be placed there. If more convenient, it could also be placed with its open end a half wavelength from the terminals of the antenna. When an antenna is fed from a flexible coaxial line, Fig. 27.4, a detuning sleeve is essential; for, if the cable is moved or coiled, its electrical length is altered and

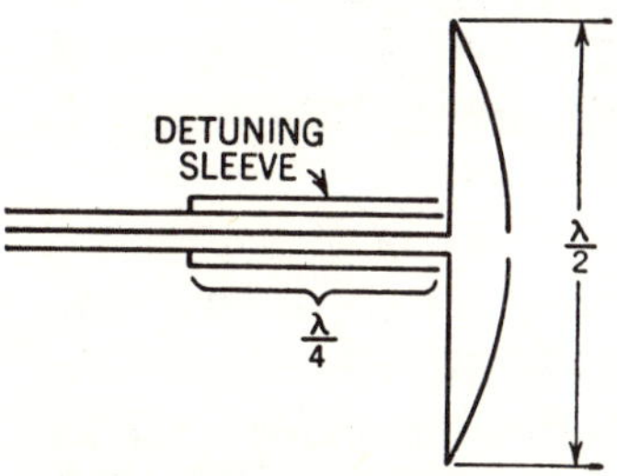

Fig. 27.4.—Detuning sleeve on a coaxial line center-driving a half-wave dipole.

resonance may be established in one position and not in another. This is a most undesirable condition, since the impedance at the terminals of the antenna and the power radiated depend on the manner in which the cable is coiled. To provide adequate coupling, detuning sleeves must be considerably larger in diameter than the coaxial line. To permit adjustment for maximum detuning, it is desirable to have them variable in length near λ/4. This is easily accomplished with a telescoping sleeve.

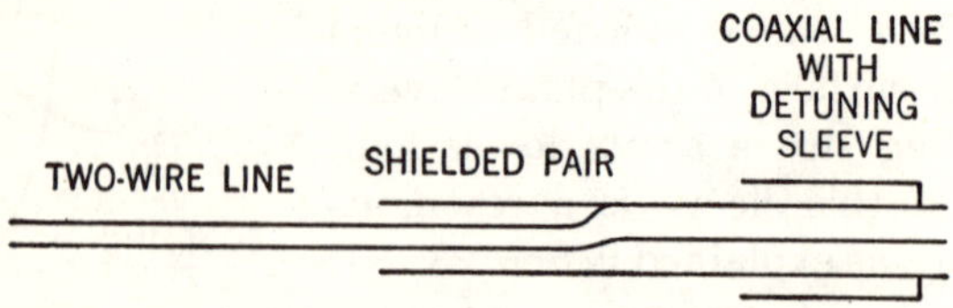

Fig. 27.5.—Simple two-wire to coaxial-line transformer consisting of a section of the shielded pair of adjustable length.

When a two-wire line and a coaxial line are connected, as in Fig. 27.5, unsymmetrical conditions exist at the junction point not unlike those discussed for a symmetrical antenna center-driven from a coaxial line. As a result, antenna currents may exist on the outer surface of the coaxial line and as unbalancing currents on the two-wire line. A simple method of reducing these is to insert a shielded pair, preferably adjustable in length, between the coaxial

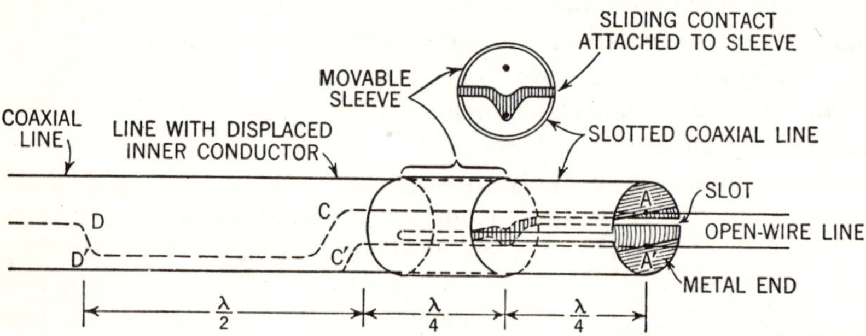

Fig. 27.6.—Slotted shielded-pair transformer.

line and the two-wire line and, if necessary, to detune the outer surface with coaxial sleeves. The shielded pair may consist of a tube or pipe around the two-wire line for the required distance, Fig. 27.5. The length of the tube must be such that resonant currents do not exist on its inner surface. A modification[1] of the shielded-pair transformer is shown in Figs. 27.6 and 27.7. The coaxial line on the left is modified from D to C by moving the inner

[1] Patent applied for.

conductor closer to the bottom of the shield. This concentrates the current on the bottom of the shield for better transfer to the lower conductor of the shielded pair that begins at *C′*. It also equalizes the over-all electrical length of the two sides of the line in the transforming section *DD′* to *AA′*. In order to transfer all the current from the shield to the lower wire of the shielded pair between *C′* and *A′*, the shield is slotted for a distance somewhat exceeding $\lambda/2$ and connected to the upper conductor of the shielded pair at *A*, to the lower conductor at *A′*, using metal end plates. By adjusting the length of the slotted section (using a movable sleeve enclosing the entire shield) to near $\lambda/4$, so that the impedance

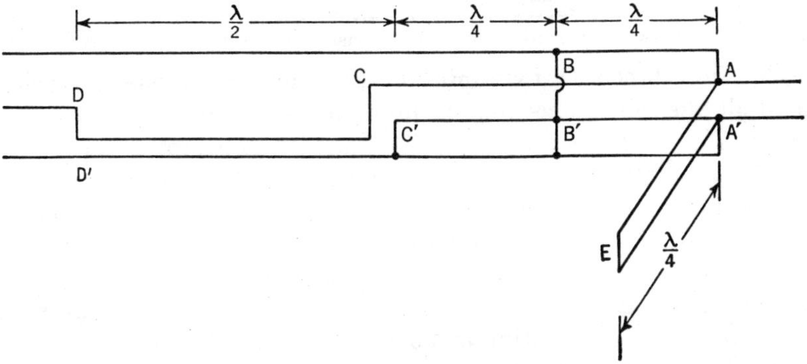

FIG. 27.7.—Circuit diagram of slotted shielded-pair transformer.

of the closed-end stub, consisting of the upper half of the slotted shield and the upper wire of the pair, is very high, any current in the upper half of the shield must turn either to the lower wire of the pair or to the lower half of the slotted shield. The path is by way of the sleeve and its connection to the lower wire of the pair. The currents on the lower half of the slotted shield combine at *A′* with those on the lower wire. Both inner and outer surfaces of the slotted section of the shield from *AA′* to the sleeve are equivalent to high impedance, $\lambda/4$ closed-end stubs across *AA′*. The slotted shielded-pair transformer is superior to the arrangement of Fig. 25.2*c* for center-driving a symmetrical antenna from a coaxial line. Detuning sleeves are usually not required on the outside of the shield if the unit is proportioned correctly. If the bends in the section *DC* are omitted for simplicity in construction, a detuning sleeve is desirable, especially if a voltage-fed antenna is connected directly at *AA′* in place of the open-wire line.

If the frequency is sufficiently high, a cavity resonator, Chap. III, may be inserted between a two-wire line and a coaxial line, as a balanced-to-unbalanced-line transformer, Fig. 27.8. Another useful device for connecting a two-wire line to a coaxial line is a T section to transfer from a single coaxial line to two coaxial lines. By having one of these $\lambda/2$ longer than the other and connecting the two-wire line to the two inner conductors, a smooth transfer results.

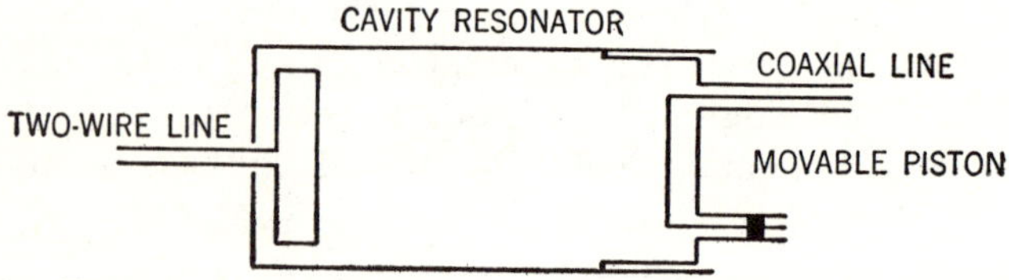

FIG. 27.8.—Two-wire line to coaxial-line transformer using a cavity resonator.

28. Impedance Transforming or Matching Sections.—In order to fulfill the conditions for optimum operation of a transmission system that includes generator, line, and antenna (or other load), matching networks must be provided at both ends of the transmission line if this is long. The matching network for the antenna serves to provide a line terminated in its characteristic impedance and hence assures minimum losses along the line. The matching network at the generator is necessary in order that the generator be correctly loaded for optimum output. Several types of network are available, some useful for both ends or for either end of the line, others serving the more special function of matching at the end connected to the load. They will now be reviewed.

a. T Section with Reactive Elements.—A T section of practically resistanceless reactive elements is useful for matching, especially at low frequencies. Such a section may be designed for use at either the input or the output end of the line. Since only two variables are necessary in order to achieve a match, it is always possible to select arbitrarily one of the three reactances forming the T. A match can be obtained with one series and one shunt element. Matching sections of this type are useful at all frequencies that are not so high that the reactance of a coil or of a capacitor becomes difficult to predetermine. In such cases, sections of transmission line are usually more convenient because their input impedance can be calculated accurately in terms of their length.

b. General Resonant Line with Adjustable Taps.—A very general form of network constructed of a short terminated section of transmission line that may be used as a matching network is shown in Fig. 28.1. The conditions for match are

$$R_{MN} \text{ (into matching network)} = R_c \tag{28.1}$$
$$X_{MN} \text{ (into matching network)} \doteq 0 \tag{28.2}$$

The impedance of the load at the terminals AB is $Z_{AB} = R_{AB} + jX_{AB}$. (In some instances, it may be more convenient for physical reasons to have Z_{AB} the input impedance of a short section of line connected to the load and treated as a part of it.) The characteristic impedance of the long line is $Z_c \doteq R_c$; that of the short matching section

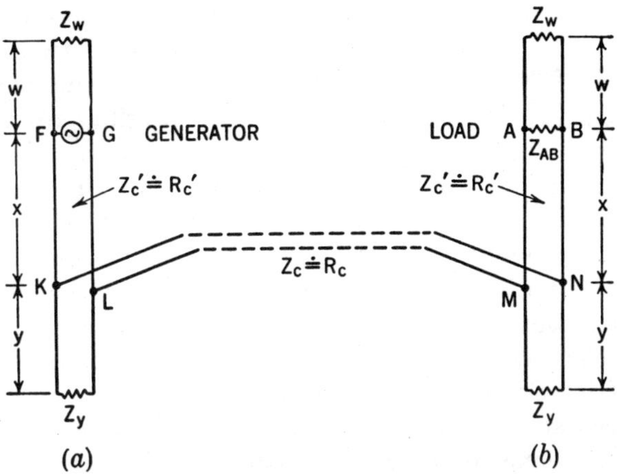

FIG. 28.1.—General matching network.

is $Z_c' \doteq R_c'$. The matching section of line is terminated at one end by Z_w and at the other end by Z_y. Both Z_w and Z_y must be high-Q impedances; usually they are the impedances of open or bridged ends.

The conditions for providing the optimum load impedance $Z_{FG} = Z_{\text{opt}}$ for the arrangement of Fig. 28.1*a* are

$$R_{FG} \text{ (into matching network)} = R_{\text{opt}} \tag{28.3}$$
$$X_{FG} \text{ (into matching network)} = X_{\text{opt}} \tag{28.4}$$

The impedance Z_{KL} is the input impedance of the line. If the line is terminated in R_c, then $Z_{KL} \doteq R_c$.

Usually only two variables are necessary to fulfill the conditions (28.1) and (28.2) or (28.3) and (28.4). There are available the three lengths w, x, y; the characteristic impedance Z_c; and the terminal impedances $Z_w = R_w + jX_w$ and $Z_y = R_y + jX_y$. By arbitrarily assigning convenient values to all except two of these, but in such a way that losses in the entire matching section are kept very small, a large variety of possible arrangements with somewhat different properties can be devised. They include the $\lambda/4$ transformer, single- and double-stub tuners, which are analyzed in Chap. I. An arrangement using a $\lambda/4$ matching transformer and a detuning sleeve with a coaxial line driving an antenna is shown in Fig. 28.2.

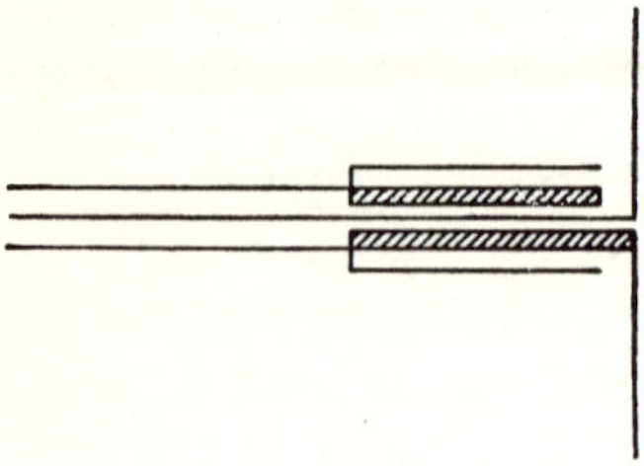

Fig. 28.2.—Antenna center-driven by coaxial line with quarter-wave transformer and detuning sleeve.

c. "Delta Match" for Antennas.—Instead of inserting a matching network consisting of a resonant section of transmission line between the input terminals of an antenna and the long feeder, it is possible to modify the antenna itself in such a way that the input impedance at its terminals is equal to the characteristic impedance of the line. The usual arrangement, Fig. 28.3, is to attach conductors at points CD along the antenna (which are not sufficiently close to be in the near zone) and then join these to the new input terminals at AB. The accurate calculation of the input impedance at AB of the modified antenna as a function of the distance CD and the lengths AC and BD has not been accomplished. Methods that treat the antenna as a highly attenuated section of transmission line give only rough approximations for small separations CD. The delta circuit permits an approximate match that is convenient experimentally. It is to be noted that the matching section is a *part of the antenna*, in that equal and opposite currents are not so close together that forces exerted at distant points are negligible compared with those exerted by the currents and charges in the antenna proper.

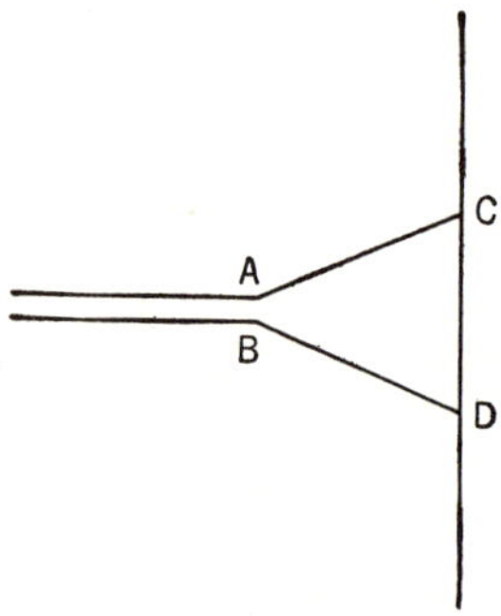

Fig. 28.3.—Delta-matched antenna.

Regardless of the particular structure of the matching section that is inserted between a transmission line (that is to be made nonresonant thereby) and an antenna or generator, the matching network is always a part of the resonant circuit and not a part of the nonresonant line. Accordingly, power dissipated in a matching network may be considerable unless the network is properly designed.

IV. THE RECEIVING ANTENNA AS A CIRCUIT ELEMENT

29. Distribution of Current and Charge along a Thin Unloaded Receiving Antenna Parallel to an Electric Field.—The distribution of current in a single parasitic antenna loosely coupled to all other conductors in the universe (including in particular a driven antenna) is known for a cylindrical and symmetrical antenna of small radius. This is the important case of a single receiving antenna in the far zone of a transmitter. The distribution of current along a cylindrical receiving antenna of moderately small radius is as intricate as that along a similar driven antenna. Much valuable information about receiving antennas as circuit elements can be derived, however, using only the relatively simple leading terms in the distribution function. These give a surprisingly good approximation for many purposes involving moderately thin antennas.

As is explained in general terms in Sec. 1, the interaction of electric charges in two separate conductors, such as a transmitting and a receiving antenna, is analyzed in two steps. The first of these is the definition of the electromagnetic field due to moving charges in the transmitting antenna. Since it is assumed here that the receiving antenna (and every other conductor) is in the far zone of the transmitter, the distributions of charge and current in the receiving antenna have a negligible effect on the current in the transmitter. Accordingly, the electromagnetic field may be defined directly in terms of the distribution of current already discussed for the driven antenna, and this is done in Sec. 35.

The second step in the analysis consists in expressing the force on the charges in the receiving antenna in terms of the electromagnetic field defined for the transmitter. Since there is no significant reaction of the receiver on the transmitter, the distributions of current and charge in the receiver may be obtained directly from the expression for the force on the charges subject to the conditions imposed by the shape and material of the antenna. Intricate simultaneous equations thus are avoided in this far-zone case.

The electromagnetic vectors in the far zone of a transmitting antenna may be taken to lie approximately in a plane insofar as a relatively short receiving antenna is concerned. The distribution of current in the receiving antenna can be expressed in terms of the electric vector or the magnetic vector. *Either one is sufficient.* For a straight receiving antenna, the electric vector is more convenient; for a loop or frame antenna, the magnetic vector usually is preferred. The distribution of current in the straight antenna depends on the direction of the electric vector, not merely on the component of the electric vector *parallel* to the antenna, as has sometimes been assumed erroneously. For optimum performance the receiving antenna is usually placed parallel to the electric field, and for the present the discussion is limited to this more important and simpler case.

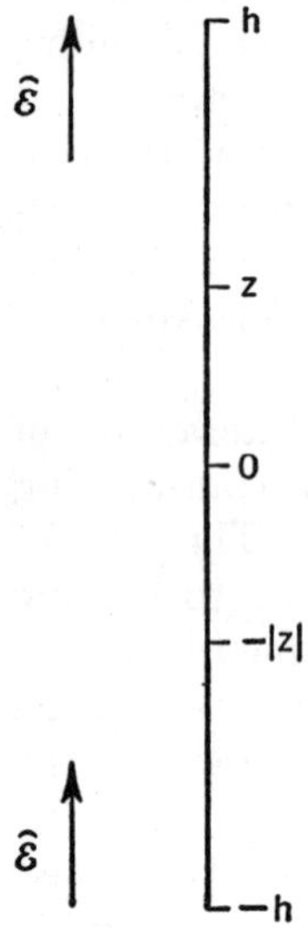

FIG. 29.1.—Unloaded receiving antenna in a uniform electric field parallel to the antenna.

The leading terms in the distributions of current and charge along an extremely thin, highly conducting thread of half length h, Fig. 29.1, are contained in the following expressions for the complex amplitudes:

$$\hat{I}_z = \hat{I}_0 \frac{\cos \beta z - \cos \beta h}{1 - \cos \beta h} \tag{29.1}$$

$$\hat{Q}_z = -j \frac{\hat{I}_0}{v_c} \frac{\sin \beta z}{1 - \cos \beta h} = \hat{Q}_{\max} \sin \beta z \tag{29.2}$$

$$\hat{I}_0 = \frac{j\hat{\mathcal{E}}}{30\beta\Omega} \frac{1 - \cos \beta h}{\cos \beta h} \tag{29.3}$$

$$\hat{Q}_{\max} = \frac{\hat{\mathcal{E}}}{30\omega\Omega \cos \beta h} \tag{29.4}$$

$$\Omega = 2 \ln \left(\frac{2h}{a}\right) \tag{29.5}$$

$\hat{\mathcal{E}}$ is the electric field due to a distant transmitter so oriented with respect to the receiver that the *electric vector is parallel to the axis of the receiving antenna.* Expressions for $\hat{\mathcal{E}}$ are given in Sec. 37 for linear radiators. It is interesting to compare (29.1) and (29.2) with corresponding formulas for the center-driven antenna, Sec. 8. These are written below.

$$\hat{I}_z = \hat{I}_0 \frac{\sin \beta(h - |z|)}{\sin \beta h} \tag{29.6}$$

$$\hat{Q}_z = -j \frac{\hat{I}_0}{v_c} \frac{\cos \beta(h - |z|)}{\sin \beta h} = \hat{Q}_{\max} \cos \beta(h - |z|) \tag{29.7}$$

Note that, whereas the distributions along the receiving antenna depend upon *the distance* $|z|$ *from the center* of the antenna, those along the transmitting antenna depend upon *the distance* $h - |z|$ *from the ends* of the antenna. They have the same form only when $\beta h = (2n + 1)\pi/2$ $(n = 0, 1, 2, \cdots)$, when both distributions reduce to

$$\hat{I}_z = \hat{I}_0 \cos \beta z \tag{29.8}$$

$$\hat{Q}_z = -j\frac{\hat{I}_0}{v_c} \sin \beta z = \hat{Q}_{\max} \sin \beta z \tag{29.9}$$

Also

$$\hat{I}_{\max} = \hat{I}_0 \qquad \text{and} \qquad \hat{Q}_{\max} = \hat{Q}_h \tag{29.10}$$

The distribution functions (29.1) and (29.2) are plotted in Fig. 29.2 for several values of h. They should be compared with

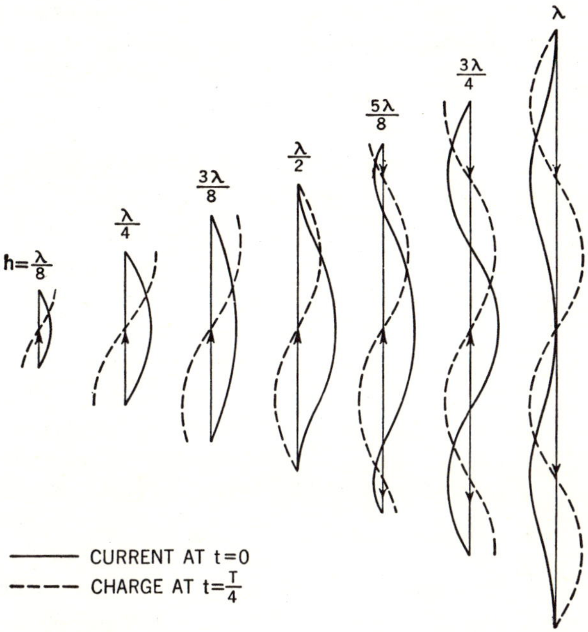

FIG. 29.2.—Distributions of current and charge in a receiving antenna without load in an electric field parallel to the antenna. Amplitudes shown for successive antennas are not to scale.

the corresponding distributions along a transmitting antenna of the same length, Fig. 8.2.

The fact that $\hat{I}_z$ becomes infinite when $\cos \beta h = 0$ in the receiving antenna or when $\sin \beta h = 0$ in the driven antenna is a consequence of retaining only the leading terms in the distribution function. In terms of the complete solution, the current may become large but never infinite. This is discussed for the driven antenna in Sec. 9. Analogous conditions exist in the receiving antenna but are not described here. However, it may be concluded that the current in the unloaded antenna is maximum (though not infinite) near $\cos \beta h = 0$ or $h = \lambda/4$. The condition for maximum current is that desired when a coupled load receives negligible power. The condition for maximum power to the load is not the same as for maximum current, as is shown in Sec. 32.

30. Distribution of Current along a Highly Conducting Symmetrical Loaded Antenna of Extremely Small Radius.—In normal operation a receiving antenna is connected to a load impedance in the form of a network of tuned circuits either directly or by means of a transmission line. In some instances, it is necessary or desirable to design the load and the antenna so that maximum power is transferred from the distant transmitter to the load. In other cases, the receiver itself is made so sensitive that it need dissipate only negligible energy as heat, and practically all the power received by the antenna is reradiated. This case then becomes approximately that of an unloaded antenna.

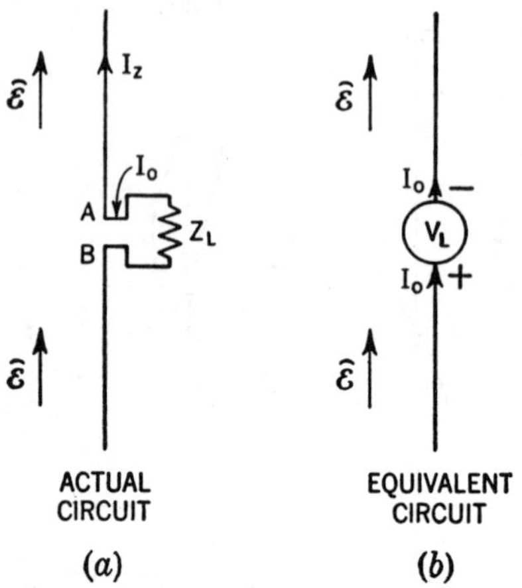

FIG. 30.1.—Loaded receiving antenna.

Consider the general case, Fig. 30.1, of a receiving antenna placed parallel to the electric field $\hat{\mathcal{E}}$ of a distant transmitter. A load impedance Z_L (which may be the input impedance of a transmission line loaded at the output end) is connected across two terminals at the center of the antenna. These are assumed to be sufficiently close together to be in the near zone with respect to each other. With the aid of the compensation theorem, the analysis of this circuit can be reduced mathematically to that of an equivalent circuit consisting of an unloaded receiving antenna with an impedanceless generator at its center. According to this theorem, any impedance Z_L that carries a current $\hat{I}_0$ may be replaced by an impedanceless generator with an emf $\hat{V}_L$ equal to the potential drop across Z_L, or

$$\hat{V}_L = \hat{I}_0 Z_L \tag{30.1}$$

If the substitution (30.1) is made, the circuit of Fig. 30.1*a* is equivalent to that of Fig. 30.1*b*.

By direct application of the principle of superposition, the total current $\hat{I}_0$ is the resultant of the current $\hat{I}_{0V}$ due to the generator and the current $\hat{I}_{0\mathcal{E}}$ due to the action of the distant transmitter as expressed in the electric field $\hat{\mathcal{E}}$, or

$$\hat{I}_0 = \hat{I}_{0\mathcal{E}} - \hat{I}_{0V} \tag{30.2}$$

The current $\hat{I}_{0V}$ due to the impedanceless generator is the same as that in a driven antenna, or

$$\hat{I}_{0V} = \frac{\hat{V}_L}{Z_0} = \hat{I}_0 \frac{Z_L}{Z_0} \tag{30.3}$$

where Z_0 is the self-impedance of a center-driven antenna that has the same dimensions as the receiving antenna under consideration. For an extremely thin antenna, the leading term over most of the range is (10.4)

$$Z_0 = jX_0 \doteq -j60\Omega \cot \beta h \tag{30.4}$$

With (30.1) and (30.3), (30.2) may be written

$$\hat{I}_{0\mathcal{E}} = \hat{I}_0 + \hat{I}_{0V} = \hat{V}_L \left(\frac{1}{Z_L} + \frac{1}{Z_0} \right) \tag{30.5}$$

At any point z along the antenna, the current $\hat{I}_z$ by the principle of superposition is also the algebraic sum of the separate currents due to the electric field and the generator, or

$$\hat{I}_z = \hat{I}_{z\mathcal{E}} - \hat{I}_{zV} = \hat{I}_{0\mathcal{E}} f_{\mathcal{E}}(z) - \hat{I}_{0V} f_V(z) \tag{30.6}$$

where $f_{\mathcal{E}}(z)$ and $f_V(z)$ are the distribution functions for an unloaded receiving antenna ($Z_L = 0$) and a driven antenna. In an extremely thin antenna, the leading terms are

$$f_{\mathcal{E}}(z) = \frac{\cos \beta z - \cos \beta h}{1 - \cos \beta h} \qquad (\beta h \text{ not near } 2\pi) \tag{30.7}$$

$$f_V(z) = \frac{\sin \beta(h - |z|)}{\sin \beta h} \qquad (\beta h \text{ not near } \pi) \tag{30.8}$$

These are rough approximations for thin antennas, provided that their lengths are not such as to make the denominators in (30.7) and (30.8) very small. With (30.3) and (30.5), (30.6) can be written in the form

$$\hat{I}_z = \hat{I}_0 \left[\left(1 + \frac{Z_L}{Z_0} \right) f_{\mathcal{E}}(z) - \frac{Z_L}{Z_0} f_V(z) \right] \tag{30.9}$$

For thin antennas, the approximate distribution of current is

$$\hat{I}_z = \hat{I}_0\left[\left(1 + \frac{Z_L}{Z_0}\right)\left(\frac{\cos\beta z - \cos\beta h}{1 - \cos\beta h}\right) - \frac{Z_L}{Z_0}\frac{\sin\beta(h - |z|)}{\sin\beta h}\right] \quad (30.10)$$

where

$$\hat{I}_0 = \frac{j\hat{\varepsilon}}{30\beta\Omega}\left(\frac{1 - \cos\beta h}{\cos\beta h}\right)\left(\frac{Z_0}{Z_0 + Z_L}\right) \quad (30.11)$$

In general, this is a complicated function, since Z_L may be complex.

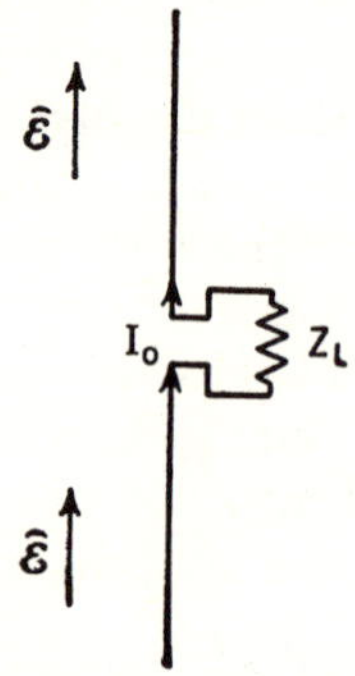

FIG. 31.1a.—Loaded receiving antenna parallel to electric field.

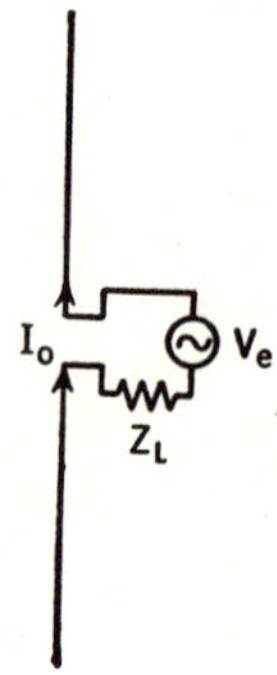

FIG. 31.1b.—Circuit equivalent to (a) insofar as current in load is concerned.

An important and simple special case is an antenna of half length $\lambda/4$, where (30.10) reduces to

$$\hat{I}_z = \hat{I}_0 \cos\beta z \quad (30.12)$$

for any load. If $\mathbf{P}_L$ is infinite, $\hat{I}_0$ is zero in (30.11).

31. The Equivalent Circuit for a Receiving Antenna in the Far Zone of a Transmitter.—The relation (30.5) for the components $\hat{I}_{0\varepsilon}$ and $\hat{I}_{0V}$ of the total current $\hat{I}_0$ may be rewritten using (30.3).

$$\hat{I}_{0\varepsilon}Z_0 = \hat{I}_0(Z_L + Z_0) \quad (31.1)$$

This equation gives the correct amplitude of the current $\hat{I}_0$ entering or leaving the load Z_L connected at the center of a symmetrical receiving antenna placed *parallel* to the electric field $\hat{\varepsilon}$ of a *distant* transmitter, Fig. 31.1*a*. Actually, however, (31.1) is the equation of the same antenna with the same load at its center when driven by an impedanceless generator, Fig. 31.1*b*. The emf of the generator is

$$\hat{V}_e = \hat{I}_{0\varepsilon}Z_0 \quad (31.2)$$

Thus the forces acting on the charges distributed along the receiving antenna in the actual circuit are replaced in the "equivalent" circuit by forces due to an impedanceless generator acting as a concentrated emf at the center of the antenna.

For the purpose of determining the current $\hat{I}_0$, the two circuits are equivalent, but they are not equivalent for all purposes. For example, the distribution of current along an extremely thin receiving antenna is given approximately by (30.9), whereas the distribution along the antenna in the equivalent circuit is given to the same approximation by (29.6). The two distributions are not similar in general, and they lead to quite different electromagnetic fields, *i.e.*, the power that would be radiated from the "equivalent" driven antenna would be transferred to different parts of the distant universe from the power radiated from the actual antenna. Whereas the fictitious emf (31.2) is chosen so that the current $\hat{I}_0$ is the same in the actual and "equivalent" circuit, the total power radiated from the driven antenna in the "equivalent" circuit is not equal to the power reradiated from the receiving antenna. *From the point of view of the power absorbed in the load Z_L,* this is immaterial, and the two circuits are equivalent.

The emf in the equivalent circuit, Fig. 31.1*b*, is evaluated readily for a sufficiently thin antenna where $\hat{I}_{0\mathcal{E}}$ is given approximately by (29.3) and Z_0 approximately by (30.4). With $H = \beta h$, and substituting (29.3) and (30.4) in (31.2),

$$\hat{V}_e = \frac{j\hat{\mathcal{E}}}{30\beta\Omega}\left(\frac{1 - \cos H}{\cos H}\right)(-j60\Omega \cot H) \tag{31.3}$$

or

$$\hat{V}_e = \frac{2\hat{\mathcal{E}}}{\beta} \tan\left(\tfrac{1}{2}H\right) \tag{31.4}$$

The quantity

$$2h_e = \frac{2}{\beta} \tan\left(\tfrac{1}{2}H\right) = \frac{\lambda}{\pi} \tan\left(\tfrac{1}{2}H\right) \tag{31.5}$$

is called the *effective length or height* of the "equivalent" antenna of length $2h$. (For a receiving antenna of length h erected vertically over a perfectly conducting horizontal plane to which one end of Z_L is attached, the effective length of the "equivalent" antenna is h_e.) Thus the emf at the center of the "equivalent" circuit of a receiving antenna of length $2h$ is

$$V_e = 2h_e\mathcal{E} \tag{31.6}$$

This can be determined experimentally by measuring $\hat{\mathcal{E}}$ with a field-intensity meter at the location of the receiving antenna (with antenna removed) if the effective length is known. It can be computed approximately from (31.5) for very thin antennas that are well below $h = \lambda/2$ in half length. If the antenna is not thin or

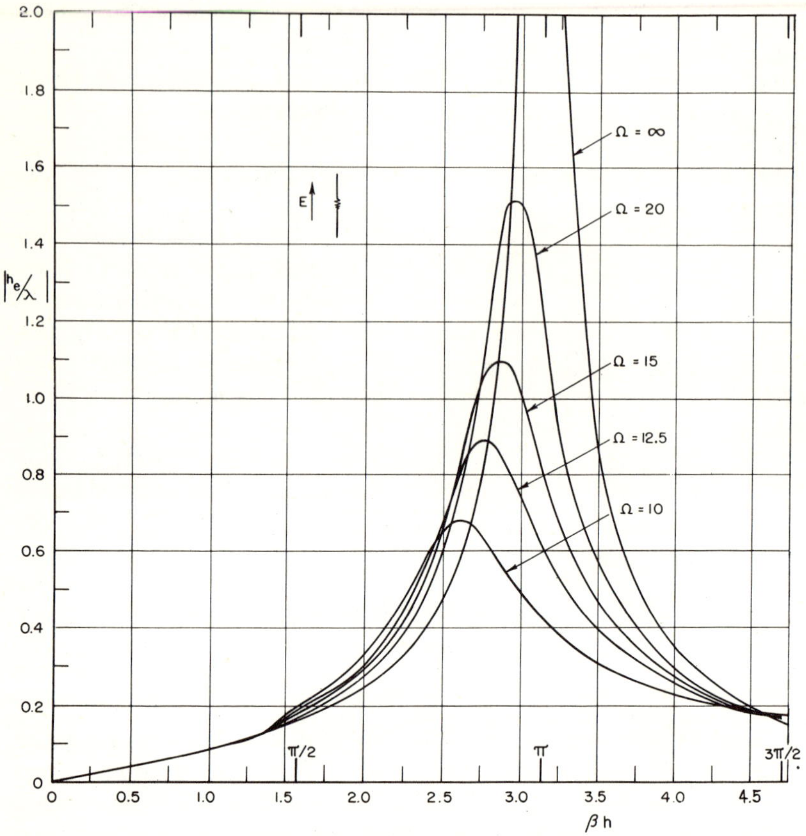

FIG. 31.2.—The normalized effective half length $|h_e/\lambda|$ for an antenna parallel to the incident electric field; $\beta h = 2\pi h/\lambda$.

if it approaches or exceeds $h = \lambda/2$ in half length, the curves of Fig. 31.2 may be used to determine h_e. They are computed from the actual distribution of current for the indicated thickness. The curve marked $\Omega = \infty$ is computed using (31.5). It is interesting to note that the curve for infinitely thin antennas is a good approximation even for thick antennas, provided that these are well below $h = \lambda/2$ in half length.

If the antenna is not parallel to the electric field but *lies in a plane of constant phase of the electric field, i.e., is perpendicular to the line joining the receiving antenna with the distant transmitter*, Fig. 31.3 *with* $\bar{\theta}_1 = 0$, the analysis applies throughout if the component of the electric field parallel to the antenna, *viz.*, $\mathcal{E} \cos \psi$, is written for $\hat{\mathcal{E}}$ in (31.3) and (31.6). The value of h_e is not changed. Thus, for the receiving antenna in the plane of the electric field, use

$$\hat{V}_e = 2h_e\hat{\mathcal{E}} \cos \psi \tag{31.7}$$

with h_e obtained from Fig. 31.2.

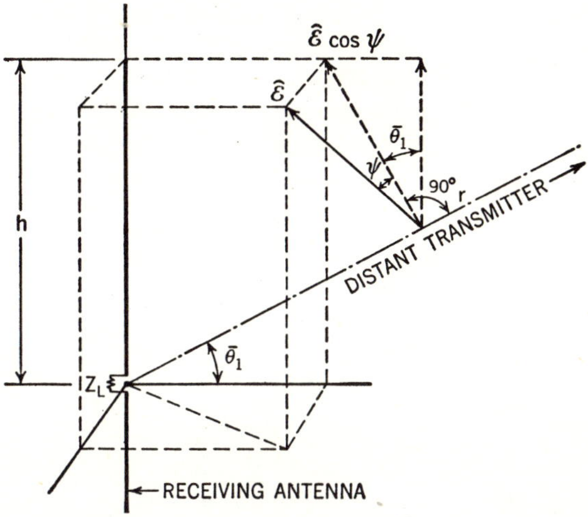

FIG. 31.3.—Receiving antenna in an electric field. In the diagram the vector representing the field is shown at a distance from the antenna. The field used in calculation is that *at the center* of the antenna.

If the receiving antenna is neither parallel to nor in the plane of the electric field but is oriented as in Fig. 31.3, *it is not correct* merely to use in (31.3) and (31.6) the component of the electric field parallel to the antenna, *viz.*, $\hat{\mathcal{E}} \cos \bar{\theta}_1 \cos \psi$. Since with $\bar{\theta}_1 \neq 0$ one end of the antenna is inclined toward and the other end away from the transmitter, the phase of the electric field is not the same at the same instant at points along the receiving antenna. The field experienced by the nearer end is due to a current in the distant transmitter at a later time than the field experienced at the same instant by the more distant end. Merely projecting $\hat{\mathcal{E}}$ on the antenna does not take into account this difference in time but assumes that the electric field at all points along the receiving antenna is due to the current in the distant transmitter at one and the same instant. Only if the antenna is quite short, $h < \lambda/4$, and $\bar{\theta}_1$ does not exceed 10° is it a moderately good approximation to write $\hat{\mathcal{E}} \cos \psi \cos \bar{\theta}_1$ for $\hat{\mathcal{E}}$ in (31.3) and (31.6).

If the receiving antenna is inclined at an appreciable angle $\bar{\theta}_1$ with the plane of the electric field, the effective length $2h_e$ for an

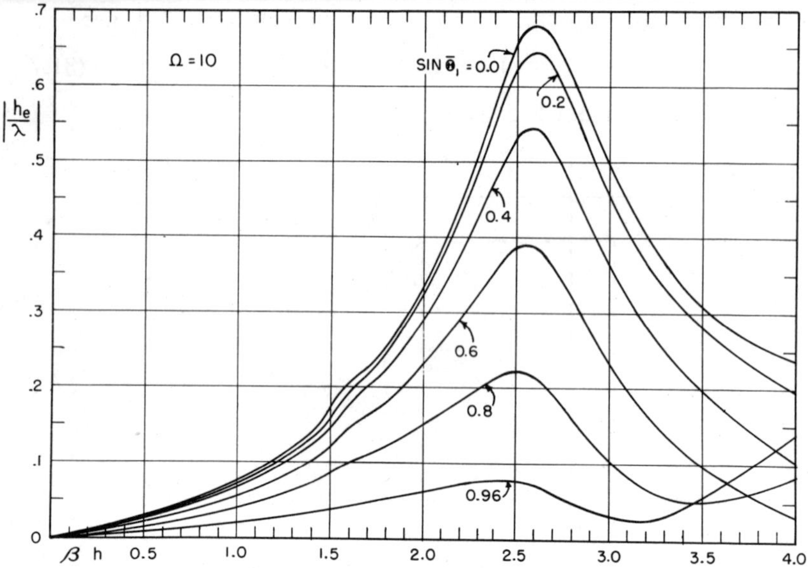

FIG. 31.4.—The normalized effective half length $|h_e/\lambda|$ for an antenna inclined at different angles with respect to the electric field; $\Omega = 10$.

infinitely thin antenna of full length $2h$ is

$$2h_e = \frac{\lambda}{\pi}\left[\frac{\cos (H \sin \bar{\theta}_1) - \cos H}{\sin H \cos \bar{\theta}_1}\right] \tag{31.8}$$

This reduces to (31.5) with $\bar{\theta}_1 = 0$. For practical antennas, the effective half length h_e may be obtained from Figs. 31.4 to 31.6. These figures give h_e/λ as a function of h/λ for three different thicknesses of antenna ($\Omega = 10, 20, 15$) and for the angles corresponding to $\sin \bar{\theta}_1 = 0.2, 0.4, 0.6, 0.8$, or $\bar{\theta}_1 = 11°.5, 23°.6, 36°.9, 53°.1$. For an arbitrary orientation of the receiving antenna, the value of h_e obtained from these curves may be substituted in (31.6) with $\hat{\mathcal{E}} \cos \psi$ written instead of $\hat{\mathcal{E}}$, *i.e.*, $\hat{V}_e$ is given by (31.7) with h_e taken from Figs. 31.4 to 31.6 for the appropriate value of $\bar{\theta}_1$ and Ω. Intermediate values may be estimated.

It is important to note that the effective half lengths h_e in all these figures are *magnitudes* of complex quantities. The phase angles are of no significance if *only a single field* is involved. On the other hand, if interest lies in the combined effect due to a distant

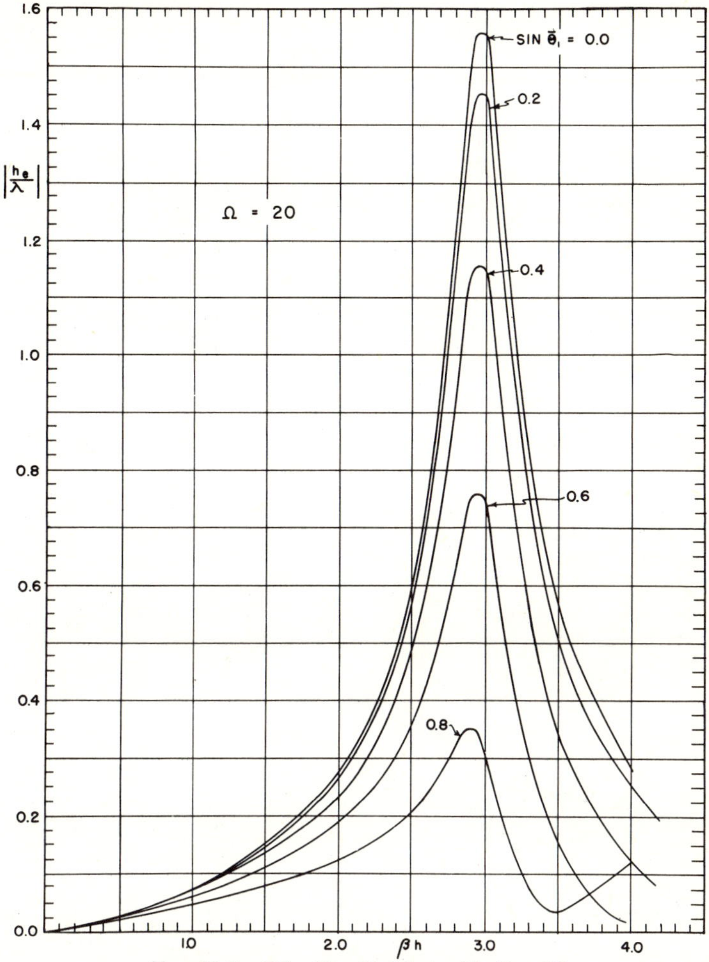

FIG. 31.5.—Like Fig. 31.4 but with $\Omega = 20$.

transmitter and a reflection from the ionosphere or due to any arbitrarily oriented, elliptically polarized electric field, the current in the load of the receiving antenna cannot be determined without the phase angles, which are available in the literature.[1]

[1] R. W. P. KING, "Theory of Linear Antennas," pp. 488–491, Harvard University Press, 1956.

If a receiving antenna is in the presence of other antennas that are *not in the far zone*, the current through its load Z_L may be determined with Thévenin's theorem applied at the terminals AB of the antenna. Thus

$$\hat{I}_0 = \frac{\hat{V}_{AB}\text{ (open)}}{Z_L + Z_{AB}} \tag{31.9}$$

where $\hat{V}_{AB}$ (open) is the open-circuit voltage at the terminals of the antenna and Z_{AB} is the input impedance of the receiving antenna

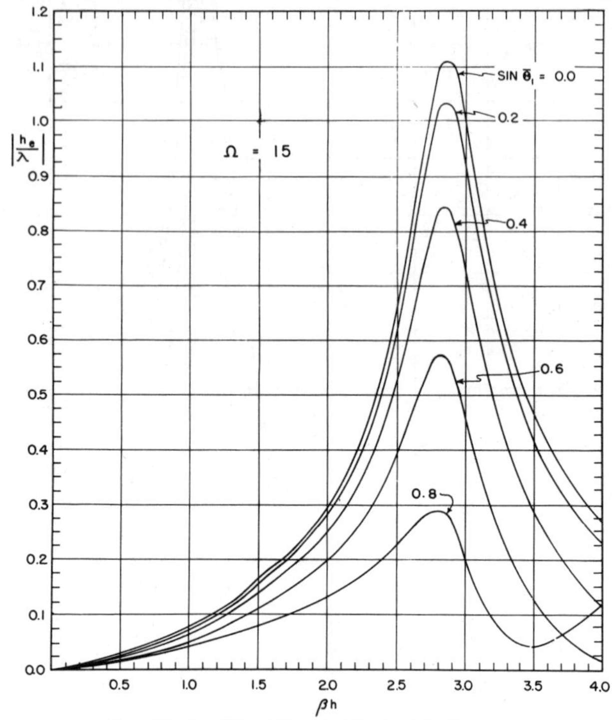

FIG. 31.6.—Like Fig. 31.4 but with Ω = 15.

in the presence of the other antennas in which all generators have been replaced by their internal impedances. In general, the computation of $\hat{V}_{AB}$ (open) and of Z_{AB} has not been accomplished. If the receiving antenna is far from all other antennas, the computation is precisely that just given, in which Z_{AB} reduces to Z_0 and $\hat{V}_{AB}$ (open) is the same as $\hat{V}_e$ of (31.6). If only two parallel antennas are involved, see Sec. 43.

32. Maximum Power Transferred to the Load.—The power transferred to the load Z_L, Fig. 31.1*b*, is a maximum when Z_L is the conjugate of the input self-impedance Z_0 of the antenna with Z_0 obtained from (10.4) or from the curves computed therefrom. The power transferred to the matched load ($R_L = R_0$; $X_L = -X_0$) of a thin antenna is one-half the total power supplied. The total power is $|\hat{V}_e|^2/4R_L$. The power to the load is

$$P_L = \frac{|\hat{V}_e|^2}{8R_L} = \frac{|\hat{\mathcal{E}}2h_e|^2}{8R_L} = \frac{|\hat{\mathcal{E}}h_e|^2}{2R_L} \tag{32.1}$$

where $\hat{V}_e$ is the emf of the generator in the "equivalent" circuit.

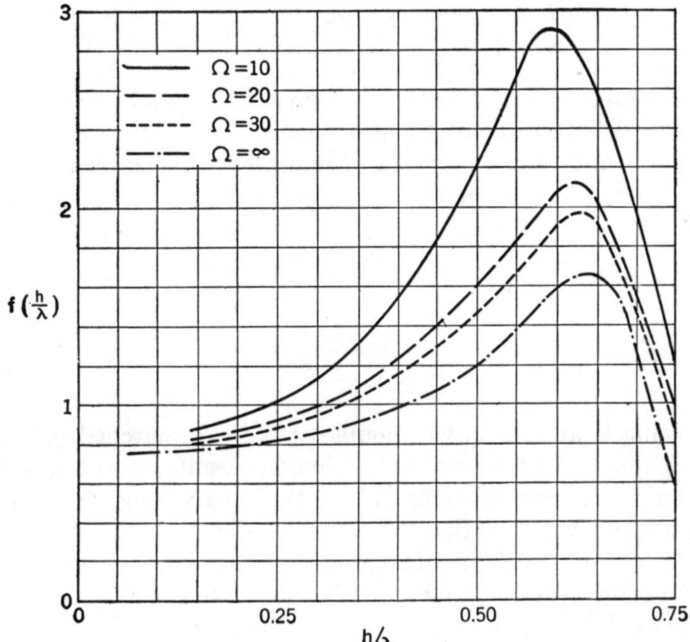

FIG. 32.1.—The function $f\left(\frac{h}{\lambda}\right)$ for a receiving antenna with matched load, placed parallel to the electric field.

The effective length $2h_e$ of a thin antenna of length $2h$ is given in (31.5). With it and using $\hat{\mathcal{E}} = \sqrt{2}\,\mathcal{E}$

$$P_L = \left(\frac{\mathcal{E}}{\beta}\right)^2 \frac{\tan^2 (\frac{1}{2}H)}{R_0} \tag{32.2}$$

By introducing the radiation resistance R^e_m referred to maximum current [as defined in (17.9) and plotted in Fig. 17.1] in place of R_0

according to

$$R_m^e = R_0 \sin^2 H \tag{32.3}$$

$$\frac{1}{R_0}\tan^2\left(\tfrac{1}{2}H\right) = \left(\frac{\sin^2 H}{R_m^e}\right)\left(\frac{1-\cos H}{\sin H}\right)^2 = \frac{(1-\cos H)^2}{R_m^e} \tag{32.4}$$

Substituting (32.4) in (32.2), with $\beta = 2\pi/\lambda$

$$P_L = \frac{(\mathcal{E}\lambda)^2}{(2\pi)^2}\frac{(1-\cos H)^2}{R_m^e} = \frac{(\mathcal{E}\lambda)^2}{240\pi^2} f\left(\frac{h}{\lambda}\right) \tag{32.5}$$

The function

$$f\left(\frac{h}{\lambda}\right) = \frac{60(1-\cos H)^2}{R_m^e}; \qquad H = \frac{2\pi h}{\lambda} \tag{32.6}$$

is represented in Fig. 32.1 by the curve $\Omega = \infty$. Corresponding curves for $\Omega = 30, 20, 10$ are also shown.

Figure 32.1 shows that, for an infinitely thin antenna, $(P_L)_{\max}$ can be made a maximum with

$$\frac{h}{\lambda} = 0.637; \qquad f\left(\frac{h}{\lambda}\right) = 1.65 \tag{32.7}$$

Thus

$$(P_L)_{\max\,\max} = 1.65\,\frac{(\mathcal{E}\lambda)^2}{240\pi^2} = 0.7 \cdot 10^{-3}(\mathcal{E}\lambda)^2 \text{ watts} \tag{32.8}$$

This value is approximately double that for a matched load with an antenna of half length $\lambda/4$. Similar results are obtained for antennas with nonvanishing radius, the maximizing value of h/λ being only slightly less than 0.637.

If the electric field is not parallel to the receiving antenna, it may be represented as in Fig. 31.3. With this orientation of the electric field

$$(P_L)_{\max} = \frac{(\mathcal{E}\lambda)^2}{240\pi^2}\cos^2\psi\, f\left(\bar{\theta}_1, \frac{h}{\lambda}\right) \tag{32.9}$$

with

$$f\left(\bar{\theta}_1, \frac{h}{\lambda}\right) = \frac{60}{R_m^e}\left[\frac{\cos(H\sin\bar{\theta}_1) - \cos H}{\cos\bar{\theta}_1}\right]^2 \tag{32.10}$$

For h/λ less than about 0.7, the maximum power is always obtained with the antenna parallel to the electric vector. For antennas longer than 1.4λ, this is not true; maximum power in the matched load is obtained when the antenna is inclined even at large angles with respect to the electric field.

33. Maximum Current through the Load—Maximum Potential Difference across the Load.—The root mean square current through the load is given by

$$I_0 = \frac{V_e}{Z_0 + Z_L} \tag{33.1}$$

This is a maximum with respect to X_L if

$$X_L = -X_0 \tag{33.2}$$

so that

$$I_0 = \frac{V_e}{R_0 + R_L} \tag{33.3}$$

If the receiving antenna is parallel to the electric field, V_e can be varied only by changing the length of the antenna. This also changes R_0, and not according to any simple relation. For any given value of R_L, the maximum value of I_0 and the maximizing half length h may be determined easily by plotting I_0 in (33.3) as a function of h, using values of h_e and R_0 obtained from the appropriate curves. In the special case in which

$$Z_L = 0 \tag{33.4}$$

$$I_0 = \frac{V_e}{Z_0} = I_{0\varepsilon} \tag{33.5}$$

This has its maximum value practically at resonance.

The voltage across the load Z_L is

$$V_L = I_0 Z_L = \frac{V_e}{1 + (Z_0/Z_L)} \tag{33.6}$$

This has its largest value if Z_L is made very large compared with Z_0,

$$|Z_0| << |Z_L| \qquad V_L \doteq V_e \tag{33.7}$$

If the receiving antenna is parallel to the electric field, V_e is a maximum with respect to the length of the antenna if this length is adjusted for *antiresonance*, *i.e.*, with the half length h somewhat less then $\lambda/2$ by an amount depending on the thickness of the antenna.

V. ELECTROMAGNETIC FIELD OF ANTENNAS AND ARRAYS

34. Vectors and Complex Numbers—General Definitions.—Before entering into a discussion of the electromagnetic fields that can be calculated for different antennas or arrays of antennas, a possible source of confusion must be clarified. It is the distinction between space vectors and complex numbers represented by rotating vectors in the complex plane.

A space vector is a quantity that has both direction and magnitude. Three space coordinates, such as rectangular, cylindrical, spherical, or ellipsoidal, are required to specify the three mutually perpendicular components of a vector. A mathematical function that assigns a vector to every point in a given region is called a vector-point function. The electric-field vector $\mathcal{E}$ and magnetic-field vector $\mathcal{H}$ are vector-point functions in that a vector $\mathcal{E}$ and a vector $\mathcal{H}$ are assigned to every point in space. Such a vector may be constant in both magnitude and direction, as in electrostatics or

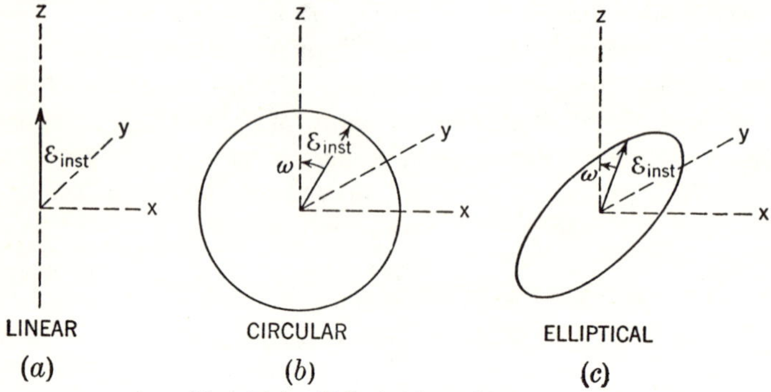

FIG. 34.1.—Polarization of a vector.

magnetostatics; or it may vary in magnitude or in direction or in both magnitude and direction, as time passes.

The direction of a vector is often referred to as its polarization. If the electric vector, for example, varies periodically in magnitude but is always directed along a fixed line such as the z axis, it is said to be *linearly polarized*, Fig. 34.1*a*. If it is constant in magnitude but rotates in a fixed plane in space, *e.g.*, in the xz plane with a constant angular velocity ω, it is *circularly polarized*, Fig. 34.1*b*. If both magnitude and direction vary periodically in the xz plane, it is *elliptically polarized* in that plane, Fig. 34.1*c*. Elliptical polarization is the general form; linear and circular polarizations are special cases. The distant field of a single straight antenna is linearly polarized, as is that of two crossed antennas with currents of equal amplitudes and in the same phase. If the currents in two crossed antennas lying along the x and z axes are equal in amplitude but differ in phase by a quarter period, the distant field for points along the y axis is circularly polarized; if the currents have different amplitudes and arbitrary phase relations, the distant field is elliptically polarized. If the electric vector is linearly polarized along an axis perpendicular to the surface of the earth, it is said to be vertically polarized; if it is parallel to the surface of the earth, it is horizontally polarized.

A *linearly polarized* vector $\mathcal{E}$ that varies periodically in time may be represented in the form

$$\mathcal{E}_{\text{inst}} = \hat{\mathcal{E}} \cos \omega t = \text{real part of } \hat{\mathcal{E}} \epsilon^{j\omega t} \tag{34.1}$$

where $\mathcal{E}$ is a vector fixed in space in the direction of the maximum of $\hat{\mathcal{E}}_{\text{inst}}$ at $t = 0$, Fig. 34.2. As time passes, the length of the vector is reduced until at $\omega t = \pi/2$ the length vanishes. It then increases in the reverse direction until at $\omega t = \pi$ it is again at maximum length. The *instantaneous length* of the vector may be looked upon as the projection on the axis of polarization of a *fictitious* "vector" of constant magnitude $|\hat{\mathcal{E}}|$ rotating with an angular velocity ω. This fictitious rotating vector is a graphical representation of the complex number $\hat{\mathcal{E}}\epsilon^{j\omega t}$ of which only the real part has a physical meaning in the present case. Great care must be exercised not to confuse a circularly polarized vector, *i.e.*, one that actually rotates in a real plane in space, with the fictitious "rotating vector" used to obtain the magnitude of a *linearly polarized* vector that varies periodically in time. This fictitious rotating vector is defined in the *complex* plane so that it is not a space vector. (The name phasor has been suggested.) In the theory of alternating currents, complex numbers are frequently called vectors, because they can be represented graphically by a directed magnitude *in the complex plane.* This terminology is not used in this chapter. It is important to note that a linearly polarized complex space vector is not a rotating "vector," and a rotating space vector is not complex. In the following discussion, only space vectors are called vectors.

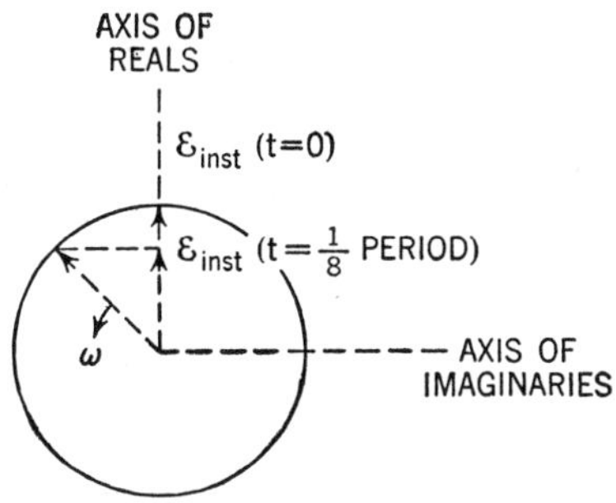

FIG. 34.2.—Complex number $\hat{\mathcal{E}}\epsilon^{j\omega t}$ of which the real part is the magnitude of the instantaneous space vector $\mathcal{E}_{\text{inst}}$.

35. Leading Term in the Instantaneous Electromagnetic Field of a Thin Center-driven Antenna near Resonance.—The electromagnetic field of an antenna of finite radius is extremely intricate because of the complexity of the distributions of current and charge from which it must be computed. On the other hand, if only the leading term in the distribution of current is retained, the problem is reduced to the idealized case of an infinitely thin conducting thread with its sinusoidal distribution of current. This permits an amazingly simple solution in terms of a highly instructive electromagnetic field that adequately portrays for most qualitative and many quantitative purposes the field due to the current in a thin antenna near resonance.

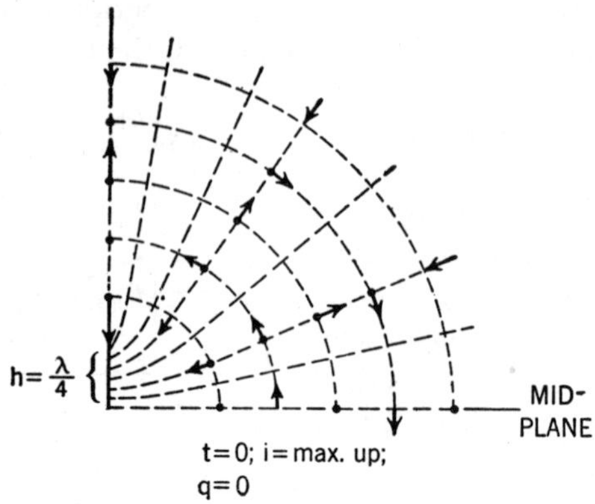

FIG. 35.1.—Instantaneous directions and relative magnitudes of the electric field of a center-driven antenna of vanishingly small radius and half length $h = \lambda/4$. Only the upper half of the antenna and the field in one quadrant are represented.

An electromagnetic field may be described in terms of mathematical surfaces of simple geometric shape on all points of which a suitably chosen component of the electric or magnetic field has the same phase. Such a surface is called an *equiphase surface.* When it moves with a finite velocity, it is called a wave front. For a conducting thread of length $2h = \lambda/2$, the equiphase surfaces are all ellipsoids of revolution with the ends of the antenna as foci. On each ellipsoid drawn around these two foci, the complete electromagnetic field consists of three mutually perpendicular components, two of the electric vector and one of the magnetic vector. The two components of the electric vector are a component $\mathcal{E}_\epsilon$ tangent to an ellipse that when rotated about the antenna as axis forms an *equiphase ellipsoid of revolution,* and a component $\mathcal{E}_h$ perpendicular to this ellipse (tangent to an orthogonal hyperbola). The entire magnetic field is represented by $\mathcal{H}_\Phi$. It is tangent to circles around the axis through the antenna, and thus tangent to the same ellipsoids of revolution as $\mathcal{E}_\epsilon$ but in a direction perpendicular to $\mathcal{E}_\epsilon$.

The components $\mathcal{E}_\epsilon$ and $\mathcal{H}_\Phi$ at any point are always in phase. The component $\mathcal{E}_h$ always lags $\mathcal{E}_\epsilon$ at the same point in space by a quarter period in time. All three components $\mathcal{E}_\epsilon$, $\mathcal{E}_h$, and $\mathcal{H}_\Phi$ may be expressed in simple form in confocal (ellipsoidal-hyperboloidal) coordinates. For the present qualitative discussion, it is sufficient

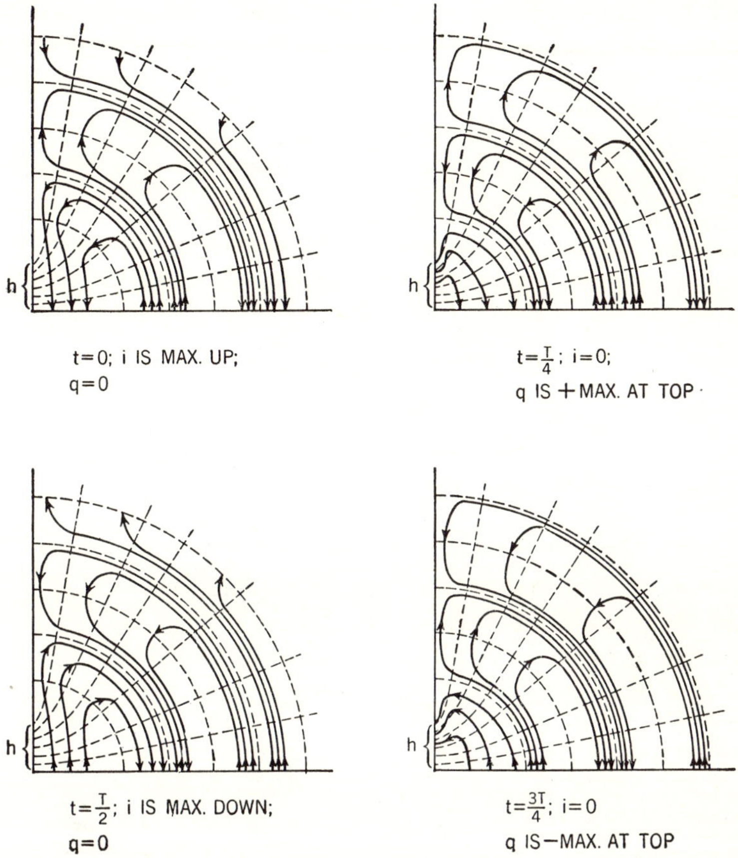

FIG. 35.2.—Instantaneous direction of the electric field of a center-driven antenna of vanishingly small radius and half length $h = \lambda/4$. Only the upper half of the antenna and the field in one quadrant are shown.

to represent them graphically, Fig. 35.1, where the instantaneous directions and relative magnitudes of the two components of $\mathcal{E}$ are shown by the arrows. (Their lengths are shown to decrease in correct sequence but not at all to scale.) The instant chosen is $t = 0$ with $i_0 = \hat{I}_0 \cos \omega t$.

In Fig. 35.2, the instantaneous directions of the electric field are shown by contours for $t = 0$, $t = T/4$, $t = T/2$, $t = 3T/4$, where T is the period. The magnitude of $\mathcal{E}$ is *not constant* along the contours. Since the electric field at every point is composed of two components $\mathcal{E}_\epsilon$ and $\mathcal{E}_h$, which are mutually perpendicular and which differ in phase by a quarter period ($\mathcal{E}_h$ lags $\mathcal{E}_\epsilon$), the resultant electric vector is elliptically polarized. The plane of polarization at any point is always the plane containing the point and the

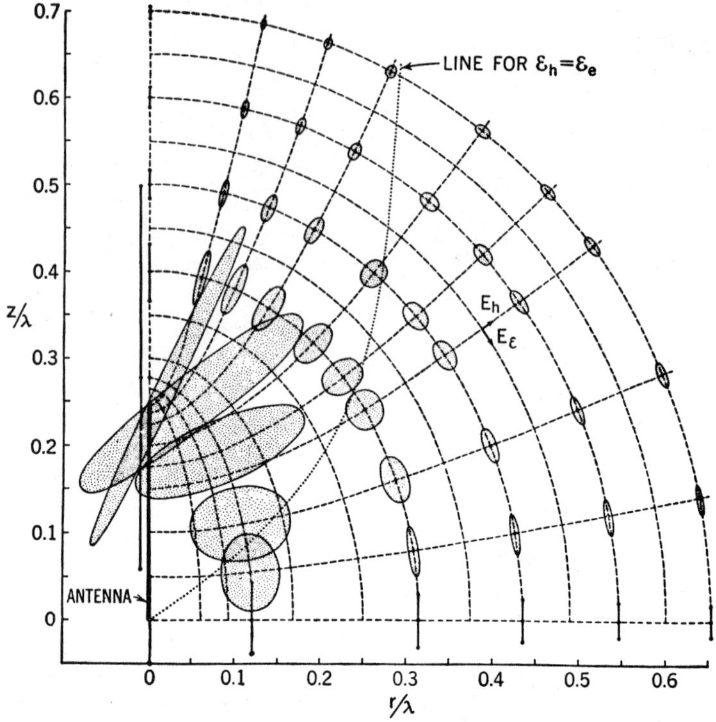

Fig. 35.3.—Ellipses of the end points of the elliptically polarized electric vector $\mathcal{E}$ at different fixed points in space about a vanishingly thin antenna of half length $h = \lambda/4$. Only the upper half of the center-driven antenna and the field in one quadrant are represented.

antenna. Since complete rotational symmetry exists, any such plane is typical. The ellipses corresponding to the loci of the end points of the vectors at different fixed points in space are drawn in Fig. 35.3. They lie in a plane passing through the antenna and are perpendicular to the ellipsoidal surfaces of constant phase. The dimensions of the ellipses are drawn to scale so that the relative magnitudes of $\mathcal{E}_\epsilon$ and $\mathcal{E}_h$ at any point are given by the major and minor semiaxes of the ellipse. $\mathcal{E}_h$ is larger than $\mathcal{E}_\epsilon$ near the z axis; $\mathcal{E}_\epsilon$ is larger than $\mathcal{E}_h$ near the mid-plane. A line of circular polarization along which $\mathcal{E}_h$ and $\mathcal{E}_\epsilon$ have the same magnitude is shown dotted in Fig. 35.3.

It is of primary importance that the electric field even very close to an antenna *cannot be described in terms of an instantaneous electrostatic field.* That this is not possible is seen perhaps most easily by comparing the *electric field* near the upper end of an antenna of length $2h = \lambda/2$ *at an instant* when this end is charged positively and the lower end charged negatively with the *electrostatic field* of a dielectric rod which *always* has the same distribution of charge as the antenna at the instant considered. Whereas the *electrostatic* field near the upper end of the charged dielectric rod is calculated in terms of positive charge at the top and negative charge at the bottom, the *electric* field of the antenna must be calculated in terms of the charge distributions along the antenna at *earlier* times. Thus the field near the upper end (at the instant when this has a maximum positive charge) due to the extreme lower tip of the antenna (one-half wavelength away) must be calculated not in terms of the maximum negative charge that is actually there at that instant, but in terms of the charge that *was there* one-half period *earlier.* This was a maximum positive charge. For other points, the charge at appropriately *retarded times* must be used. Thus the electric field of the antenna is *entirely different* from the electrostatic field of the charged dielectric rod even though, at the *instant in question,* the charge distributions are exactly the same, simply because the field of the antenna is *nowhere determined exclusively by the charge distribution at that instant.* The charge distribution at different points and different times along the antenna is involved.

36. Phase and Group Velocities; Wavelength; Ellipsoidal, Spherical, and Plane Waves.—If attention is focused on any one of the ellipsoidal surfaces on which at a given instant the components $\mathcal{E}_\epsilon$ and $\mathcal{H}_\Phi$ of the electromagnetic field of a center-driven antenna of negligible radius are a maximum, while the component $\mathcal{E}_h$ is zero, the following is observed: First it is noted that, at the given instant, there exists an infinite family of confocal ellipsoidal surfaces on each of which $\mathcal{E}_\epsilon$ and $\mathcal{H}_\Phi$ simultaneously are a maximum while $\mathcal{E}_h$ is zero. These ellipsoids differ in size in such a way that the distance between successive ones *as measured along the z axis* is exactly one wavelength λ, where by definition

$$\lambda = \frac{v_c}{f} \tag{36.1}$$

The *radial* distance between them *as measured along the mid-plane* cannot be a constant. Very near the antenna, *i.e.*, within several wavelengths, the radial distances along the mid-plane between successive ellipsoids are much greater than λ. At greater and greater distances from the antenna, the intervals approach λ asymptotically.

After having noted the family of confocal ellipsoidal surfaces on each and all of which *at a given instant* $\mathcal{E}_\epsilon$ and H_Φ are maximum and $\mathcal{E}_h$ is zero, let the behavior of these surfaces be observed at successive instants. As time passes, these surfaces expand in such a way that their *points of intersection with the z axis* all travel outward in the z direction with the constant velocity $v_c = 3 \cdot 10^8$ m/sec. Since the ellipsoidal surfaces in expanding become more nearly spherical, the outward velocity (along a hyperboloid) at any point other than that traveling along the z axis must *exceed* v_c. The highest velocity is along any radial line in the mid-plane. Since the velocity described in this way is the velocity of a particular phase in a continuous periodic phenomenon (the phase for the maximum of $\mathcal{E}_\epsilon$ and $\mathcal{H}_\Phi$, and the zero value of $\mathcal{E}_h$ have been chosen for convenience), this is a *phase velocity* v_p.

The ellipsoidal surfaces associated with constant phases of the components of the electromagnetic field are called *expanding electromagnetic waves*. They may be regarded as mathematical envelopes traveling through space, carrying with them particular phases and components of the electromagnetic vectors. Near the antenna, the waves are ellipsoidal; at great distances (in the far zone), they become practically spherical. Over any sufficiently small part of an ellipsoidal or spherical surface, the electromagnetic waves may be treated as approximately *plane*. In the far zone, the length of a receiving antenna is the chord for a sufficiently small arc to permit treating the passing spherical electromagnetic waves at the antenna as plane waves.

In order to transmit a signal in the form of a change in current observed in the loudspeaker of a distant receiver, it is necessary to vary the current induced in the receiving antenna. Since this current is proportional to the tangential electric field, and this in turn depends on the current in the transmitting antenna, a variation in the transmitter current is required, *i.e.*, the steady carrier must be modulated in some way.

The delayed time at which a modulation sequence (*e.g.*, in the form of a change in amplitude) is observed in the receiver after being applied to the transmitter is not necessarily the same as the delayed time computed for a given phase of the carrier. Much as a particular crest in a train of circular wavelets set up by a stone dropped into a smooth pool travels and dies out more rapidly than the group of wavelets as a whole, so a given phase in the train of carrier electromagnetic waves of an antenna may travel more rapidly than the envelope of a modulation. The velocity of such an envelope is called the *group velocity* and is denoted by v_g. For the antenna of length $2h = \lambda/2$ in free space, the phase and group velocities are equal along the z axis and are

$$(v_p)_z = (v_g)_z = v_c = 3 \cdot 10^8 \text{ m/sec} \tag{36.2}$$

Along the mid-plane, on the other hand,

$$(v_g)_r = \frac{v_c}{\sqrt{1 + (h^2/r^2)}} \tag{36.3}$$

$$(v_p)_r = v_c \sqrt{1 + \frac{h^2}{r^2}} \tag{36.4}$$

so that

$$v_p v_g = v_c^2 \tag{36.5}$$

where r is the radial distance out from the center of the antenna. Note that h is written for $v_c/4f$ in (36.3) and (36.4). Thus $(v_p)_r$ approaches v_c asymptotically from larger values, and $(v_g)_r$ approaches v_c asymptotically from smaller values as $(h/r)^2$ diminishes with increasing distance. The term $(h/r)^2$ may usually be neglected at distances from the antenna exceeding a few wavelengths.

The electromagnetic field of a base-driven antenna erected vertically over a highly conducting half-space, Sec. 12, is the same in the upper half-space as the field of a symmetrical center-driven antenna the upper half of which is identical with the base-driven antenna. In Fig. 35.3, for example, the lower half of the antenna (not shown) may be disregarded and all space below the mid-plane may be supposed filled with a good conductor. Along the surface of this conductor, the electric field is perpendicular, the magnetic field tangent to the surface. By applying appropriate boundary conditions, Chap. III, (5.4) and (5.5), radially expanding concentrations of surface charge are found to exist where the perpendicular electric field $\mathcal{E}_\epsilon$ and the tangential magnetic field $\mathcal{H}_\Phi$ are nonvanishing. Each ring of charge concentration on the surface travels outward with the same radial phase velocity as the field just above the surface. Rings of outwardly moving concentrations of positive charge, constituting radially *outward* surface currents, alternate with rings of outwardly moving concentrations of negative charge constituting radially *inward* surface currents. Radially expanding *circles of constant phase* for the surface current and charge are called *traveling waves of surface current and charge or surface waves.* At sufficiently great distances from the antenna and on a perfectly conducting surface, their radial phase velocity is $v_c = 3 \cdot 10^8$ m/sec, and the distance between successive circles of constant phase is λ.

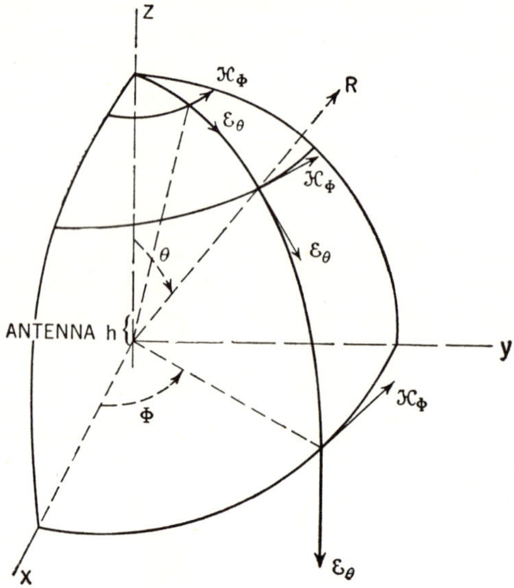

FIG. 37.1.—Components $\mathcal{E}_\theta$ and $\mathcal{H}_\Phi$ of the far-zone field of a center-driven linear radiator; polar coordinates R, θ, Φ.

Near the antenna, the phase velocity v_p exceeds v_c, and a constant wavelength cannot be defined.

37. Leading Term in the Distant Field of a Thin Center-driven Antenna near Resonance.—The far zone with respect to a transmitting antenna of half length h may be defined by

$$\beta R >> 1; \qquad \frac{h}{R} << 1 \tag{37.1}$$

where R is the distance from the center of the antenna to a point in the far zone. When these conditions are satisfied, the ellipsoids of revolution in Figs. 35.1 to 35.3 practically become spheres, and the hyperboloids become radial lines. The confocal system of coordinates approaches a spherical system R, θ, Φ, Fig. 37.1 and

$\mathcal{E}_\epsilon$ and $\mathcal{E}_h$ approach $\mathcal{E}_\theta$ and $\mathcal{E}_R$. The formulas for the field with $h = \lambda/4$ have the following simple structure for the time dependence:

$$i_0 = \hat{I}_0 \cos \omega t \tag{37.2}$$

$$(\mathcal{E}_\theta)_{\text{inst}} = \mathcal{R}_c(\mathcal{H}_\Phi)_{\text{inst}} = -\frac{60\hat{I}_0}{R} \frac{\cos\left(\frac{\pi}{2}\cos\theta\right)}{\sin\theta} \sin(\omega t - \beta R) \tag{37.3}$$

$$(\mathcal{E}_R)_{\text{inst}} = \frac{60\hat{I}_0 h}{R^2} \frac{\sin\left(\frac{\pi}{2}\cos\theta\right)}{\sin\theta} \cos(\omega t - \beta R) \tag{37.4}$$

The numerical factor 60 is the approximate value of

$$\frac{\mathcal{R}_c}{2\pi} = \frac{376.7}{2\pi} = 59.96 \text{ ohms} \tag{37.5}$$

Except for very small or vanishing values of θ, or those near π (near the z axis) where $\mathcal{E}_\theta$ becomes small or actually zero, the second condition in (37.1) permits writing

$$\hat{\mathcal{E}}_R << \hat{\mathcal{E}}_\theta \tag{37.6}$$

for most of the range of θ from zero to π. The larger R becomes as compared with h, the greater is the range of θ over which (37.6) is true. At very great distances (for which $h/R << 1$), it is usually justifiable to neglect $\mathcal{E}_R$. (When $\mathcal{E}_\theta$ becomes as small as $\mathcal{E}_R$ near $\theta = 0$, $\mathcal{E}_R$ likewise may be neglected.) In this important and useful case, the entire distant electromagnetic field is given by the simple formula (37.3). In verifying the magnitude of (37.3) experimentally it is important to note that the *form* (37.3) for $\mathcal{E}_\theta$ is correct subject only to $h^2 << R^2$ instead of (37.1), but $\mathcal{E}_R$ is not negligible. If a receiving antenna is oriented to respond only to $\mathcal{E}_\theta$, it is immaterial whether $\mathcal{E}_R$ is negligible or not.

The function

$$F(\theta) = \frac{\cos\left(\frac{\pi}{2}\cos\theta\right)}{\sin\theta} \tag{37.7}$$

is called the "vertical" characteristic or factor of the field, the "vertical" direction always being taken along the axis of the antenna however this may be oriented in space. $F(\theta)$ is plotted in polar form in Fig. 37.2 for the range $\theta = 0$ to $\theta = \pi/2$. The lower half from $\theta = \pi/2$ to $\theta = \pi$ is not shown, because it is an image of the upper half. Since the antenna is rotationally symmetrical about its axis, this pattern is correct for all values of Φ, and the "horizontal" field pattern is a circle.

If the antenna of height h is erected vertically over a perfectly conducting half-space, the pattern of Fig. 37.2 applies directly.

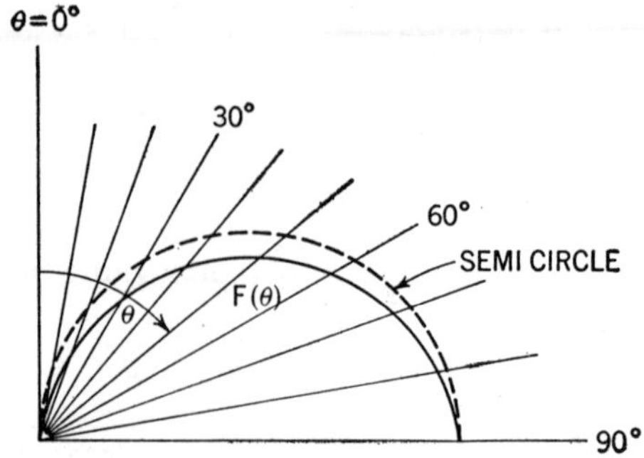

FIG. 37.2.—Polar plot of the "vertical" field factor $F(\theta) = \dfrac{\cos\left(\frac{\pi}{2}\cos\theta\right)}{\sin\theta}$ in one quadrant.

If the lower half-space is not perfectly conducting, the field pattern is modified, Fig. 37.3.

The physical significance of the field patterns of Figs. 37.2 and 37.3 is easily visualized. Consider a straight receiving antenna placed at a great distance from the transmitter with its axis parallel to the distant electric field $\mathcal{E}_\theta$ of the driven antenna, *i.e.*, tangent to a meridian on a great sphere drawn around the driven antenna. Let the receiving antenna be moved always tangent to the meridian beginning at a point directly above the antenna ($\theta = 0$) and continuing through an angle of 90° to the mid-plane. The current induced in this antenna when held at any angle θ will be proportional to the length of the radial line drawn from the origin in Fig. 37.2 or 37.3 to the curve at the angle θ. For Fig. 37.2, for example, the induced current will increase from zero when the receiving antenna is directly above the driven antenna to a maximum when it is parallel to it in the mid-plane.

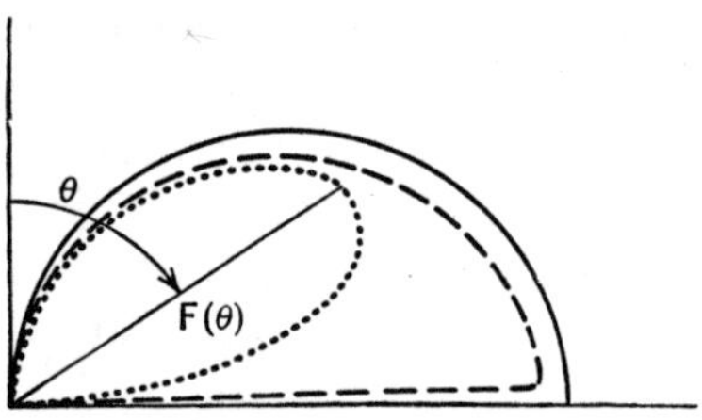

FIG. 37.3.— "Vertical" field pattern of a vanishingly thin antenna of length $h = \lambda/4$ over variously conducting earths: solid line, perfect conductor; broken line, good conductor; dotted line, poor conductor.

In terms of the convenient complex notation in which

$$i_0 = \hat{I}_0 \epsilon^{j\omega t} \tag{37.8}$$

with the understanding that only the real part is physically significant, the instantaneous electric field in the far zone is given by the real part of

$$(\mathcal{E}_\theta)_{\text{inst}} = \hat{\mathcal{E}}_\theta \epsilon^{j\omega t} \tag{37.9}$$

where

$$\hat{\mathcal{E}}_\theta = j\,\frac{60\hat{I}_0}{R}\,F(\theta)\epsilon^{-j\beta R} \tag{37.10}$$

38. Directivity and Gain.—The field patterns of Figs. 37.2 and 37.3 show that the degree of coupling between a very thin transmitting antenna for which $h \doteq \lambda/4$ and a distant receiving antenna placed parallel to the electric field of the driven antenna depends upon the spherical coordinate θ measured along any meridian. The driven antenna has directional properties such that the induced voltage in the receiving antenna is not the same when this is placed successively so that θ has different values, even though all other possible variables are maintained unchanged. (In other types of antenna which are not rotationally symmetrical, the degree of coupling between transmitter and distant receiving antenna depends upon the angle Φ as well as the angle θ.) This directional characteristic of an antenna is of great practical importance. Indeed, the purpose of most antenna arrays is to modify and make use of it so that receiving antennas in one or more directions receive a strong signal, while those in all other directions are able to detect no signal at all or merely an extremely weak one. A quantitative measure of the directional properties of an antenna is given by its absolute directivity D. This may be defined by

$$D = \frac{P_N}{P} \tag{38.1}$$

where P is the power radiated from the antenna of directivity D when it maintains an electric field of amplitude $(\hat{\mathcal{E}}_\theta)_{\text{max}}$ at a great distance R in that particular direction in which $\hat{\mathcal{E}}_\theta$ is a maximum. (The direction is specified by the spherical coordinate θ_m, or more generally by the two coordinates θ_m and Φ_m. In Fig. 37.2, $\theta_m = 90°$; in Fig. 37.3, θ_m is approximately 60° for the antenna over a poorly conducting earth.) P_N is the power that would be radiated from a physically fictitious *omnidirectional* antenna that maintained an electric field of the same amplitude $(\hat{\mathcal{E}}_\theta)_{\text{max}}$ at the same distance R

as the actual antenna but in *all directions, i.e.*, for all possible values of both θ and Φ. A simple method of calculating the absolute directivity of an antenna in terms of its distant electromagnetic field is described in Sec. 44. The directivity of the dipole for which

$$F(\theta) = \frac{\cos\left(\frac{\pi}{2}\cos\theta\right)}{\sin\theta}$$

is 1.64 for $\theta_m = 90°$. Neglecting power dissipated in heating the antenna, this means that 1.64 times the power actually supplied to the dipole would have to be delivered to a completely nondirectional antenna in order to maintain the same field $\hat{\mathcal{E}}_\theta$ at $\theta = \pi/2$.

In comparing the directional properties of antennas and arrays, it is often convenient to use the simple $\lambda/2$ dipole (the thin center-driven antenna with $h = \lambda/4$) as a standard of comparison. This is done by defining a relative directivity D_r by

$$D_r = \frac{D}{1.64} \tag{38.2}$$

The gain in decibels of an antenna of absolute directivity D referred to a half-wave dipole is

$$G(db) = 10 \log_{10} D_r \tag{38.3}$$

39. Distant Field of Linear Radiators.—If the half length h of a very thin center-driven antenna (or the full length h of a base-driven antenna erected vertically over a perfectly conducting plane) is not restricted to $h \doteq \lambda/4$, the leading term in the distant electromagnetic field differs from (37.3) only in the form of $F(\theta)$. Thus

$$(\mathcal{E}_\theta)_{\text{inst}} = \mathcal{R}_c(\mathcal{H}_\Phi)_{\text{inst}} = \frac{-60\hat{I}_0}{R} F_0(\theta) \sin(\omega t - \beta R) \tag{39.1}$$

$$F_0(\theta) = \frac{\cos(H\cos\theta) - \cos H}{\sin\theta \sin H} \tag{39.2}$$

and $H = \beta h = 2\pi h/\lambda$. The formula (39.2) reduces to (37.7) with $H \doteq \pi/2$. Since both $\hat{I}_0$ and $\sin H$ vanish for $H = n\pi$, an indeterminate form results. This is due to the fact that only the leading term in the distribution of current has been retained. The difficulty can be avoided by introducing the maximum amplitude I_m given by

$$\hat{I}_m = \frac{I_0}{\sin H} \tag{39.3}$$

and writing

$$(\mathcal{E}_\theta)_{\text{inst}} = \mathcal{R}_c(\mathcal{H}_\Phi)_{\text{inst}} = -\frac{60\hat{I}_m}{R} F_m(\theta) \sin(\omega t - \beta R) \tag{39.4}$$

$$F_m(\theta) = F_0(\theta) \sin H = \frac{\cos(H\cos\theta) - \cos H}{\sin\theta} \tag{39.5}$$

The function $F_m(\theta)$ is plotted in Fig. 39.1 for four values of h including $\lambda/4$. As long as h does not exceed $\lambda/2$, there is only a single ear in the field pattern, with its maximum value occurring always at $\theta = 90°$. The absolute directivity when $h = \lambda/2$ for $\theta_m = 90°$

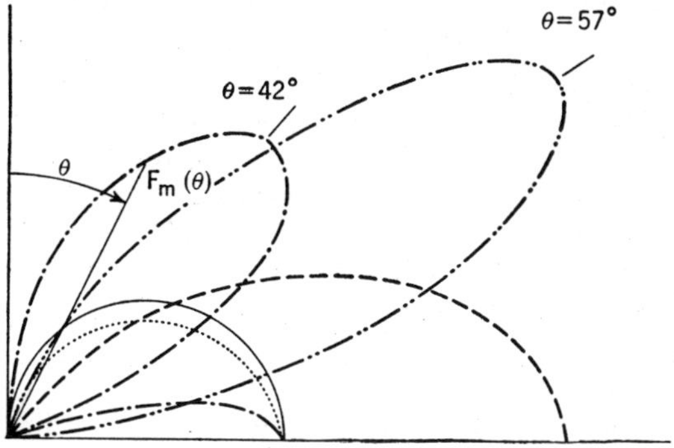

FIG. 39.1.—The "vertical" field factor $F_m(\theta) = \dfrac{\cos(H\cos\theta) - \cos H}{\sin\theta}$ with $H = \dfrac{2\pi h}{\lambda}$ for the following values of h: $h = \lambda/4$, dotted curve; $h = \lambda/2$, curve in broken line; $h = 3\lambda/4$, curve in dot dash; $h = \lambda$, curve in dot dot dash. The solid curve is a semicircle of unit diameter for reference.

is 2.41; the relative directivity referred to the dipole with $h = \lambda/4$ as standard is

$$D_r = \frac{2.41}{1.64} = 1.47 \tag{39.6}$$

The gain in decibels referred to the dipole ($h = \lambda/4$) is

$$G(db) = 10 \log_{10} D_r = 1.67 \tag{39.7}$$

The absolute directivity when $h = 3\lambda/4$ for $\theta_m = 42°$ is 2.11; the relative directivity is 1.29, and the gain in decibels over a dipole is 1.1.

If an antenna is very short compared with the wavelength, (39.2) reduces to the form

$$F_0(\theta) = \tfrac{1}{2}H \sin\theta; \qquad (H^2 << 1) \tag{39.8}$$

This applies to a short antenna of half length h with vanishing current at the ends. If the antenna is so end-loaded that the current is sensibly uniform over the entire length $2h$,

$$F_0(\theta) = H \sin \theta; \qquad (H^2 << 1) \tag{39.9}$$

The absolute directivity in both cases is 1.5, and this value, instead of the value 1.64, is sometimes used as the standard in defining relative directivity.

An interesting case is that of an antenna with $h = 0.581\lambda$. Approximately this value for the half length was recommended for use at broadcast frequencies to minimize the field at small angles θ from the vertical and so to reduce ionospheric reflection.

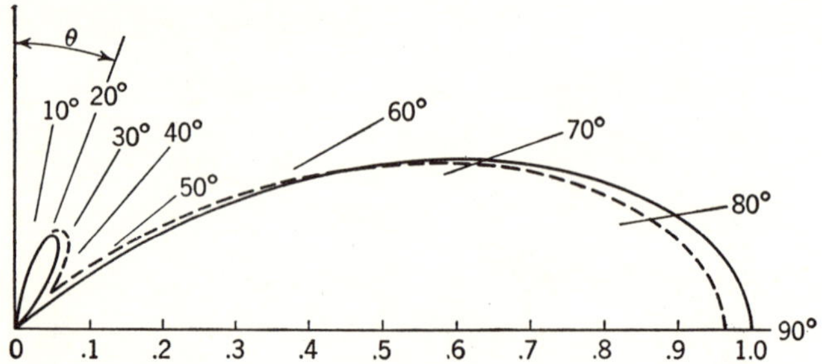

FIG. 39.2.—"Vertical" field patterns computed for an antenna of half length $h = 0.581\lambda$.

For $h = 0.581\lambda$, the principal ear at $\theta = 90°$ is a little flatter than that for $h = \lambda/2$, and the secondary ear is still so small that its contribution to the field is insignificant. The computed field pattern is shown in solid line, Fig. 39.2. So long as the antenna is sufficiently thin so that the leading term in the distribution of current is sufficient, the pattern shown in solid line in Fig. 39.2 is a fair approximation. For most practical antennas, the component $\hat{I}_z''$, Sec. 9, is not completely negligible compared with $\hat{I}_z'$, as can be seen from Fig. 9.3. Its effect on the field pattern is shown in Fig. 39.2 for a moderately thick cylindrical antenna. The principal effect of the nonvanishing radius is to increase the size of the secondary ear while the sharp zero between the two ears is rounded off. In a still thicker antenna or an equivalent tower antenna, the field at high angles increases so that stronger ionospheric reflections may be expected with thick than with thin antennas.

If a base-driven antenna of full length h is erected over an imperfectly conducting earth, the field patterns of Fig. 39.1 or

39.2 bend in to zero near $\theta = 90°$ much as in Fig. 37.3 for the special case $h = \lambda/4$. In addition, the sharp zero values between ears are very much rounded off in a way similar to Fig. 39.2. This rounding off of zero values between ears is due to components of current a quarter period out of phase with the principal component. In a thick antenna, this component $\hat{I}_z''$ is in the antenna itself; in a thin antenna over an imperfectly conducting earth, the out-of-phase component is in the earth. If a thick antenna is erected over an imperfectly conducting earth, out-of-phase components of current are in both antenna and earth, and the components of fields due to the two combine to round off still further the sharp zero values between ears that exist in the idealized case of an infinitely thin antenna over a perfectly conducting half-space.

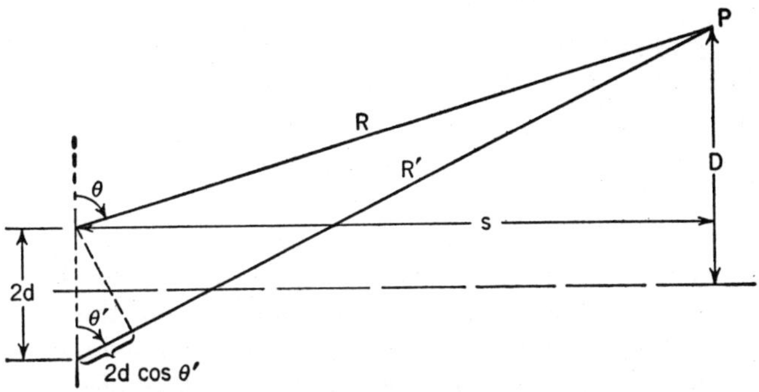

FIG. 40.1.—Collinear antennas separated at a distance of $2d$ between centers.

40. Distant Field of Collinear Arrangements of Antennas; Vertical Antenna over the Earth; Collinear Array.—The distant field of two center-driven antennas erected one above the other, Fig. 40.1, is readily obtained by direct combination of the individual fields if it is assumed that to a first approximation the distribution of current along each unit is the same as when it is far from the other. Assume that the point P where the field is to be calculated is so distant from both antennas that the vertical distance $2d$ between centers is very small compared with R and R'. Under this condition, no significant error is introduced by setting θ' equal to θ and R' equal to R, except in the phase factors βR and $\beta R'$. Although $(R' - R)$ may be negligible compared with R, $\beta(R' - R)$ may be an appreciable fraction of 2π and therefore significant. Accordingly

$$\beta R' \doteq \beta(R + 2d \cos \theta) \tag{40.1}$$

in phase angles.

If it is assumed that the currents in the two identical antennas are equal in amplitude and phase, the resultant complex electric field at a distant point is

$$\hat{\mathcal{E}}_\theta = j\,\frac{60\hat{I}_0}{R}\,F(\theta)(1 + \epsilon^{-j2\beta d\cos\theta})\epsilon^{-j\beta R} \tag{40.2}$$

or

$$\hat{\mathcal{E}}_\theta = j\,\frac{120\hat{I}_0}{R}\,F(\theta)\cos(\beta d\cos\theta)\epsilon^{-j\beta(R+d\cos\theta)} \tag{40.3}$$

Multiplying both sides of (40.3) by $\epsilon^{j\omega t}$ and retaining only the real part, the instantaneous value of the distant field due to both antennas is

$$(\mathcal{E}_\theta)_{\text{inst}} = -\frac{120\hat{I}_0}{R}\,F(\theta)\cos(\beta d\cos\theta)\sin[\omega t - \beta(R + d\cos\theta)] \tag{40.4}$$

Comparing (40.4) with (37.3), using (37.7), it is seen that the *amplitude* of the distant field due to both antennas differs from that due to a single antenna by the factor

$$A(\theta) = 2\cos(\beta d\cos\theta) \tag{40.5}$$

A more restricted but often useful expression for $A(\theta)$ may be derived in terms of the height D (Fig. 40.1) of the point P (where the field is calculated) above the mid-plane and the horizontal distance s along the mid-plane between the points directly below the antenna and below P. Instead of writing $(R' - R) \doteq 2d\cos\theta$ as in (40.1), it is possible to write

$$R' - R = \sqrt{s^2 + (D+d)^2} - \sqrt{s^2 + (D-d)^2} \tag{40.6}$$

If the condition

$$s > (D + d) \tag{40.7}$$

is satisfied

$$R' - R \doteq \frac{2dD}{s} \tag{40.8}$$

so that

$$A(\theta) \doteq 2\cos\left(\frac{\beta dD}{s}\right) \tag{40.9}$$

$A(\theta)$ is called the array factor of the two-antenna array. The combined vertical field pattern is given by the product $F(\theta)\,A(\theta)$ of the vertical characteristic $F(\theta)$ of a single antenna and the array factor $A(\theta)$ of the two antennas. Depending upon the length of the antennas, $F(\theta)$ may be given by (37.7) or more generally by (39.5).

An important application of the analysis of two collinear antennas is the approximate determination of the distant field of a vertical linear radiator with its center at a height d (not $2d$) above the

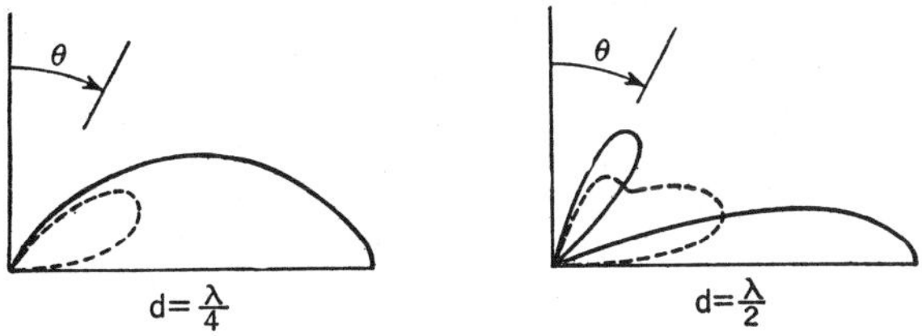

FIG. 40.2.—"Vertical" field pattern of vertical, center-driven antenna of half length $h = \lambda/4$ with its center at height d over conducting earth. Solid lines, perfectly conducting earth; broken lines, poorly conducting earth.

surface of the earth. If the frequency is not too high and the conductivity of the earth is large, a rough estimate is obtained by treating the earth as a perfectly conducting half-space, in which case the field at points above the earth is given exactly by (40.4). This is true because a perfectly conducting half-space can always be replaced by an image antenna which has its current reversed with respect to corresponding image points. For a vertical antenna, this means that currents are in the *same* direction in the image and the antenna, since the image point corresponding to the upper tip of the antenna is at the extreme bottom of the image, and that corresponding to the bottom of the antenna is at the top of the image. Vertical field patterns for the antenna at a height d over a perfectly conducting half-space, or for two isolated antennas separated a distance $2d$ between centers are shown in solid lines for two values of d, Fig. 40.2. The corresponding patterns for the same antenna over a poorly conducting earth are shown in broken lines.

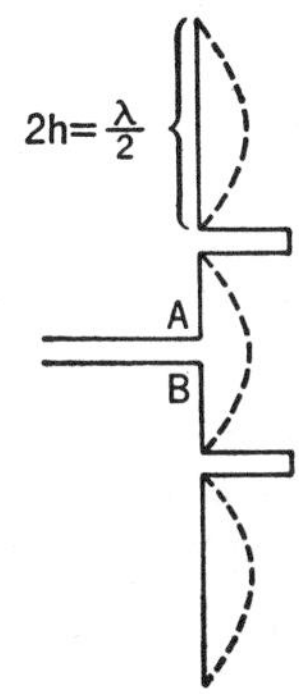

FIG. 40.3. Center-driven collinear array with sinusoidal current in broken lines. Phase-reversing sections are shown as closed-end open-wire stubs of length $\lambda/4$.

If more than two antennas are arranged along a common axis, their resulting field is determined as with two antennas, by direct combination. Especially simple expressions are obtained for a collinear array of N identical end-to-end units each of half length $h = \lambda/4$, Fig. 40.3, for $N = 3$. The construction of such an array is discussed in detail in Secs. 22 and 23. If the array is perfectly adjusted so that each unit carries the same current in both magnitude and phase, the array factor reduces to

$$A(\theta) = \frac{\sin [N(\pi/2) \cos \theta]}{\sin [(\pi/2) \cos \theta]} \tag{40.10}$$

Along the mid-plane, $\theta = \pi/2$, the pattern $F(\theta)$ of a single unit is multiplied by the number of units N; in other directions, it is not magnified in proportion to N, although small sharp ears do appear. The result is a considerable flattening of the vertical field pattern with a corresponding increase in the directivity for $\theta_m = \pi/2$. Typical vertical characteristics are shown in Fig. 40.4 for $N = 1$, 4, 8, assuming the same current amplitude maintained in each

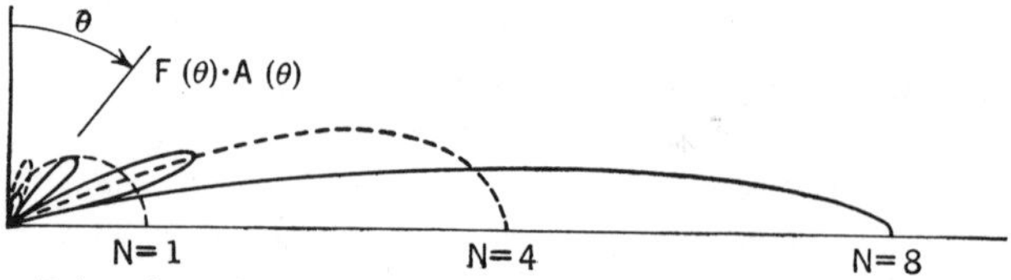

FIG. 40.4.—Polar plots of $F(\theta)A(\theta)$ for collinear arrays of N units each of half length $h = \lambda/4$. All units have equal currents in phase.

extremely thin unit irrespective of the number of units. This does not mean the same power for each unit. Since the input resistance at the terminals AB, Fig. 40.3, of an array of N units is greater than N times the input resistance of a single unit, *more* than N times as much power must be supplied to an array of N units in order to obtain the same current in each unit and N times the magnitude of the electromagnetic field for $\theta = \pi/2$. The relative directivity of N units as referred to a single unit is not N but roughly only about $N/2$. A great increase in "vertical" directivity for $\theta_m = 90°$ may be achieved with the collinear array. Note that "vertical" refers to the axis of the array regardless of its orientation in space.

41. Distant Field of Parallel Arrays—All Units Driven.—If two identical driven antennas are parallel to each other, Fig. 41.1, both the "vertical" and "horizontal" field patterns in general differ from those of a single unit. This is primarily a result of the superposition of the fields due to the currents in the two units; it is also partly due to a changed distribution of current along each antenna when each is near and hence quite closely coupled to the other. The leading term in the distribution of current in each of the two antennas when these are in proximity is the sinusoidal distribution of an infinitely thin isolated antenna. For calculating the far-zone field, this simple approximation is adequate for most purposes and will be used.

The array, Fig. 41.1, is not rotationally symmetrical (as was the stacked or collinear arrangement), and the distant electromagnetic field must depend therefore on both spherical coordinates, θ and Φ. The angle θ is measured from the "vertical" axis along any meridian; the angle Φ is measured in the mid-plane from a line drawn through the array. At very distant points such as P where the electromagnetic field is to be determined, it is possible to write θ for θ' and R for R' (except in phase angles), provided that R and R' are large compared with both the "horizontal" separation b and the half length h of the antennas. In phase angles, it is necessary to write

$$\beta R' = \beta(R - b \sin \theta \cos \Phi) \tag{41.1}$$

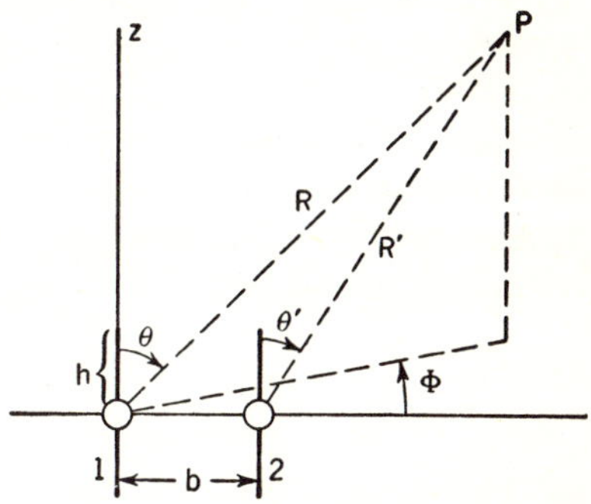

FIG. 41.1.—Two parallel antennas. The left-hand unit (No. 1) has its center at the origin. (R and R' should be very much greater relative to h and b for correct proportions with P in the far zone.)

If the complex current at the center of antenna 2, Fig. 41.1, is related to that at the center of antenna 1 by $\hat{I}_{02} = \hat{I}_{01} k \epsilon^{-j\delta}$, where k is an arbitrary amplitude factor and δ is an arbitrary phase angle, the complex amplitude of the distant field of both antennas is obtained using (37.10) and is

$$\hat{\mathcal{E}}_\theta = j \frac{60 I_{01}}{R} F(\theta) \epsilon^{-j\beta R}[1 + k\epsilon^{-j(\delta - \beta b \sin \theta \cos \Phi)}] \tag{41.2}$$

If the currents in the two antennas have the same amplitude so that $k = 1$, (41.2) may be written

$$\hat{\mathcal{E}}_\theta = j \frac{60 I_{01}}{R} F(\theta)\ A(\theta,\Phi) \epsilon^{-j\left(\beta(R - \frac{1}{2} b \sin \theta \cos \Phi) + \frac{\delta}{2}\right)} \tag{41.3}$$

with the array factor defined by

$$A(\theta,\Phi) = 2 \cos [\tfrac{1}{2}(\delta - \beta b \sin \theta \cos \Phi)] \tag{41.4}$$

Thus the "vertical" characteristic $F(\theta)$ of a single antenna is multiplied by $A(\theta,\Phi)$ to give the general field characteristic of the array of two antennas. Interest is usually in the "horizontal" field factor, which is obtained from the general relation $F(\theta)\ A(\theta,\Phi)$

with $\theta = \pi/2$. Thus the "horizontal" field factor of the two antennas is given by $F(\pi/2)\ A(\pi/2,\ \Phi)$. For $h = \lambda/4$, $F(\pi/2) = 1$, and the "horizontal" factor is

$$F\left(\frac{\pi}{2}\right) A\left(\frac{\pi}{2}, \Phi\right) = 2 \cos\left[\tfrac{1}{2}(\delta - \beta b \cos \Phi)\right] \tag{41.5}$$

The "vertical" field factor for the particular value of Φ for which the horizontal factor is a maximum is important and is given by $F(\theta)\ A(\theta,\Phi_m)$.

The meaning of these rather general formulas may be illustrated by special cases.

CASE 1: Broadside arrangement: $\delta = 0$; $\beta b = \pi$. In the so-called "broadside array" the currents in the two antennas are equal in phase and amplitude. Choosing $b = \lambda/2$ and $\delta = 0$, the general field function for an array of dipoles (extremely thin resonant antennas of half length $h \doteq \lambda/4$) is

$$F(\theta)\ A(\theta,\Phi) = \frac{\cos\left(\frac{\pi}{2}\cos\theta\right)}{\sin\theta}\ 2\cos\left(\frac{\pi}{2}\sin\theta\cos\Phi\right) \tag{41.6}$$

The "horizontal" factor with $\theta = \pi/2$ is

$$F\left(\frac{\pi}{2}\right) A\left(\frac{\pi}{2}, \Phi\right) = 2\cos\left(\frac{\pi}{2}\cos\Phi\right) \tag{41.7}$$

and is plotted in Fig. 41.2. It has maxima at $\Phi_m = \pm\pi/2$ given by

$$F\left(\frac{\pi}{2}\right) A\left(\frac{\pi}{2}, \frac{\pi}{2}\right) = 2 \tag{41.8}$$

The "vertical" factor in the direction for maximum horizontal field is

$$F(\theta)\ A\left(\theta, \frac{\pi}{2}\right) = 2\,\frac{\cos\left(\frac{\pi}{2}\cos\theta\right)}{\sin\theta} = 2F(\theta) \tag{41.9}$$

Thus, in the broadside array, the "vertical" field factor *in the direction* Φ_m is the same in form as that for a single antenna.

CASE 2: Double-end-fire arrangement: $\delta = \pi$; $\beta b = \pi$. If the conditions of the broadside array are all maintained except that the currents in the two antennas are opposite in phase, $\delta = \pi$ instead of $\delta = 0$, the general characteristic of the array is

$$F(\theta)\ A(\theta,\Phi) = \frac{\cos\left(\frac{\pi}{2}\cos\theta\right)}{\sin\theta}\ 2\sin\left(\frac{\pi}{2}\sin\theta\cos\Phi\right) \tag{41.10}$$

The "horizontal" field factor with $\theta = \pi/2$ is

$$F\left(\frac{\pi}{2}\right) A\left(\frac{\pi}{2}, \Phi\right) = 2 \sin\left(\frac{\pi}{2} \cos \Phi\right) \tag{41.11}$$

and is plotted in Fig. 41.3. It differs from the broadside factor plotted in Fig. 41.2 in having its maxima rotated an angle Φ of 90° in space, reversing the phases in the two antennas and shifting the maxima from the double-broadside positions at $\Phi = \pm\pi/2$ to the double-end-fire positions at $\Phi = 0, \pi$. The vertical field factor

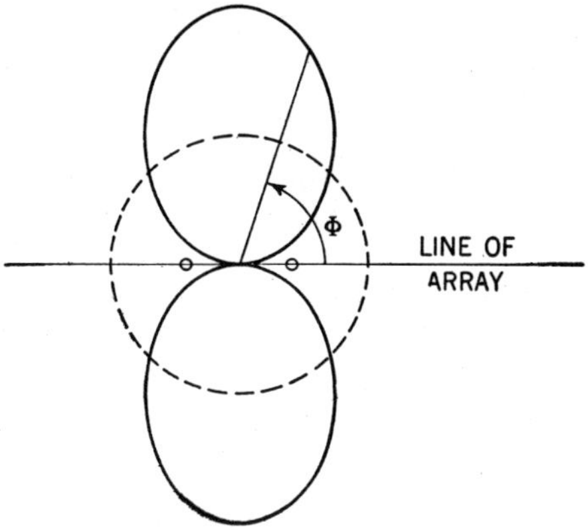

FIG. 41.2.—"Horizontal" field pattern of a two-element "broadside array" with half-wavelength spacing. The circle in broken line is the "horizontal" pattern of a single unit.

in the direction of the maximum horizontal field is the same in form as for a single unit.

CASE 3: End-fire arrangement: $\delta = \pi/2$; $\beta b = \pi/2$. If the two antennas are separated horizontally by $b = \lambda/4$ and the phases are adjusted so that current I_{02} in antenna 2 lags current I_{01} in antenna 1 at the origin by a quarter period, the general field function is

$$F(\theta)\, A(\theta,\Phi) = \left[\frac{\cos\left(\frac{\pi}{2}\cos\theta\right)}{\sin\theta}\right]\left\{2\cos\left[\frac{\pi}{4}(1 - \sin\theta\cos\Phi)\right]\right\} \tag{41.12}$$

The "horizontal" pattern with $\theta = \pi/2$ is plotted in Fig. 41.4 from

$$F\left(\frac{\pi}{2}\right) A\left(\frac{\pi}{2}, \Phi\right) = 2 \cos\left[\frac{\pi}{4}(1 - \cos \Phi)\right] \tag{41.13}$$

It is seen to be broadly unidirectional in the sense that there is only a single maximum in the direction $\Phi = 0$ (in the direction from the antenna with leading current toward the antenna with lagging current) with a zero value in the opposite direction, $\Phi = \pi$.

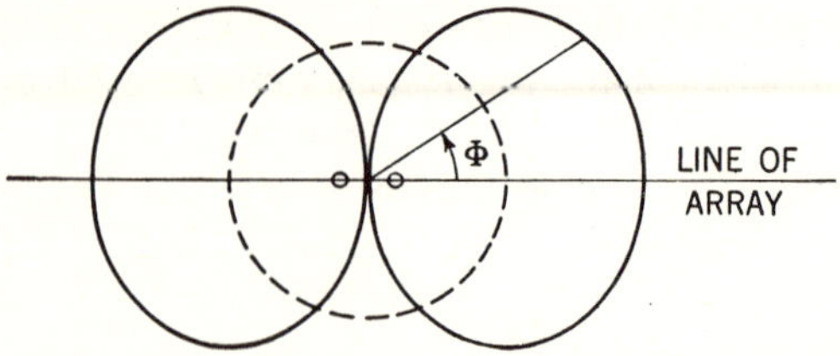

FIG. 41.3.—"Horizontal" field pattern of a two-element, double-end-fire array with half-wavelength spacing. The circle in broken line is the horizontal pattern of a single unit at the center.

The "vertical" pattern in the direction $\Phi = 0$, in which the "horizontal" factor has its maximum, is plotted in Fig. 41.5 from

$$F(\theta)\,A(\theta,\pi) = \frac{\cos\left(\frac{\pi}{2}\cos\theta\right)}{\sin\theta}\,2\cos\left[\frac{\pi}{4}(1 - \sin\theta)\right] \qquad (41.14)$$

It differs only slightly from the "vertical" pattern of a single

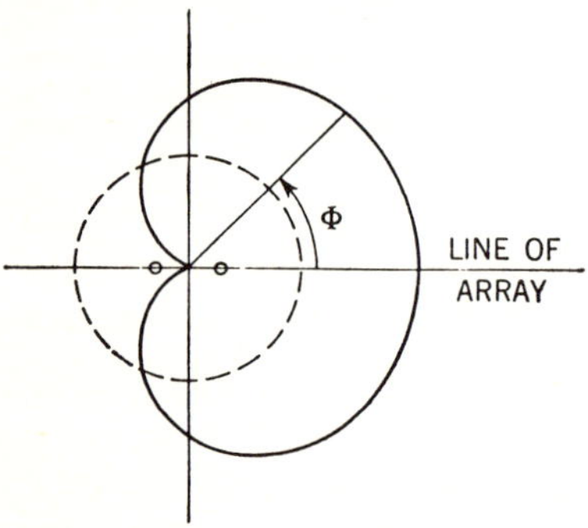

FIG. 41.4.—"Horizontal" field pattern of two-element end-fire array with quarter-wavelength spacing. The current in the right-hand unit lags by a quarter period.

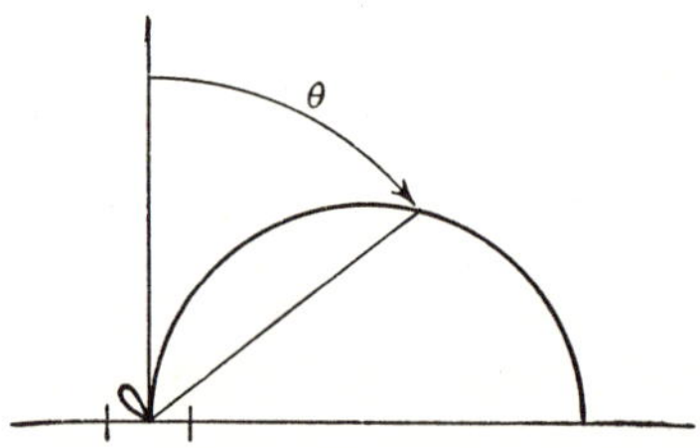

FIG. 41.5.—Upper half of the "vertical" field pattern of the array of Fig. 41.4 along the line of the array. $\Phi = 0$ to the right; $\Phi = \pi$ to the left.

antenna but is entirely different and of negligible magnitude in the opposite direction, $\Phi = \pi$.

The analytically derived facts regarding the horizontal field patterns of the three special cases may be illustrated graphically if it is recalled that at distant points surfaces of constant phase for either the electric or the magnetic field are spheres that differ in radius by λ. (This is not true near the antenna.) Thus, in two antennas driven with currents in phase and with the same amplitude, two great circles of the same radius, one drawn around each antenna, represent contours along which the electric field is everywhere in the same phase. Points of intersection therefore are points at which the fields due to the two antennas add directly and give a maximum. Circles drawn around the same antennas with radii that differ from that of the first pair of circles by $\pi/2$ represent contours along which the electric field is opposite in direction to that of the first pair and nearly equal in magnitude as in the first pair of circles. Hence, intersections of a circle of the first pair with a circle of the second pair locate points of almost complete cancellation of the field. In Fig. 41.6, the first pair of circles is drawn in solid lines and their intersections give the directions M of maxima; intersections of solid lines with broken lines give zeros at O.

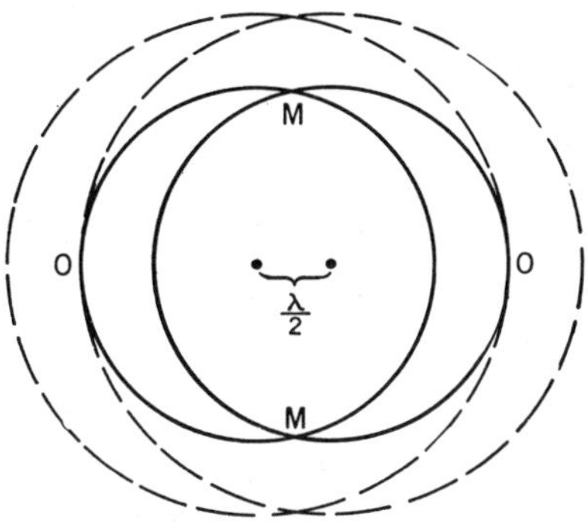

FIG. 41.6.—Circles of constant phase for broadside array. The field at the circles in broken line is 180° out of phase with the field at the solid circles.

If the two antennas are separated by a half wavelength, $b = \lambda/2$, and are driven with currents of the same amplitude and opposite phase, two circles of the same radius drawn around the two antennas must be drawn, one with a solid line and the other with a broken line. The same rules apply as before for locating maxima and zeros in Fig. 41.7.

If the two antennas are separated by a quarter wavelength, $b = \lambda/4$, and are driven so that the two units carry currents of the same amplitude but with that in the right-hand unit in Fig. 41.7 lagging that in the left-hand unit by a quarter period, the phases of the electric field due to each current on two circles of the same radius R drawn around the two antennas differ by a quarter period. On the other hand, the phase of the field on a circle of radius $R + \lambda/4$ drawn around the left-hand unit is the same as that on a circle of radius R drawn around the right-hand unit, and points of intersection of these two circles indicate directions for a maximum of the "horizontal" field pattern, M in Fig. 41.8. If circles are drawn with broken lines to represent loci of fields that differ by 180° from those represented by the circles with solid lines, then points of intersection O of a circle in solid line with one in broken line indicate cancellation of the field due to the two units.

The directional properties of the broadside and of the end-fire arrangements can be increased greatly by using more than two antennas. It is assumed for simplicity that the distribution of current along each unit is to a first approximation the same as

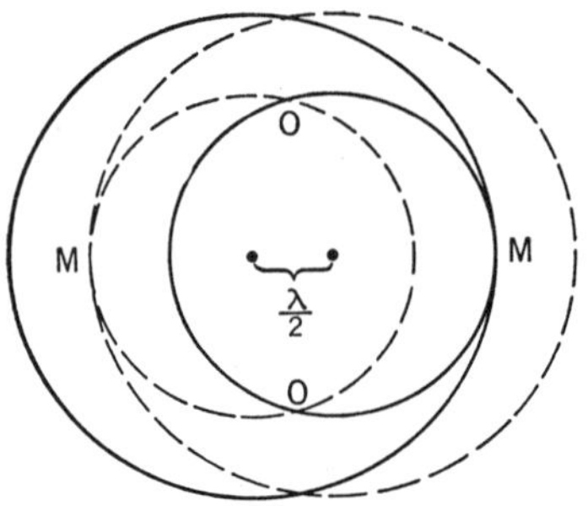

FIG. 41.7.—Circles of constant phase for double-end-fire array.

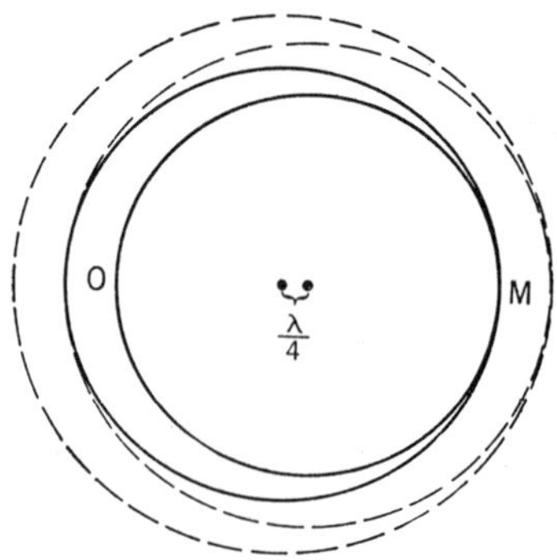

FIG. 41.8.—Circles of constant phase for an end-fire array with lagging current on the right.

when it is isolated. This assumption certainly is justified for units of half length h near $\lambda/4$.

The array characteristic of N parallel identical antennas with all currents the same in amplitude but differing progressively in phase by δ is

$$A(\theta,\Phi) = \frac{\sin\left[\frac{1}{2}N(\delta - \beta b \sin\theta \cos\Phi)\right]}{\sin\left[\frac{1}{2}(\delta - \beta b \sin\theta\cos\Phi)\right]} \tag{41.15}$$

Principal maxima of magnitude N occur in directions for which both numerator and denominator in (41.15) vanish. Zeros occur in directions for which the numerator in (41.15) vanishes but the denominator does not. Minor maxima in the "horizontal" plane ($\theta = \pi/2$) occur for values of the argument that are located about midway between the values that locate zero. The general formula (41.15) is of the form $\dfrac{\sin N\chi}{N \sin\chi}$ with $\chi = \frac{1}{2}(\delta - \beta b \sin\theta\cos\Phi)$, and is tabulated in Table 41.1 for a range of values of χ. It is specialized for a broadside array, in which all currents are in the same phase and have the same amplitude, by writing $\delta = 0$.

TABLE 41.1.—$\frac{\sin N\chi}{N \sin \chi}$

χ Radians	χ Degrees	N = 2	3	4	5	6	8	10	12
0	0.0	1.0000	1.000	1.000	1.000	1.000	1.000	1.000	1.000
π/24	7.5	0.9914	0.9778	0.9578	0.9502	0.9027	0.8298	0.7402	0.6383
π/20	9.0				0.9040			0.6393	0.5067
π/12	15.0	0.9659	0.9107	0.8268	0.7465	0.6441	0.4184	0.1932	0.000
π/10	18.0				.6472			0.000	−0.1585
3π/24	22.5	0.9239	0.8048	0.6532	.4829	.3078	0.0000	−0.1847	−0.2177
3π/20	27.0				.3114			−0.2202	−0.1485
π/6	30.0	0.8660	0.6667	0.4280	0.2000	0.0000	−0.2166	−0.1732	0.0000
π/5	36.0				0.0000			0.0000	0.1348
5π/24	37.5	0.7934	0.5059	0.2053	−0.0429	−0.1934	−0.1778	0.0425	0.1369
π/4	45.0	0.7071	0.3333	0.000	−0.2000	−0.2358	0.000	0.1414	0.000
7π/24	52.5	0.6088	0.1608	−0.1575	−0.2509	−0.1485	0.1365	0.0326	−0.1050
3π/10	54.0				−0.2472			0.000	−0.0980
π/3	60.0	0.5000	0.000	−0.2471	−0.2000	0.000	0.1251	−0.1000	0.000
7π/20	63.0				−0.1586			−0.1122	0.0757
9π/24	67.5	0.3827	−0.1381	−0.2705	−0.0828	0.1275	0.000	−0.0765	0.0902
2π/5	72.0				0.000			0.000	0.0515
5π/12	75.0	0.2588	−0.2440	−0.2215	0.0536	0.1725	−0.1121	0.0518	0.000
9π/20	81.0				0.1432			0.1012	−0.0802
11π/24	82.5	0.1305	−0.3107	−0.1261	0.1600	0.1188	−0.1092	0.0974	−0.0840
π/2	90.0	0.0000	−0.3333	0.000	0.2000	0.000	0.000	0.000	0.000
13π/24	97.5	−0.1305	−0.3107	0.1261	0.1600	−0.1188	0.1092	0.0974	0.0840
11π/20	99.0				0.1432			−0.1012	0.0802
7π/12	105.0	−0.2588	−0.2440	0.2215	0.0536	−0.1725	0.1121	−0.0518	0.000
3π/5	108.0				0.0000			0.000	−0.0515
15π/24	112.5	−0.3827	−0.1381	0.2705	−0.0828	−0.1275	0.000	0.0765	−0.0902
13π/20	117.0				−0.1586			0.1122	0.0757
2π/3	120.0	−0.5000	0.0000	0.2471	−0.2000	0.0000	−0.1251	0.1000	0.0000
7π/10	126.0				0.2472			0.0000	0.0980
17π/24	127.5	−0.6088	0.1608	0.1575	−0.2509	0.1485	−0.1365	−0.0326	0.1050
3π/4	135.0	−0.7071	0.3333	0.0000	−0.2000	0.2358	0.000	−0.1414	0.000
19π/24	142.5	−0.7934	0.5059	−0.2053	−0.429	0.1934	0.1778	−0.0425	−0.1369
4π/5	144.0				0.000			0.0000	−0.1348
5π/6	150.0	−0.8660	0.6667	−0.4280	0.2000	0.0000	0.2166	0.1732	0.0000
17π/20	153.0				0.3114			0.2202	0.1485
21π/24	157.5	−0.9239	0.8048	−0.6532	0.4829	−0.3078	0.0000	0.1847	0.1766
9π/10	162.0				0.6472			0.0000	0.1585
11π/12	165.0	−0.9659	0.9107	−0.8268	0.7465	−0.6441	−0.4184	−0.1932	0.000
19π/20	171.0				0.9040			−0.6393	−0.5067
23π/24	172.5	−0.9914	0.9778	−0.9578	0.9502	−0.9027	−0.8298	−0.7407	−0.6383
π	180.0	−1.000	1.000	−1.000	1.000	−1.000	−1.000	−1.000	−1.000

$$A_B(\theta,\Phi) = \frac{\sin\left(\frac{1}{2}N\beta b \sin\theta \cos\Phi\right)}{\sin\left(\frac{1}{2}\beta b \sin\theta\cos\Phi\right)} \tag{41.16}$$

A typical "horizontal" field pattern for $\theta = \pi/2$ and with $\beta b < 2\pi$ is shown in Fig. 41.9. (If βb approaches or exceeds 2π, some of the minor lobes become large, and the array ceases to be bidirectional.) The maximum value for $\theta = \pi/2$, $\Phi = \pm\pi/2$, is N. Since the input resistance of each antenna in the presence of the others when all are driven in phase with the same input current is somewhat less than the input resistance when isolated, the power required to

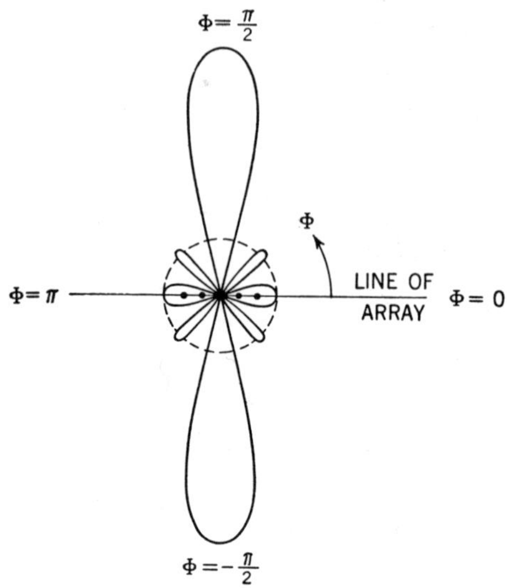

FIG. 41.9.—"Horizontal" field pattern of a broadside array of five units spaced one-half wavelength. The circle in broken line is the "horizontal" pattern of the central unit alone with the same current.

maintain N times the field of one antenna in the directions $\theta = \pi/2$, $\Phi = \pm\pi/2$, is not N times as great as that required for the single unit alone. The optimum spacing of units is not critical but lies between $\lambda/2$ and $3\lambda/4$.

The general characteristic of the end-fire array of N units driven so that all units have currents of the same amplitude but with phases such that each lags its neighbor on the left by δ and leads its neighbor on the right by δ, with $\delta = \beta b$, is

$$A_E(\theta,\Phi) = \frac{\sin\left[\frac{1}{2}N\beta b(1 - \sin\theta\cos\Phi)\right]}{\sin\left[\frac{1}{2}\beta b(1 - \sin\theta\cos\Phi)\right]} \tag{41.17}$$

A typical horizontal pattern for $\theta = \pi/2$ and with $\beta b < 2\pi$ is shown in Fig. 41.10. The vertical pattern in the direction $\Phi = 0$ is given in Fig. 41.11. The true end-fire characteristic with only a single direction in which the field attains a maximum comparable with N is achieved only with spacings that are in any case

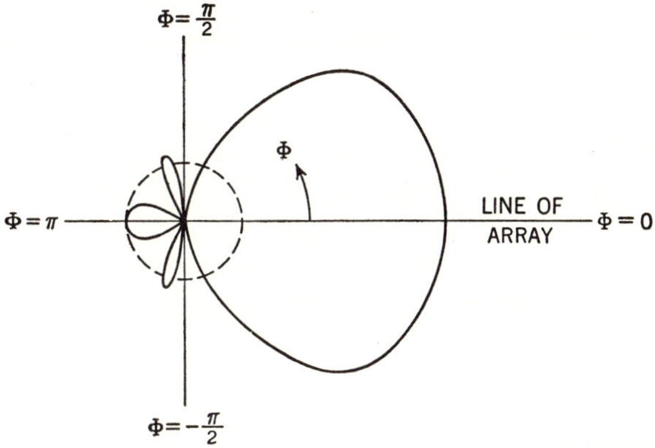

FIG. 41.10.—"Horizontal" field pattern of an end-fire array of five units spaced one-quarter wavelength with lagging currents on the right.

less than $\lambda/2$. The optimum spacing is between $\lambda/4$ and $3\lambda/8$. If b becomes as large as a half wavelength, the pattern is that of a double end fire. The relative directivity of an end-fire array of N units with optimum spacing of elements is somewhat less than N. The "vertical" field pattern of the end-fire array in the direction Φ_m, in which the horizontal pattern has its maximum, resembles the "horizontal" pattern if N is large. For small values of N, the "vertical" pattern is more sharply directive than the "horizontal" pattern, as can be seen for two units by comparing Figs. 41.4 and 41.5 or for five units by comparing Figs. 41.10 and 41.11.

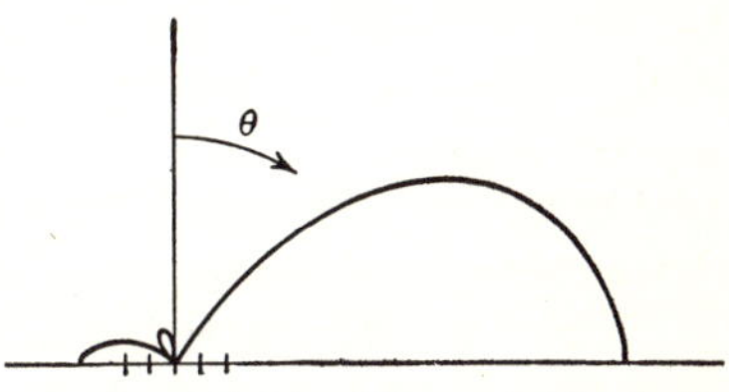

FIG. 41.11.—Upper half of the "vertical" field pattern of the array of Fig. 41.10 in the vertical plane containing the array.

The distant field of an antenna placed horizontally at a distance $b/2$ over a highly conducting half-space is closely approximated by the distant field of the same antenna with the entire conducting half-space replaced by an image antenna carrying a current opposite to that in the actual antenna. The image antenna is at a distance b below the actual antenna. The distant field of the antenna with its image is simply that of two parallel antennas driven 180° out of phase except that the antennas are horizontal instead of vertical. With this change in orientation, the field of a horizontal antenna over a conducting plane may be derived directly from that of two parallel antennas. With $b = \lambda/2$, *i.e.*, an an enna placed horizontally over the conducting earth at a height $\lambda/4$, the entire field pattern is roughly that of a sphere in contact with the earth directly below the center of the antenna. The shape of the pattern changes greatly as the antenna height is increased with several ears appearing at heights of a half wavelength or more. But, for all heights, the value of the field at the surface of the earth is vanishingly small.

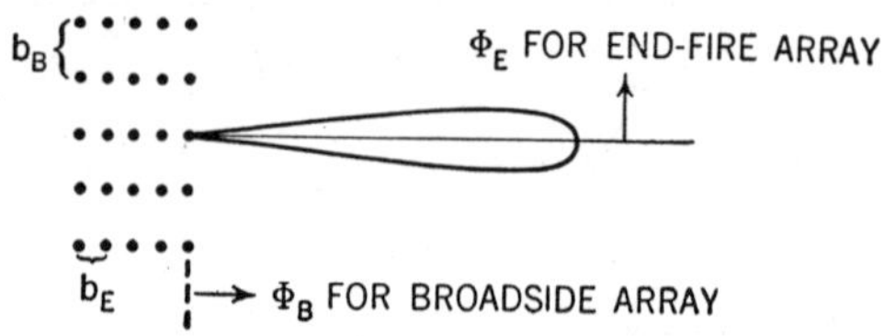

FIG. 41.12.—Array of arrays consisting of five broadside rows of collinear antennas in end fire. Horizontal plane and principal ear of horizontal field pattern are shown.

Highly directional arrays of arrays of antennas that have almost cigar-shaped, unidirectional field patterns in both horizontal and vertical planes may be constructed by combining the properties of broadside, end-fire, and collinear arrays, as shown in the horizontal plane in Fig. 41.12. Each dot is a collinear array of several $\lambda/2$ dipoles arranged with suitable phase-reversing sections. Each widely spaced row (spacing b_B) consists of antennas in broadside; each more closely spaced row (spacing b_E) at right angles to the broadside rows consists of antennas in end fire with lagging currents at the right. Appropriate spacings are $b_B = \lambda/2$, $b_E = \lambda/4$. The characteristic of the array of arrays is obtained by multiplying together the characteristics of the broadside, end-fire, and collinear arrays. Since the angle Φ in all parallel arrays has been measured from the line of antennas, the angle Φ_B for the broadside rows in Fig. 41.12 differs from Φ_E for the end-fire rows by $\pi/2$ or 90°. Accordingly, if only one angle, Φ_E, that for the end-fire array, is to be used, Φ in (41.16) for the broadside array must be replaced by $\Phi_E - \pi/2$. The characteristic of the array of arrays is

$$A_A(\theta,\Phi) = A_C(\theta)\, A_B\left(\theta, \Phi_E - \frac{\pi}{2}\right) A_E(\theta,\Phi_E) \qquad (41.18)$$

Since $\cos \Phi_B = \cos (\Phi_E - \pi/2) = \sin \Phi_E$, it is merely necessary to write sin Φ for cos Φ in (41.16) in order to use it as a factor in a general formula using the angle Φ_E.

For identical units of length $h = \lambda/4$, (41.18) becomes

$$A_A(\theta,\Phi) = \frac{\sin\left[N_c \frac{\pi}{2}\cos\theta\right]\sin\left(\frac{1}{2}N_B\beta b_B \sin\theta\sin\Phi_E\right)}{\sin\left(\frac{\pi}{2}\cos\theta\right)\sin\left(\frac{1}{2}\beta b_B\sin\theta\sin\Phi_E\right)} \frac{\sin\left[\frac{1}{2}N_E\beta b_E(1-\sin\theta\cos\Phi_E)\right]}{\sin\left[\frac{1}{2}\beta b_B(1-\sin\theta\cos\Phi_E)\right]} \tag{41.19}$$

The subscripts C, B, E refer to collinear, broadside, end-fire arrays; the subscript A stands for the complete array of arrays. The magnitude of the distant field of the array of arrays is obtained by multiplying the magnitude of (37.10) for one unit by $A_A(\theta,\Phi)$.

The problem of driving all the antennas in an array so that the relative amplitudes and phases of the currents are as desired is usually not a simple one if practical difficulties are considered. Theoretically, it is certainly possible to drive all units from the same generator if individual transmission lines and amplifiers are provided. If a matching section is arranged for each antenna so that all lines are nonresonant, the relative phases can be fixed in terms of the relative lengths of the feeders, and the amplitudes can be adjusted with the amplifiers. Since most arrays require the currents in the several units to have the same amplitude, the amplifiers are not necessary if the lines do not differ greatly in length. At very high frequencies, amplifiers are difficult to provide. For a broadside array, all nonresonant feeders must have the same length; for an end-fire array, they must differ in length by the same fraction of a wavelength as the fraction of a period that the currents differ in phase, the shortened lines going to the units with leading currents. If coaxial feeders are used, these are easily coiled if their correct length for phasing exceeds that required physically. Two- and four-wire open lines can be zigzagged. Practical difficulties are encountered because of the presence of insulators, which always reduce the actual length of a line below its electrical length by an amount that is difficult to compute accurately and may vary owing to moisture. In addition, there may be appreciable antenna currents on the outer surface of a coaxial line or superimposed on the line current on an open-wire line. Detuning sleeves or stubs are usually necessary, therefore, if the entire system of transmission lines is not to act as part of the antenna array.

Instead of providing individual nonresonant lines for the several antennas, it is possible to arrange the antennas as loads along a resonant line. This is most easily accomplished for a half-wave-

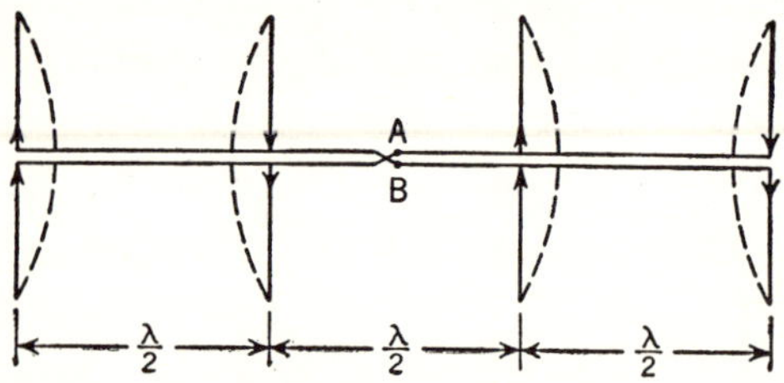

FIG. 41.13.—Center-driven resonant antennas in double-end-fire array. Approximate distribution of current is in broken lines. *AB* are high-impedance terminals.

length spacing of antennas with phase differences of a half period between units, Fig. 41.13, or (by transposing the line) with all units in phase, Fig. 41.14. By bending, coiling, or zigzagging the

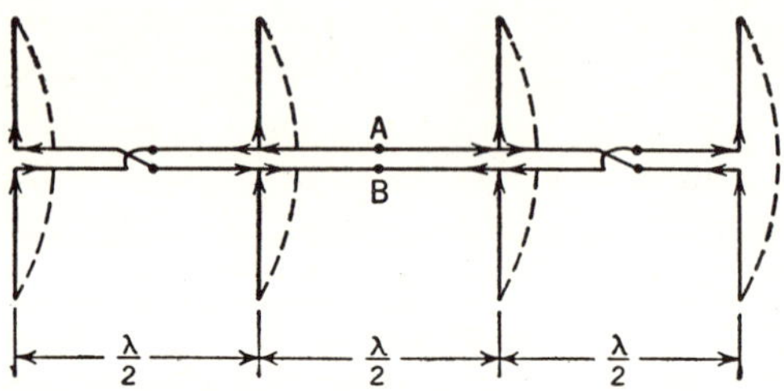

FIG. 41.14.—Center-driven resonant antennas in broadside array. Approximate distribution of current is in broken lines. *AB* are high-impedance terminals.

half-wavelength sections of line between antennas, the spacing may be reduced. The input terminals should be at the center and should be provided with a matching network for properly terminat-

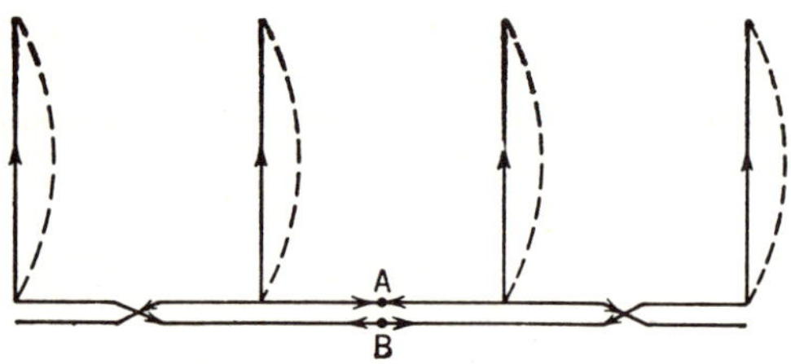

FIG. 41.15.—Broadside array of end-driven units.

ing the long line. Single end-driven units (like the two vertical units in Fig. 22.5*b*) may be used with open-wire and coaxial lines, Figs. 41.15 and 41.16. Note that a resonant line coupled to resonant antennas constitutes a complicated network of closely coupled circuits. The antennas are all coupled; they are low Q circuits because of the radiation load. The sections of line are high Q. Arrays may be constructed also by appropriately bending or zigzagging a long antenna; one possible broadside arrangement is shown in Fig. 41.17. This is known as a Bruce array. The distribution of current is not the same as along the same wire when

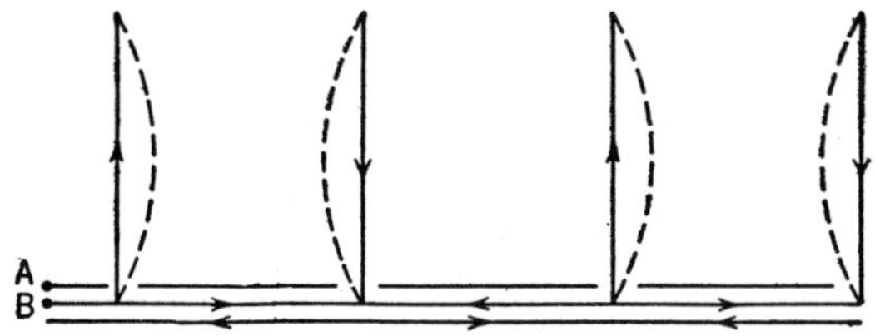

FIG. 41.16.—Double-end-fire array end-driven from a coaxial line. (A broadside array is obtained by omitting every second unit and folding the line so that the remaining units are spaced $\lambda/2$.)

straight, because the forces acting on the charges in the several parts of the conductor are changed. A very rough approximation for sufficiently thin wires is given by (8.3) with z and h measured *along* the conductor. If the array is long, a very large component of current corresponding to I_z'' in (9.1) exists even for the thinnest useful wire. Accordingly, the distant field as calculated under the assumption of a sinusoidal current is only a rough approximation. The principal contribution to the field is made by the currents

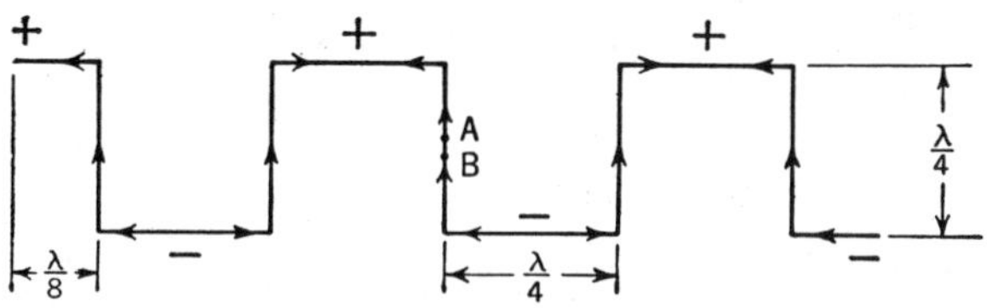

FIG. 41.17.—Broadside (Bruce) array of five vertical units each of length $\lambda/4$. Maximum currents are at the centers of the vertical elements. Location of maxima of charge, a quarter period after maximum upward current, is indicated with + and −. AB are low-impedance terminals.

in the vertical elements of Fig. 41.17. The currents in the horizontal elements are small and are in opposite directions in the halves of each element.

42. Distant Field of Parallel Arrays with Parasitic Elements; Reflectors.—The parallel arrays described in the preceding sections consist entirely of driven antennas in that each is connected to a generator either directly or by a transmission line. In such an array of driven elements, both the amplitude and the phase of the input current in each unit are individually adjustable to any desired values. It is by suitably fixing the relative amplitudes and phases of the units as well as their spacing that various directional field patterns are obtained.

In a parasitic array, only one antenna is driven directly by a generator, and all the power is supplied to it. All other antennas have no generator and have minimum load. They are placed sufficiently close to the driven unit and to each other for large forces to act to induce currents in them. The distant field is that due to all the currents in both driven and parasitic units. Since the spacing of the antennas in the array is neither so small that the condition for the near zone is satisfied nor so great that they are in the far zone with respect to one another, general electromagnetic coupling exists. Because the relative amplitudes and phases of the currrents in the parasitic elements are not individually variable by controllable driving voltages, as are the currents in the units of a driven array, the variety of available distant field patterns is much more restricted unless other means for adjustment are provided.

If all antennas in the array have the same length and if the parasitic elements consist merely of straight conductors, the only variable is the spacing of the elements. The amplitudes and phases of the currents are determined completely by the spacing and the relative location of the elements. If, on the other hand, the parasitic elements are adjustable in length or if they are provided with variable reactances at their centers, an additional parameter is available for controlling the relative phases of the currents. By adjusting the spacing and either the length or the tuning reactance (which at very high frequencies may consist of a short section of transmission line of variable length), a variety of directional field patterns can be obtained.

A single parasitic antenna placed parallel to a driven antenna, Fig. 42.1, may be analyzed in terms of (18.5) and (18.6) for two coupled antennas.

$$V_1 = I_{01}Z_{11} + I_{02}Z_{12} \tag{42.1}$$
$$0 = I_{01}Z_{21} + I_{02}Z_{22} \tag{42.2}$$

where $Z_{11} = Z_1 + Z_{s1}$; $Z_{22} = Z_2 + Z_{s2}$; Z_{s1} and Z_{s2} are the input self-impedances of the driven antenna and the parasitic antenna for the actual distributions of current; and Z_{12} and Z_{21} are the corresponding mutual impedances and are equal if the two antennas are immersed in a medium for which μ and ϵ are constant, as is usual. If it is assumed that both antennas are the same in length and radius,

$$Z_{s1} = Z_{s2} \tag{42.3}$$

and Z_{s1} and Z_{12} are given in Figs. 19.2 to 19.5; for indefinitely thin antennas $Z_{s1} = Z_0$, and Z_{12} is given by the curves of Fig. 19.1 for $h = \lambda/4$. On solving (42.2) for the current at the center of the

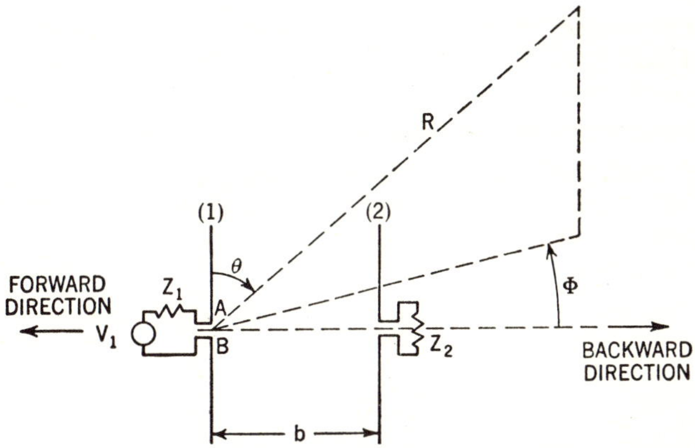

FIG. 42.1.—Center-driven antenna (1) with parallel parasitic antenna (2).

parasitic antenna,

$$\hat{I}_{02} = -\hat{I}_{01}\frac{Z_{21}}{Z_{22}} = -\hat{I}_{01}\frac{|Z_{21}|}{|Z_{22}|}\epsilon^{-j(\theta_{22}-\theta_{21})} \tag{42.4}$$

where $\theta_{22} = \tan^{-1} X_{22}/R_{22}$ and $\theta_{21} = \tan^{-1} X_{21}/R_{21}$. (42.5)
Using the notation of (41.2),

$$k = -\frac{|Z_{21}|}{|Z_{22}|}; \qquad \delta = \theta_{22} - \theta_{21} \tag{42.6}$$

The distant field of the two parallel antennas is given by (41.2), using (42.5) and with $F(\theta)$ given by (37.7). It is

$$\hat{\mathcal{E}}_\theta = j\frac{60\hat{I}_{01}}{R}F(\theta)\epsilon^{-j\beta R}\left[1 - \frac{|Z_{12}|}{|Z_{22}|}\epsilon^{-j(\theta_{22}-\theta_{21}-\beta b\sin\theta\cos\Phi)}\right] \tag{42.7}$$

This is an approximate general expression for the distant field of a center-driven antenna at a distance b from a parallel parasitic antenna. It is approximate because $F(\theta)$ applies rigorously only to the infinitely thin antenna. The "horizontal" field factor is defined by setting $\theta = \pi/2$. Three typical theoretically determined patterns for infinitely thin antennas with $h = \lambda/4$ are shown in

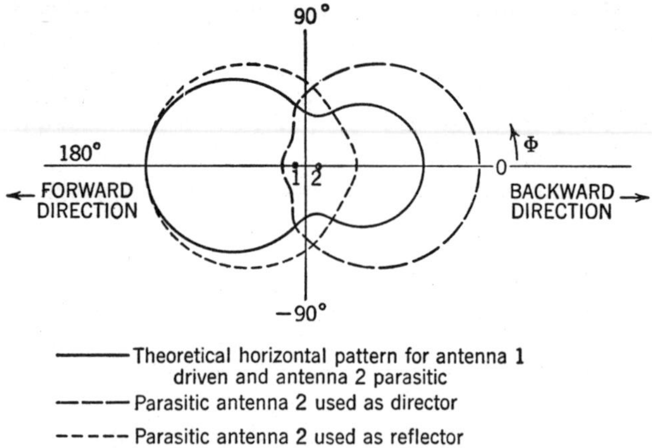

FIG. 42.2.—Possible "horizontal" field patterns of a driven antenna with single parasitic antenna. Both antennas are assumed to be infinitely thin.

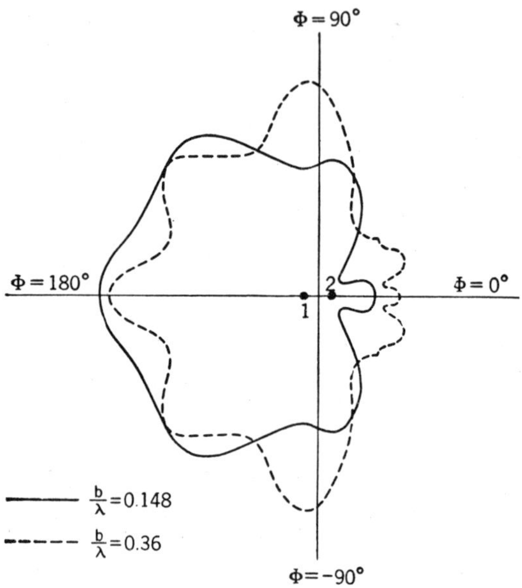

FIG. 42.3.—Experimental "horizontal" field patterns of a driven antenna (1) with a single parallel parasitic antenna (2). For both antennas $h = \lambda/4$.

Fig. 42.2. Many other types are possible. Two experimentally determined horizontal patterns for thick antennas with $h = \lambda/4$ are shown in Fig. 42.3. The "vertical" field pattern in the "forward" direction (*i.e.*, from the parasitic toward the driven antenna) is obtained by setting $\Phi = \pi$ or 180°. The "vertical" field pattern in the "backward" direction (*i.e.*, from the driven toward the parasitic antenna) is obtained with $\Phi = 0$. In practice, it is usually important to maximize the ratio of forward to backward field or the ratio of backward to forward field. If the ratio of forward to backward field is maximized, the parasite is called a reflector; if the ratio of backward to forward field is maximized, the parasite is called a director. The field in the forward direction ($\Phi = \pi$, $\theta = \pi/2$) is

$$\hat{\mathcal{E}}_\theta = j\,\frac{60\hat{I}_{01}}{R}\,\epsilon^{-j\beta R}\left[1 - \frac{|Z_{12}|}{|Z_{22}|}\,\epsilon^{-j(\theta_{22}-\theta_{21}+\beta b)}\right] \tag{42.8}$$

The field in the backward direction ($\Phi = 0$, $\theta = \pi/2$) is

$$\hat{\mathcal{E}}_\theta = j\,\frac{60\hat{I}_{01}}{R}\,\epsilon^{-j\beta R}\left[1 - \frac{|Z_{12}|}{|Z_{22}|}\,\epsilon^{-j(\theta_{22}-\theta_{21}-\beta b)}\right] \tag{42.9}$$

Since there are several parameters and variables involved, these expressions may be maximized in a number of different ways by holding one or more parameters constant. The most common variable is the spacing b. Since the input impedance of the driven antenna varies with the spacing, the input current $\hat{I}_{01}$ is itself a function of the spacing *for constant power P_1 supplied to the driven antenna*, and hence to the array. Thus,

$$|\hat{I}_{01}| = \sqrt{\frac{2P_1}{R_{AB}}} \tag{42.10}$$

where R_{AB} is the input resistance of the driven antenna in the presence of the parasite, as given by (18.12). If (42.10) is substituted in the magnitude of (42.7),

$$|\hat{\mathcal{E}}_\theta| = \frac{60F(\theta)}{R}\left\{\frac{2P_1}{R_{AB}}\left[1 + \left|\frac{Z_{12}}{Z_{22}}\right|^2 - 2\left|\frac{Z_{12}}{Z_{22}}\right|\cos\left(\theta_{22} - \theta_{21} - \beta b\sin\theta\cos\Phi\right)\right]\right\}^{\frac{1}{2}} \tag{42.11}$$

The corresponding expression for the magnitude of the field due to the driven unit when isolated and supplied the same power is

$$|\hat{\mathcal{E}}_\theta|_{1\,\text{unit}} = \frac{60F(\theta)}{R}\sqrt{\frac{2P_1}{R_0}} \tag{42.12}$$

where R_0 is the self-resistance of the isolated antenna. The ratio

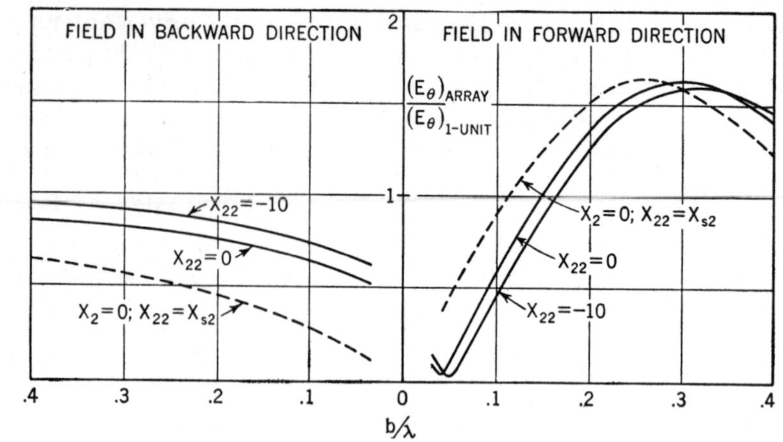

FIG. 42.4*a*.—Calculated ratio of the magnitude of the far-zone field of parasitic array of two antennas to the field of a single antenna with *constant current* in the driven unit. For each unit $h = \lambda/4$, $\Omega = 20$.

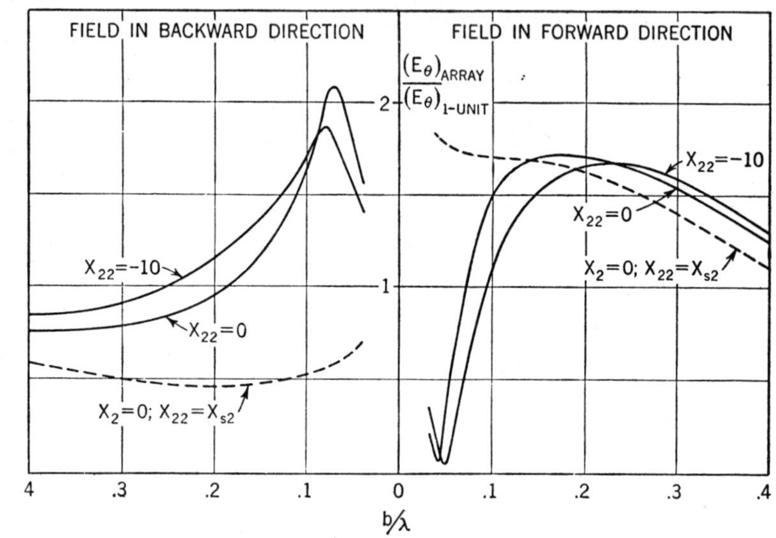

FIG. 42.4*b*.—Like Fig. 42.4*a* but with *constant power* to the driven unit.

of (42.11) to (42.12) is

$$\left|\frac{\mathcal{E}_\theta\ (\text{array})}{\mathcal{E}_\theta\ (1\ \text{unit})}\right| = \left\{\frac{R_0}{R_{AB}}\left[1 + \left|\frac{Z_{12}}{Z_{22}}\right|^2 - 2\left|\frac{Z_{12}}{Z_{22}}\right| \cos\,(\theta_{22} - \theta_{21} - \beta b \sin\theta \cos\Phi)\right]\right\}^{\frac{1}{2}} \tag{42.13}$$

In the mid-plane, $\theta = \pi/2$; in the forward direction, $\Phi = \pi$; in the backward direction, $\Phi = 0$. The front-to-back ratio is

$$\left|\frac{\mathcal{E}_\theta(\Phi = \pi)}{\mathcal{E}_\theta(\Phi = 0)}\right| = \left[\frac{1 + \left|\dfrac{Z_{12}}{Z_{22}}\right|^2 - 2\left|\dfrac{Z_{12}}{Z_{22}}\right| \cos\,(\theta_{22} - \theta_{21} + \beta b)}{1 + \left|\dfrac{Z_{12}}{Z_{22}}\right|^2 - 2\left|\dfrac{Z_{12}}{Z_{22}}\right| \cos\,(\theta_{22} - \theta_{21} - \beta b)}\right]^{\frac{1}{2}} \tag{42.14}$$

This ratio is independent of the power supplied to or the current in the driven element. For thin antennas with $h = \lambda/4$, maximum forward field with constant power to the driven antenna is obtained with b near 0.15λ and with X_{22} near +10 ohms. Maximum forward field with constant current in the driven antenna is obtained with b between 0.25λ and 0.3λ with $|X_{22}|$ small. Maximum backward field with constant power is obtained with b between 0.05λ and 0.1λ and with X_{22} between zero and 10 ohms negative. There is no maximum in the backward field with constant current in the driven antenna for small spacings. The forward and backward fields for constant current and for constant power in the driven antenna, as calculated for two antennas each of half length $h = \lambda/4$ with $\Omega = 20$, are shown in Fig. 42.4*a* and *b*. Curves for the front-to-back ratio of the field are shown in Fig. 42.5.

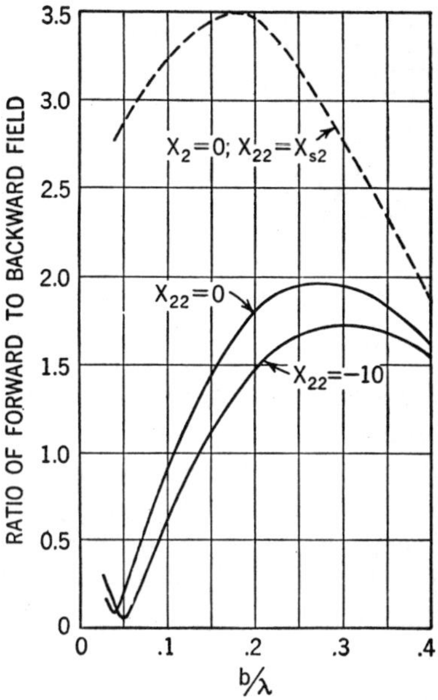

FIG. 42.5.—Ratio of forward to backward field for a driven antenna with one parasite. For each unit, $h = \lambda/4$, $\Omega = 20$.

Since the circuit reactance $X_{22} = X_{s2} + X_2$ of the parasitic antenna with its tuning reactance differs little from zero for either maximum ratio of forward to backward or backward to forward field, the required values actually are obtained easily without an external tuning reactance, *i.e.*, with $X_2 = 0$, by adjusting the length of the parasitic antenna. Although the mutual impedance is considerably changed from that for $h = \lambda/4$, even for small adjustments in length, so that Figs. 19.1 to 19.5 do not apply, the following conclusions are certainly qualitatively justified: A parasitic antenna that is to be used as a reflector to maximize the forward field by varying length and spacing of the antenna (employing no auxiliary tuning reactance) should have a half length h that is somewhat longer than the self-resonant half length obtained from Fig. 10.14 but is usually less than $\lambda/4$. A parasitic antenna that is to be used in the same manner as a director to maximize the backward radiation should have a half length h that is somewhat *shorter* than the self-resonant half length and hence considerably shorter than $\lambda/4$ in all practical cases involving antennas of reasonable thickness.

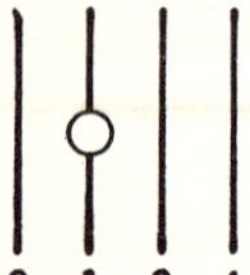

FIG. 42.6.—Side view of a driven antenna (1) with one reflector (2) and two directors (3, 4) for maximum field to the right. If no tuning reactances are available, (2) should be longer, (3, 4) shorter than (1).

A single row of parasitic reflectors is often used with a broadside array of driven antennas (a reflector behind each driven antenna) in order to make the array unidirectional.

The "horizontal" directivity of a driven antenna with a single parasitic director or reflector may be increased considerably by using both director and reflector, or several of each. They may be arranged in a single line as in Fig. 42.6; the reflectors may be arranged in the shape of a slightly flattened cylindrical parabola with the driven antenna near the focus, Fig. 42.7a; and parabolic metal sheet also may be used instead of the parallel parasitic antennas, Fig. 42.7b. Such an arrangement corresponds roughly to a unidirectional broadside array of length equal to the width of the opening of the parabola.

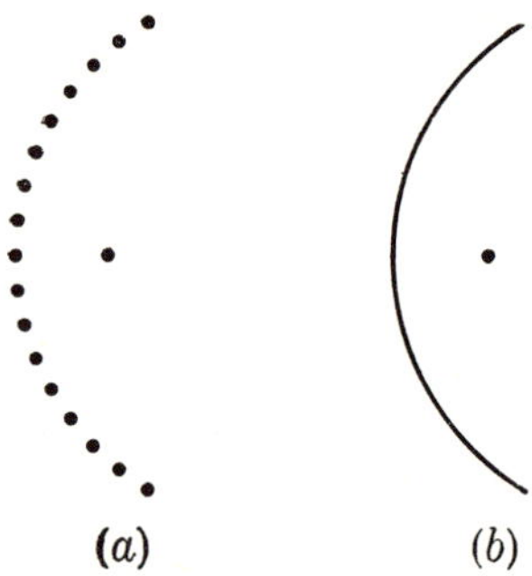

FIG. 42.7.—Section through an array consisting of a driven antenna (a) with parasitic antennas arranged in a slightly flattened parabola; (b) with a flattened parabolic cylinder of sheet metal.

The parabola must be slightly flattened in order to obtain a plane surface of constant phase for the electromagnetic field a short distance in front of the array, because the driven antenna is not sufficiently far from all parts of the conductors forming the parabola for the expanding ellipsoid of constant phase to have become sensibly spherical with a phase velocity equal to v_c. This is illustrated in Fig. 35.3. If the parabola is made so large that the distance from the focus to the nearest part of the reflecting array is so great that surfaces of constant phase expanding from the driven element are practically spherical, the reflecting parabola need not be flattened, and the driven unit should be placed at the focus. It is clear from Fig. 35.3 that an approximately spherical equiphase surface obtains even as short a distance as $\lambda/2$ from the center of a half-wave dipole, a distance not approaching the far zone. For wavelengths exceeding a few meters, such large cylindrical parabolas are inconvenient, and relatively smaller ones must be used. These may either be flattened, or, in the array of Fig. 42.7*a*, the phases of the currents in the outer units may be adjusted by changing their lengths or by using tuning reactances. For some purposes, much flatter or deeper arrangements, which do not produce plane surfaces of constant phase in front of the array, are advantageous.

The vertical directivity of any parasitic array of parallel elements can be increased by stacking several identical arrays to form a collinear array.

It is possible to obtain great directivity in both horizontal and vertical planes if a metal paraboloidal reflector (parabola of revolution) is used instead of the cylindrical parabola. For effective operation, the diameter of the paraboloid should approximate 10 wavelengths, and it is therefore not practically useful except at very high frequencies. (The cylindrical parabola needs to be only a half wavelength long.) The driven antenna is placed at the optimum point in the paraboloid; this differs only slightly from the optical focus if the focal distance of the paraboloid is considerably greater than a half wavelength. Experimentally determined field patterns are shown in Fig. 42.8. An additional parasitic reflector in front of the driven antenna, so spaced that it *minimizes* the forward direct field *due to the driven antenna,* increases the directivity still further. Such a reflector may consist of a straight conductor that is longer than the self-resonant length and is placed parallel to the driven antenna at a distance near 0.15λ, Fig. 42.9, or it may consist of a smaller paraboloid with opening toward the larger one.

The distant field of an arrangement like that in Fig. 42.9 is due almost entirely to the sheet of surface current maintained on the inside of the paraboloid. If the paraboloid extends out so far that an axis through the antenna intersects it in two points, the current at these points due to the field of the antenna is zero, because the magnetic field due to the antenna vanishes there. If it is assumed that these points are centers of maximum charge density, of opposite sign, the surface current may be visualized as in the direction from one of these points toward the other with maximum amplitude crossing a line equidistant from both. This sheet of current is primarily up and down (if the antenna is vertical as in Fig. 42.9*a*), and its effect at distant points resembles that of a parallel row of vertical antennas. If the reflector were perfectly parabolic and sufficiently large compared with the wavelength, the currents in the equivalent array of antennas would be all in phase, and the surfaces of constant phase of the electromagnetic field just outside the reflector would be plane. If, on the other hand, the reflector were flattened, the currents in the

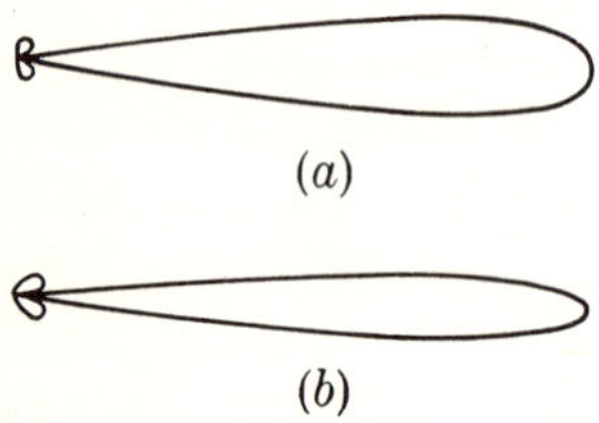

Fig. 42.8.—(*a*) "Vertical," (*b*) "horizontal" field pattern of a driven antenna with a paraboloidal reflector.

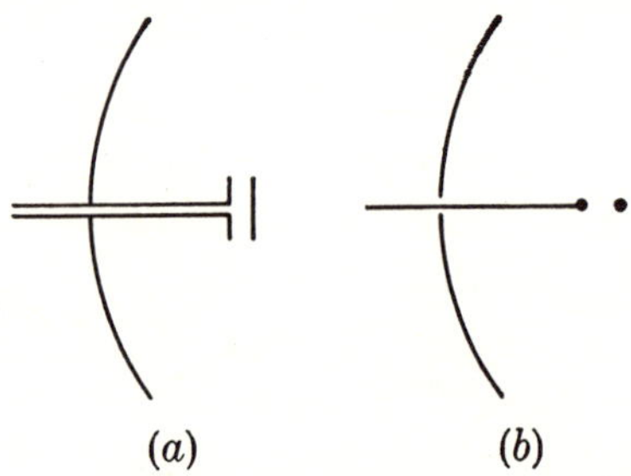

Fig. 42.9.—(*a*) Section, with antennas vertical, of paraboloidal reflector of sheet metal with the driven antenna near the focus and one parasitic antenna as reflector; (*b*) section with antennas horizontal.

middle units of an equivalent array would have to lead the current in the outer units in phase, because the surface of constant phase just outside the reflector would be convex outward. If the reflector were made more concave than a parabola, an equivalent row of antennas would have to have lagging currents in the central units, because a surface of constant phase must be concave outward. By adjusting the curvature of the reflector, the shape of the pattern and the relative magnitudes of minor lobes can be varied.

Metal reflectors may have shapes other than those of cylindrical parabolas and paraboloids, and sheets of current or layers of charge separated an appreciable fraction of a wavelength may be produced in other ways than by an antenna placed in front of the reflector. The antenna may be arranged, for example, inside an extension of a conical or horn-shaped reflector, a so-called "electromagnetic horn," Fig. 42.10*a*, provided that the diameter of this extension is not too small. (This is discussed in detail in Chap. III.) By suitably adjusting the position of the antenna and of the piston behind it, large resonant currents can be set up on the inside walls of the horn-shaped reflector and its extension. These currents have large components parallel to the antenna in the side walls of the horn and establish concentrations of alternately positive and

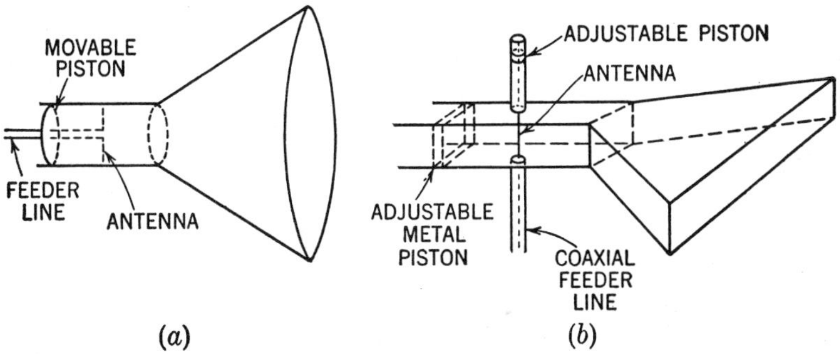

FIG. 42.10.—Electromagnetic horns.

negative charges along the upper and lower edges of the horn, Fig. 42.10*a*. If these are separated an appreciable fraction of a wavelength, a unidirectional distant field is set up resembling that of a row of antennas all driven in phase. An alternative method of driving the antenna is shown in Fig. 42.10*b* with reflector and horn of rectangular rather than circular cross section. Whatever the shape of the horn and its extension, the distant field always is due primarily to the distributions of current and charge on the horn. There may be some current on the outer surface behind the mouth of the horn. Unless a condition of resonance exists on the outside, owing to an unfortunate choice of dimensions, this current will not be large.

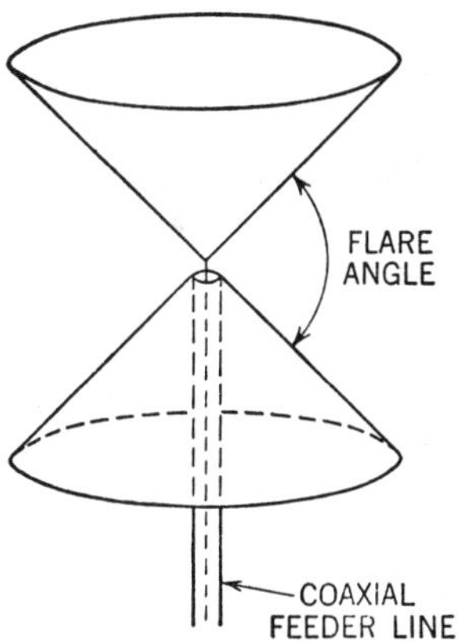

FIG. 42.11.—Biconical horn.

An antenna that is useful because of its broad-band charcteristics may be constructed by flaring and folding back the outer conductor of a coaxial line and enlarging the inner conductor to a cone that must be an appreciable fraction of a wavelength long, Fig. 42.11. This arrangement is called a biconical horn. If the flare angle is large, its characteristics somewhat resemble those of a thick center-driven antenna. For small flare angles, the field patterns are similar to those of a circular ring of antennas all driven in phase.

43. Rayleigh-Carson Reciprocal Theorem.—The distant field of an array consisting of a single driven antenna and any number of parallel parasitic antennas may be determined experimentally by moving a receiver in such a way that the receiving antenna (which is assumed to be center-loaded and symmetrical) is always tangent to a meridian on a great sphere around the transmitting array as center. The "vertical" axis of the array coincides with the radius $\theta = 0$. The current I_1'' through the load in the receiver is proportional to and hence a measure of the electric field due to the transmitter when driven by an emf V_2'' for any point θ,Φ on the sphere and may be used to determine the directional properties of the array.

Let the load in the receiver be replaced by a generator of emf V_2' with the same impedance as the load; also let the generator in the single driven antenna of the transmitting array be replaced by a load with the same impedance as the generator. The current in the load is I_2'. In this way, the fixed transmitting array has been transformed into an identical receiving array, whereas the movable receiver has become a similarly movable transmitter. The Rayleigh-Carson reciprocal theorem states that, if an emf V_1' in array 1 produces a current I_2' in array 2, and if an emf V_2'' in array 2 produces a current I_1'' in array 1, the following relation is true:

$$\frac{V_1'}{I_2'} = \frac{V_2''}{I_1''} \qquad \text{or} \qquad I_1''V_1' = I_2'V_2'' \tag{43.1}$$

In particular, if the same emf is applied to the transmitters in the two cases so that $V_1' = V_2''$, the currents in the loads are likewise the same, $I_1'' = I_2'$, provided that the movable antenna is in the same position relative to the fixed array. Thus the directional properties of the fixed array are the same when used as a receiver as when used as a transmitter. Accordingly, the directional properties, including the magnitude of the current in the load, of a receiving array may be obtained conveniently from the directional properties (measured or calculated) of the same array used as a transmitter with the load replaced by a generator having the same impedance as the load. It may be concluded that parallel directors and reflectors, cylindrical parabolic reflectors, paraboloids, and electromagnetic horns, Sec. 42, are equally useful and in the same way to obtain sharply directive receivers.

Used as a receiver, a unidirectional array responds strongly only to a signal from a given direction; used as a transmitter, the same array maintains a strong electric field only in that same direction.

As a specific example of the use of the reciprocal theorem, consider the directional properties as transmitter and as receiver of an array, designated by the number 2, consisting of a symmetrical center-driven or center-loaded antenna with one parallel parasitic antenna. Let antenna 1 be a single center-loaded or center-driven antenna. When used as a transmitter, the parasitic array 2 is driven by a generator of emf V_2''; the resulting current in the load Z_{L1} of antenna 1 used as receiver in the far zone is I_1''. When the generator and load are interchanged, an emf V_1' in the single antenna produces a current I_2' in the load Z_{L2} of antenna 2 used as receiver. Since each array is to have the same impedance in the circuit when transmitting or receiving, it is assumed that the impedance of the generator Z_G is the same as the impedance of the load Z_L, so that

$$Z_{G1} = Z_{G2} = Z_{L1} = Z_{L2} = Z_L \tag{43.2}$$

If $2h_{e1}$ is the effective length of antenna 1 when used as receiver parallel to the far-zone electric field $\hat{\mathcal{E}}_2''$ of array 2, the current in the load with V_2'' applied to array 2 is

$$\hat{I}_1'' = \frac{2h_{e1}\hat{\mathcal{E}}_2''}{Z_0 + Z_L} \tag{43.3}$$

where Z_0 is the self-impedance of the single isolated antenna. [Use (31.2) and (31.6) in (31.1) to obtain (43.3).] The far-zone electric field $\hat{\mathcal{E}}_2''$ due to array 2 is given by (42.7). The subscript θ is omitted.

$$\hat{\mathcal{E}}_2'' = \frac{j60\hat{V}_2''}{Z_{AB} + Z_L} F_2(\theta) \frac{\epsilon^{-j\beta R}}{R}\left[1 - \frac{|Z_{12}|}{|Z_{22}|}\epsilon^{-j(\theta_{22}-\theta_{21}-\beta b \sin\theta\cos\Phi)}\right] \tag{43.4}$$

Z_{AB} is the input impedance of the driven antenna in array 2 in the presence of the parasite as defined in Sec. 18. The substitution

$$\hat{I}_2'' = \frac{\hat{V}_2''}{Z_{AB} + Z_L} \tag{43.5}$$

is made in obtaining (43.4) from (42.7). With (43.4) in (43.3),

$$\hat{I}_1'' = \hat{V}_2'' \frac{j120h_{e1}F_2(\theta)}{(Z_0 + Z_L)(Z_{AB} + Z_L)} \frac{\epsilon^{-j\beta R}}{R} \left[1 - \frac{|Z_{12}|}{|Z_{22}|}\epsilon^{-j(\theta_{22}-\theta_{21}-\beta b\sin\theta\cos\Phi)}\right] \tag{43.6}$$

If the same voltage is applied to the two antennas when driven, so that

$$\hat{V}_1' = \hat{V}_2'' \tag{43.7}$$

the reciprocal theorem (43.1) permits writing

$$\hat{I}_2' = \hat{I}_1'' \tag{43.8}$$

where $\hat{I}_1''$ is given by (43.6).

If the same power P is supplied to the driven antenna in each case and for all orientations and spacings, the applied voltage is not the same, and it is therefore necessary to use the reciprocal theorem in the more general form (43.1). The applied voltages V_2'' and V_1' may be expressed in terms of the constant power and the impedance of the circuit. Since

$$P = \frac{1}{2}\left|\frac{\hat{V}_2''}{Z_{AB} + Z_L}\right|^2 R_{AB} = \frac{1}{2}\left|\frac{\hat{V}_1'}{Z_0 + Z_L}\right|^2 R_0 \tag{43.9}$$

$$|\hat{V}_2''| = |Z_{AB} + Z_L| \sqrt{\frac{2P}{R_{AB}}} \tag{43.10}$$

$$|\hat{V}_1'| = |Z_0 + Z_L| \sqrt{\frac{2P}{R_0}} \tag{43.11}$$

[If (43.10) is substituted in (43.4), this becomes (42.11).]

Upon substituting (43.10) and (43.11) in (43.1) written for magnitudes, *i.e.*, in

$$|\hat{I}_1''|\,|\hat{V}_1'| = |\hat{I}_2'|\,|\hat{V}_2''| \tag{43.12}$$

and solving for $|\hat{I}_2'|$,

$$|\hat{I}_2'| = |\hat{I}_1''| \left|\frac{Z_0 + Z_L}{Z_{AB} + Z_L}\right| \sqrt{\frac{R_{AB}}{R_0}} \tag{43.13}$$

If (43.6) and (43.10) are used in (43.13),

$$|\hat{I}_2'| = \frac{120 h_{e1} F_2(\theta)}{R|Z_{AB} + Z_L|} \left\{\frac{2P}{R_0}\left[1 + \left|\frac{Z_{12}}{Z_{22}}\right|^2 - 2\left|\frac{Z_{12}}{Z_{22}}\right| \cos\,(\theta_{22} - \theta_{12} - \beta b \sin\theta \cos\Phi)\right]\right\}^{\frac{1}{2}} \tag{43.14}$$

The power in the load Z_L of the parasitic array 2 used as receiver, if this is always kept conjugate matched so that

$$Z_L = Z_{AB}^* \qquad \text{or} \qquad R_L = R_{AB} \qquad \text{and} \qquad X_L = -X_{AB} \tag{43.15}$$

is given by

$$P_L = \tfrac{1}{2}|\hat{I}_2'|^2 R_L = \frac{h_{e1}^2}{R_0}\left(\frac{60F_2(\theta)}{R}\right)^2\left(\frac{P_0}{R_{AB}}\right)$$
$$\left[1 + \left|\frac{Z_{12}}{Z_{22}}\right|^2 - 2\left|\frac{Z_{12}}{Z_{22}}\right| \cos\left(\theta_{22} - \theta_{21} - \beta b \sin\theta \cos\Phi\right)\right] \qquad (43.16)$$

This differs only by the constant factor h_{e1}^2/R_0 from the expression $|\mathcal{E}_\theta|^2$ obtained from (42.11). This verifies, under the special conditions of (43.14), the general reciprocal relation that the directional pattern $|\mathcal{E}_\theta|^2$ for the driven array 2 for constant power is the same as the directional pattern of the same array used as receiver in terms of the power transferred to the load. In particular, the curves of Fig. 42.4*b* giving $\mathcal{E}_\theta$ for constant power give the correct variation for $\sqrt{P_L}$ if the same array is used for reception with the load conjugate matched for each spacing.

44. Poynting Vector and Effective Cross Section.—It is pointed out in general terms in Sec. 1 that, because of the complexity of the general law of force between moving charges in widely separated conductors, it is convenient to define an electromagnetic field as a mathematically useful intermediate step in calculating such forces. Specific use is made of this fact in determining the current (or more specifically the forces producing a deflection of an ammeter calibrated in terms of current) in a receiver in terms of the electric field parallel to the receiving antenna. Since the electromagnetic field is defined at all points in space directly in terms of distributions of moving charge in transmitting and receiving antennas, it is mathematically possible to express the transfer of power from a transmitting antenna to a receiving antenna, or to the distant universe as a whole, directly in terms of the electromagnetic fields due to the currents in the transmitting and receiving antennas. In fact, it is possible to define energy functions in terms of the electromagnetic vectors from which the entire transfer of power can be calculated. It can be shown that the real part of the integral over *any surface completely enclosing an antenna* of the component perpendicular to the surface of a complex vector point function $\mathcal{S}$ is a measure of the time-average power transferred from a driven antenna within the surface to antennas and to the universe outside, or from antennas and from the universe outside the surface to the antenna within the surface if this is a receiving antenna. The vector $\mathcal{S}$ is called the complex Poynting vector. It is always directed perpendicular to the plane containing the electric and magnetic vectors, and in the direction of advance of a right-hand screw when $\mathcal{E}$ is turned into $\mathcal{H}$ through the shortest arc. Its complex magnitude is

$$\mathcal{S} \text{ (watts/m}^2) = \tfrac{1}{2}\mathcal{E}\mathcal{H}^* \sin \phi \tag{44.1}$$

where $\mathcal{H}^*$ is the complex conjugate of $\mathcal{H}$, and ϕ is the angle in space between the vectors $\mathcal{E}$ and $\mathcal{H}$. If the closed surface is a great sphere in the far zone around a driven antenna where $\mathcal{E}$ and $\mathcal{H}$ have only the components $\mathcal{E}_\theta$ and $\mathcal{H}_\Phi$ that are mutually perpendicular and in phase, and that differ in magnitude only by the factor $\mathcal{R}_c$, $\mathcal{S}$ is real and directed radially outward. Its magnitude $\mathcal{S}_R$ is given by

$$\mathcal{S}_R \text{ (watts/m}^2) = \tfrac{1}{2}\hat{\mathcal{E}}_\theta\hat{\mathcal{H}}_\Phi = \frac{\hat{\mathcal{E}}_\theta^2}{2\mathcal{R}_c} = \frac{\mathcal{E}_\theta^2}{\mathcal{R}_c} \tag{44.2}$$

The average power radiated (transferred from the moving charges within the sphere to those outside) is

$$P \text{ (watts)} = \int_{\text{sphere}} \mathcal{S}_R \, d\sigma \tag{44.3}$$

where $d\sigma$ is an element of the spherical surface, and $\mathcal{S}_R$ is the value of the radial Poynting vector at that element. Note that, whereas the real part of the *integral* of the normal component of the complex Poynting vector $\mathcal{S}$ over the surface of a great sphere (or any other completely closed surface) correctly and uniquely gives the average power lost by the moving charges on one side and gained by the moving charges on the other side of the surface, the conclusion cannot be drawn that the real part of the complex Poynting vector at some point on the surface is itself a measure of the *fraction* of the total power that is transferred across a unit area at that point. There is nothing in the mathematical analysis to justify such a conclusion, and experimental verification is available only for the *total* power as measured by the complete integral. In fact, it is analytically possible with equal correctness to define a variety of energy-flow functions that are entirely different from the Poynting vector. Each of these functions gives a *different* value to the power transferred across a particular unit area, but they all give exactly the same total power when integrated over the same closed surface. It is the integral of the normal component of the Poynting vector over a closed surface that is important, not the Poynting vector itself. Thus, with the aid of this integral, it is possible to estimate the total power radiated from an intricate array of antennas if an approximate (sinusoidal) distribution of current is assumed along each conductor. By determining the far-zone electric field $\mathcal{E}_\theta$ for this distribution of current, the total power radiated is given by

$$P \text{ (watts)} = \frac{1}{\mathcal{R}_c} \int_{\substack{\text{spherical} \\ \text{surface}}} \mathcal{E}_\theta^2 \, d\sigma \tag{44.4}$$

By dividing this power by I_{max}^2, the radiation resistance R_m^e referred to maximum sinusoidal current is obtained (17.9). This calculation is usually simpler than the accurate analysis of the distribution of current and the determination from it of the input impedance. The result obtained is, however, only approximate, because an assumed and only roughly correct distribution of current is involved. If the accurate distribution of current is known, the current at the input terminals is also known and hence the input impedance, and there is no point in evaluating the integral of the normal component of the Poynting vector.

It is frequently assumed in practice (without analytical or experimental justification) that the Poynting vector is unique among all the possible power-flow functions that could be defined, in that it gives the actual direction and magnitude of the flow of energy through space, rather than being merely a mathematically convenient intermediate step in calculating the total power lost from one antenna or gained by another. Some consequences of such an assumption are considered in the following examples.

Consider first a transmitting antenna containing a generator at its center. If the Poynting vector is assumed to specify correctly the direction of flow of energy, the following conclusions are unavoidable. First, no power leaves the antenna. On the contrary, the small amount of power that is dissipated in heating the antenna enters it radially from space. Second, all the power that is radiated plus that which is dissipated in heat in the antenna is transferred directly from the generator to space. However, if the same potential difference were maintained across the same terminals of the generator without the antenna attached, no power would be transferred to space. It seems difficult to coordinate the conclusion that the antenna has no direct part in the radiation with the original postulate that it is the moving charges in the antenna that exert uncanceled forces and hence presumably do work on charges elsewhere in the universe.

As a second example, consider a receiving antenna parallel to the far-zone electric field of a transmitting antenna. The antenna is itself tangent to the surface of a great sphere on which the Poynting vector due to the field of the transmitting antenna is outwardly directed. If the Poynting vector were a measure of the total power transferred across unit area perpendicular to the vector, it would be reasonable to suppose that the normal component of the Poynting vector integrated over the area cut from the sphere by the receiving antenna, *viz.*, its length times its diameter, would be a measure of the maximum total power that could be transferred to the load of the receiving antenna. Actually, this area has nothing to do with this power. It follows from (37.3) that the distant field of a very thin center-driven antenna 1 of half length h near $\lambda/4$ has an amplitude in the mid-plane, $\theta = \pi/2$, of

$$\hat{\mathcal{E}}_{\theta 1}\left(\frac{\text{volts}}{\text{m}}\right) = \frac{60\hat{I}_{01}}{R} = \frac{2\hat{I}_{01}}{R}\frac{\mathcal{R}_c}{4\pi} \tag{44.5}$$

where R is the distance between transmitter and receiver, $\mathcal{R}_c = 376.7$ ohms, and $\hat{I}_0$ is the input current of the driven antenna. If a thin receiving antenna 2 of half length $h = \lambda/4$ and with a load R_L at its center is placed parallel to the far-zone electric field as given by (44.5), the power delivered to the load may be calculated by treating the receiving antenna as a driven antenna with a generator of emf

$$\hat{V}_2 = 2h_{e\theta}\hat{\mathcal{E}}_1 \tag{44.6}$$

in series with the load and the antenna. If the antenna is adjusted in length to be self-resonant as a center-driven antenna and the load R_L is made equal to the input resistance of the antenna, the total power transferred to the load is

$$P_L\ (\text{watts}) = \frac{(\frac{1}{2}\hat{V}_2)^2}{2R_L} = \frac{(h_e\hat{\mathcal{E}}_{\theta 1})^2}{2R_L} \tag{44.7}$$

The magnitude of the Poynting vector at the receiving antenna, as given by (44.2), is

$$\mathcal{S}_R\ (\text{watts/m}^2) = \frac{\hat{\mathcal{E}}^2_{\theta 1}}{2\mathcal{R}_c} \tag{44.8}$$

The ratio $P_L/\mathcal{S}_R$ in square meters is

$$\frac{P_L}{\mathcal{S}_R} = \left(\frac{\mathcal{R}_c}{R_L}\right)h_e^2 \tag{44.9}$$

For an antenna of half length h near $\lambda/4$,

$$h_e = \frac{\lambda}{2\pi} \tag{44.10}$$

Hence the area C_e in square meters defined by (44.8) is

$$C_e = \left(\frac{\mathcal{R}_c}{R_L}\right)\left(\frac{\lambda}{2\pi}\right)^2 \tag{44.11}$$

This is evidently not the area of cross section of the antenna, nor is it related to it. In fact, if the antenna is taken to be indefinitely thin so that its cross section *vanishes*,

$$R_0 = R_L = 73.13 \text{ ohms} \tag{44.12}$$

and

$$C_e = \left(\frac{376.7}{73.13}\right)\frac{\lambda^2}{4\pi^2} = 0.13\lambda^2 \tag{44.13}$$

This is given roughly by $\lambda^2/8$. For a practical antenna for which, for example, with $\Omega = 10$ and Figs. 31.2 and 10.4,

$$|h_e/\lambda| = 0.20, \qquad R_0 = R_L = 87 \text{ ohms} \tag{44.14}$$

$$C_e \doteq 0.17\lambda^2 \tag{44.15}$$

Although the power received by an antenna clearly has nothing to do with its actual cross section, it is still possible to insist on retaining the arbitrary assumption that the Poynting vector does measure the actual flow of power per unit area by introducing a fictitious "effective cross section" for the receiving antenna. Such an "effective cross section" for a matched load in the "equivalent" circuit for the receiver is defined, if it must be defined, by (44.9).

A physically crude "equivalent" for a center-driven antenna of half length $h = \lambda/4$ is often used. This is based on the incorrect assumption that such an antenna is equivalent to a physically impossible antenna of half length $h/2\pi$ with a *uniform* distribution of current along its entire length. The radiation resistance of such a fictitious antenna is 80 ohms; the corresponding area C_e is

$$C_e = \left(\frac{376.7}{80}\right)\left(\frac{\lambda}{2\pi}\right)^2 = \frac{3}{8\pi}\lambda^2 = 0.118\lambda^2 \tag{44.16}$$

While somewhat in error compared with (44.13) and (44.15), this value is frequently adequate for engineering purposes.

A different load with the same antenna requires the definition of a different "effective cross section" for the identical antenna.

The assumption that the Poynting vector itself specifies the direction and magnitude of the rate of flow of energy, which is then concluded to be spatially distributed, is an interesting illustration of the fact that, in physical theory, it is always possible to introduce arbitrarily defined parameters or variables together with theories to make them plausible, provided that equally arbitrary correction factors are also defined wherever necessary to obtain correct results.

VI. CLOSED CIRCUITS AS ANTENNAS

45. Frame or Loop Antenna for Transmission.—An antenna or array need not be constructed of straight and parallel conductors

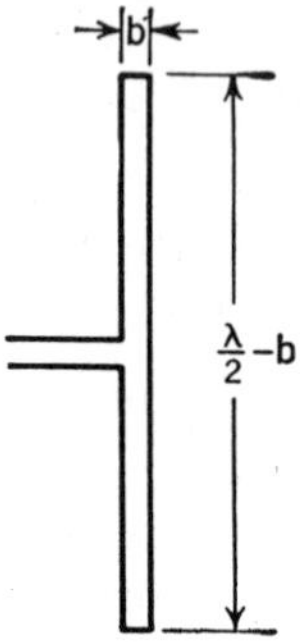

FIG. 45.1.—Folded dipole driven from a two-wire line.

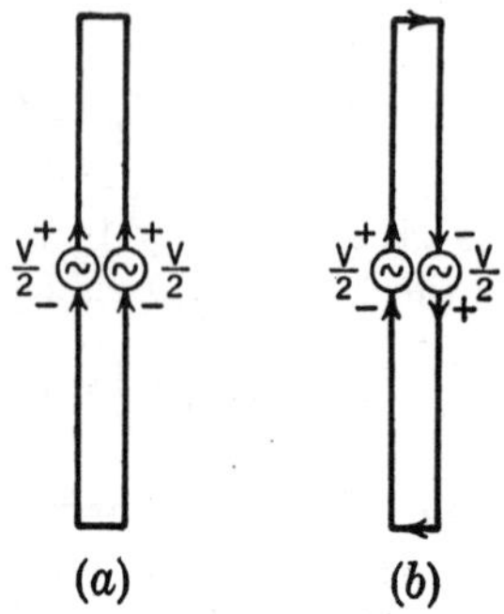

FIG. 45.2.—Circuits which, when combined, are equivalent to the folded dipole of Fig. 45.1.

each of which is an open circuit. Any closed loop of wire that is not confined to the near zone and that does not carry equal and opposite currents very close together radiates at least a fraction of the power supplied at its terminals. The distribution of current around the conductor and hence also the input impedance at the driving terminals vary greatly with the size of the loop and its shape and with the thickness of the conductor. A rigorous analysis for the distribution of current and the impedance of a loop unrestricted in size and shape is not available. Even an approximate analysis that assumes a sinusoidal distribution of current and employs the Poynting-vector method does not lead to a simple integrated formula for the radiation resistance referred to maximum current. In practice, two special forms of the loop are important, both of which can be analyzed approximately by relatively simple methods. The first form is a narrow rectangle called the folded dipole; the second is a loop of any shape that is small compared with the wavelength. They will be considered in turn.

The folded dipole is shown in Fig. 45.1. It consists of a resonant section of two-conductor transmission line that is driven at the center of one of the long sides instead of at one end. Its length is $\lambda/2 - b$, and the two conductors are separated a distance b between centers that is so small that it is possible to write, as for a transmission line,

$$b <<< \lambda \tag{45.1}$$

It may be analyzed in two steps, as follows: Consider first the circuit of Fig. 45.2*a*, in which identical *impedanceless* generators with emf $\frac{1}{2}V$, driven in phase, are connected at the centers of the long sides. Except near the ends, these sides are like parallel identical antennas driven in phase with equal antenna currents in the same direction.

With identical sides, $Z_{s1} = Z_{s2}$, $Z_{12} = Z_{21}$ and $I_{02} = I_{01}$. Hence, (18.2) reduces to

$$\tfrac{1}{2}V = I_{01}(Z_{s1} + Z_{12}) \tag{45.2}$$

$$I_{01} = \frac{V}{2(Z_{s1} + Z_{12})} \tag{45.3}$$

where Z_{s1} and Z_{12} are given in Fig. 19.2.

Consider next the circuit of Fig. 45.2*b*, which differs from Fig. 45.2*a* only in having one of the generators reversed so that the two sides have equal and opposite transmission-line currents. In this case, the circuit is equivalent to two generators in series with two extremely high impedances—several hundred thousand ohms. Accordingly, the transmission-line currents are small compared with the antenna currents in Fig. 45.2*a*. Let the two circuits, Fig. 45.2*a* and *b*, be combined so that on the left side there are two generators in series, in the same phase, and equivalent to a single generator of emf V, while on the right equal generators in series and in phase opposition are equivalent to a conductor. By the principle of superposition, each pair of generators continues to maintain the same current, so that the resultant current at the *center* of each side consists of the antenna current I_{01} in (45.3) and very small equal and opposite line currents. The latter are so small that they may be neglected, and the entire current at the center of each side and hence the input current on the left side is given therefore by (45.3). It is to be noted that the fact that the transmission-line currents at the centers of the long sides are extremely small does not mean that they are small in the short sides. Whereas the antenna currents have their maxima at the centers of the long sides and vanish at the centers of the short sides, the line currents are smallest at the centers of the long sides and maximum at the centers of the short sides. The radiation due to the line currents is small, as discussed in Sec. 48. The impedance seen by the single generator of emf V at the center of the left side is

$$Z_{\text{in}} = \frac{V}{I_{01}} = 2(Z_{s1} + Z_{12}) \tag{45.4}$$

Although Fig. 19.2 does not give values of Z_{1s} and Z_{12} for as small values of b/λ as are involved in the folded dipole except for infinitely thin conductors, it is clear nevertheless that, for thin conductors, Z_{1s} and Z_{12} do not differ greatly from each other or from Z_0, the value for an isolated center-driven dipole. As a rough estimate, therefore,

$$Z_{\text{in}} \doteq 4Z_0 \tag{45.5}$$

Thus the folded dipole has an input impedance approximately four times the simple dipole, so that a two-wire-line feeder requiring no matching section may be constructed. By connecting two or more rectangles in parallel, the input impedance can be multiplied by factors greater than 4. The distant field of the folded dipole is the same as that of the simple dipole, since the two parallel antenna currents are equivalent to a single current equal to their sum in determining the distant field.

The loop antenna of small size compared with the wavelength is of importance particularly at low frequencies. The distance around a loop of this type is usually only a very small fraction of a wavelength, so that near-zone circuit analysis and experimental techniques (which assume a current of *uniform* amplitude at all points around the loop) give good approximations for the input impedance, which is equivalent to an inductive reactance in series with a resistance. Formulas for the inductance of single-turn or several-turn loops of different shapes are given, for example, in *Bulletin* 74 of the Bureau of Standards. Although the input impedance computed from near-zone formulas is a good approximation for sufficiently small loops, the value so obtained does not and cannot take into account radiated power. When, as in the present case, the effect of radiation on the distribution of current is negligible, the small radiation resistance R^e may be computed independently by general electromagnetic methods and simply added to the ohmic resistance. Thus the contribution to the input resistance of a single-turn loop *which carries a current of sensibly uniform amplitude at all points* of the loop has the simple form

$$R^e = 20\beta^4 A^2 \tag{45.6}$$

where A is the area enclosed by the loop and $\beta = 2\pi/\lambda$.

Formula (45.6) is a good approximation for *a loop of any shape* (which encloses a singly connected area) that satisfies the inequality

$$\beta s \leqq 1 \tag{45.7}$$

If the loop is driven at one pair of terminals, s is the distance around the loop; if it is driven at several points around its periphery, s may be taken to be the longest distance measured along the wire between any pair of terminals. If the loop has N turns and one pair of terminals, s is roughly the total length of wire. In the specific case of the single-turn loop commonly used as aircraft beacons in the frequency range 200 to 400 kcps, s may be of the order of magnitude of 150 m, and βs is therefore approximately 0.6. The radiation resistance of such loops may be below 0.01 ohm, and very large currents must be used to radiate appreciable power. Usually more energy is dissipated in heat than is radiated, and it is obvious that loop antennas of this type are selected not because they are efficient radiators—which they are not—but because of other properties, in particular their directional characteristics and constructional features.

For use at ultra-high frequencies, transmitting loops with much larger radiation resistance may be constructed by means of suitably placed phase-reversing stubs, or by using the loop both as antenna and tank circuit of a triode oscillator that has several tubes appropriately spaced around its periphery (*e.g.*, at the corners of a square).

The distant field of a rectangular loop is computed readily by combining four terms of the form (37.10) to determine the characteristic of the array consisting of four straight antennas. If the length of each side of the rectangle is a small fraction of a wavelength, $F_0(\theta)$ for each has the form (39.5) multiplied by 2. The factor 2 appears because in the sides of the rectangle the current does not vanish at the ends [as assumed for (39.5)] but has approximately the same value as at the center. The average amplitude thus is doubled. If the systems of polar coordinates used for the several sides are all referred to a single system R, θ_F, Φ_F, Fig. 45.3, with origin at the center of the frame and vertical axis perpendicular to its plane, the far-zone electromagnetic field has the following complex form, also valid for loops of other shapes,

$$\hat{\mathcal{E}}_{\Phi_F} = -\mathcal{R}_c\hat{\mathcal{H}}_{\theta_F} = \frac{\mathcal{R}_c\hat{I}_F\beta^2 A}{4\pi R}\sin\theta_F\epsilon^{-j\beta R} \tag{45.8}$$

The subscript F will be used where convenient to distinguish quantities and coordinates for the frame or loop antenna.

Formula (45.8) may be compared with the corresponding expression for a short vertical antenna of half length $h << \lambda$, given by (37.10) using (39.5). An antenna of this type is also shown in Fig. 45.3*a*. Its distant field is

$$\hat{\mathcal{E}}_\theta = \mathcal{R}_c\hat{\mathcal{H}}_\Phi = j\frac{\mathcal{R}_c\hat{I}_0 H}{4\pi R}\sin\theta\,\epsilon^{-j\beta R} \tag{45.9}$$

Clearly, the electromagnetic fields in the far zone of a small horizontal loop and of a short vertical antenna differ essentially only in having electric and magnetic fields interchanged. Thus, in the loop, the radiation-zone electric field is directed along great circles in vertical planes. If Fig. 37.1 were to be changed to apply to a horizontal-frame antenna instead of to a vertical antenna along the z axis, it would be necessary merely to write $\mathcal{E}_F$ for $\mathcal{H}$ and $-\mathcal{H}_F$ for $\mathcal{E}$. On the other hand, since electric and magnetic fields in the radiation zone differ in magnitude only by a constant factor, the shapes of the vertical and horizontal field patterns for a small horizontal loop and for a short vertical antenna are identical, Fig. 45.3*c*. Both types of antennas have sharp nulls along their axes.

Several loop antennas may be combined into directional arrays in the same way as straight antennas are combined if the axes perpendicular to the planes of the loops are oriented just as are the axes of the straight antennas in the arrays already described, and phase and amplitude relations in the several loops are adjusted appropriately. In each case, corresponding field patterns result but with electric and magnetic fields interchanged. The field of a loop antenna oriented in any way over a perfectly conducting plane may be obtained by constructing an exact geometrical image of the loop below the plane, reversing the currents in all conductors of the image, and computing the field due to the loop and the image.

Loop antennas may be combined also into arrays with straight antennas. An interesting array often used in direction finders is that of a loop antenna with its axis horizontal and a straight antenna with its axis vertical, Fig. 45.4. The electric fields due to both antennas are perpendicular to the xy or $y_F z_F$ plane. In this plane,

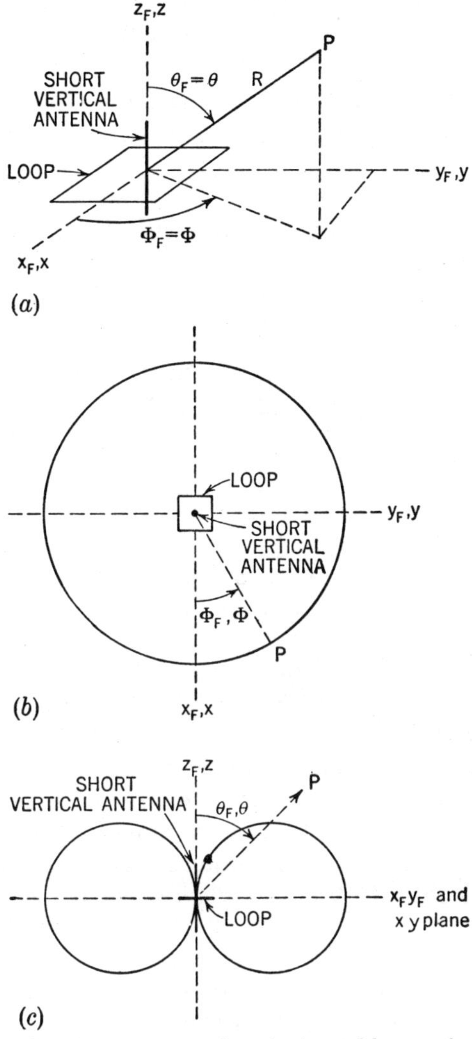

FIG. 45.3.—(a) Coordinate systems for a horizontal loop or frame antenna (subscript F) and for a short vertical antenna. (b) "Horizontal" field pattern for a horizontal loop antenna. (c) "Vertical" field pattern for a horizontal loop antenna.

$\theta = \pi/2$, and $\Phi = \pi \pm \theta_F$. If a unidirectional horizontal pattern is desired, it is necessary merely to require that

$$I_F = j\hat{I}_0 \qquad H = \beta^2 A \tag{45.10}$$

By combining (45.8) with (45.9) subject to (45.10), the resultant electric field in the xy plane is

$$\hat{\mathcal{E}}_z = j\,\frac{\mathcal{R}_c \hat{I}_0 H}{4\pi R}\,\epsilon^{-j\beta R}(1 + \sin\Phi) \tag{45.11}$$

The electric vectors due to the loop and the straight antenna are shown in Fig. 45.4. The horizontal characteristic of the array is

$$A(\Phi) = 1 + \sin\Phi \tag{45.12}$$

This gives a cardioid like Fig. 41.4 but with the zero line of Φ shifted backward 90°.

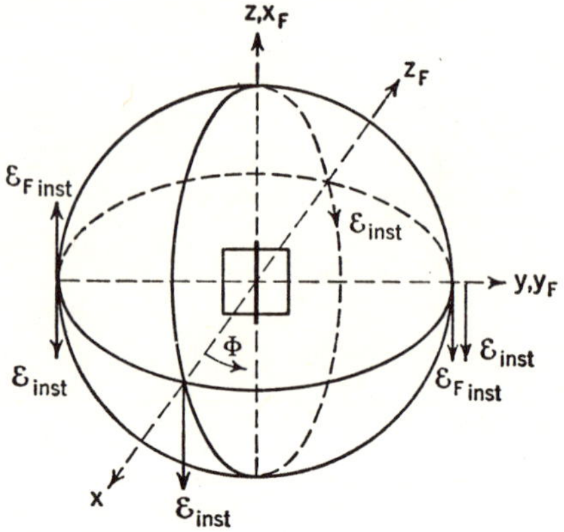

FIG. 45.4.—Coordinates and directions of electric field for a loop or frame antenna (subscript F) and a straight antenna.

46. Loop or Frame Antenna for Reception.—The loop antenna may be used for receiving as well as for transmitting. The fundamental equation for the loaded loop antenna immersed in an electromagnetic field is the same in form as for the straight receiving antenna, (31.1) with (31.2),

$$I_0(Z_0 + Z_L) = V_e \tag{46.1}$$

where I_0 is the current entering and leaving the load Z_L; Z_0 is the self-impedance of the loop; V_e is the open-circuit voltage at the load terminals of the antenna. For a loop unrestricted in size, Z_0 and V_e have been determined analytically for only the single-turn circular loop, in a linearly polarized electric field, but the results have not been translated into a practically useful form. For the "small loop," which satisfies the condition (45.7), V_e is readily evaluated by direct analysis or by application of the reciprocal theorem. Referring to Fig. 46.1, V_e may be written

$$V_e = h_e \mathcal{H} \mathcal{R}_c \cos \psi = h_e \mathcal{E} \cos \left(\psi + \frac{\pi}{2} \right) \tag{46.2}$$

with

$$h_e \equiv \beta \; A \cos \bar{\theta}_1 \tag{46.3}$$

where $\mathcal{H} \cos \psi$ is the component of the magnetic field in the plane containing the axis of the loop and a line drawn from the center of the loop to the transmitter (or a line perpendicular to the planes

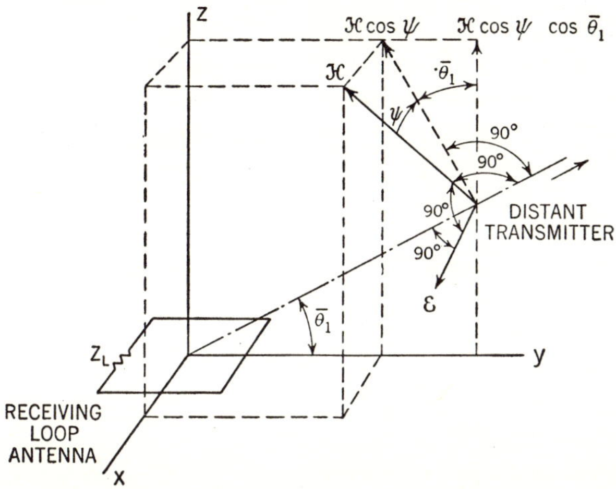

FIG. 46.1.—Horizontal loop antenna in an arbitrarily oriented, linearly polarized magnetic field $\mathcal{H}$.

containing the $\mathcal{E}$ and $\mathcal{H}$ vectors), and h_e is the effective length of the loop and has exactly the same form as the effective length of a short straight antenna *insofar as the trigonometric factor is concerned.* If the axis of the loop is in the plane of constant phase for both $\mathcal{E}$ and $\mathcal{H}$, $\theta_1 = 0$, and

$$h_e = \beta \; A \tag{46.4}$$

If the loop consists of N turns instead of one turn, Z_0 is changed. In addition, V_e and hence h_e are multiplied by N.

The loop antenna has the same directional characteristics for receiving as for transmitting. They bear the same space relations to the *magnetic* component as those of a short straight antenna do to the *electric* component of a plane-polarized electromagnetic field.

47. Unbalanced Loop Antenna—Shielding.—In describing the properties of the loop or frame antenna for both transmission and reception, all consideration of the construction and the location of the generator or load as well as of the transmission line that may lead to one or the other was omitted in order to simplify the discussion. Accordingly, the transmitting and receiving properties of the frame antenna as described are correct only if the presence of the generator or load or of the connecting transmission line does not produce a condition of unbalance in the loop. This presupposes that the loop itself as well as an attached transmission line and the load or generator *are symmetrical* with respect to a plane perpendicular to the plane of the loop and passing midway between its symmetrically located terminals. Furthermore, if the transmitting or receiving loop is near a conducting surface such as the earth, this surface must be perpendicular to the plane of symmetry across the loop. When all these requirements are fulfilled, the forces acting on the charges in the loop are distributed symmetrically with respect to the input or load terminals, and the ideal conditions and characteristics described in the preceding sections obtain. Symmetrical systems of this type are not always easy to design, because all unbalanced elements such as coaxial lines, and loads or generators not symmetrical with respect to a center tap (which may be grounded) must not be used. If a system is unsymmetrical, the same difficulties arise that are encountered in an unbalanced transmission line. For example, in a driven loop with horizontal axis, supported at some distance above the earth and using a connecting coaxial transmission line, the entire outer surface of the system may be excited as a single vertical antenna with the loop serving as a top load. The resulting field pattern is a superposition of that of a loop and that of a vertical antenna. Obviously the sharp nulls in the distant field of the loop will not be observed.

If a receiving loop is placed well above the surface of the earth (with its axis horizontal), and a two-wire line runs vertically down from it to a receiver, Fig. 47.1, an electromagnetic field with horizontal magnetic and vertical electric component induces currents in the loop and in the line that can be resolved into two parts. There is a circulating current around the loop as a consequence of the small *difference*, $\mathcal{E}_{inst}$ (right) $-$ $\mathcal{E}_{inst}$ (left), in the values of the electric field along the two vertical sides of the loop due to the phase difference resulting from their separation in space. Half of this difference in field is considered conveniently to be acting downward on the left and half upward on the right. The remaining components of the electric field on the two sides are both equal to

$\frac{1}{2}[\mathcal{E}_{inst}$ (right) + $\mathcal{E}_{inst}$ (left)] and are directed upward. They induce equal and codirectional up-and-down currents in *both* sides of the loop. If the spacing of the transmission line is assumed to be practically zero, the same field $\frac{1}{2}[\mathcal{E}_{inst}$ (right) + $\mathcal{E}_{inst}$ (left)] acts on both conductors simultaneously and induces equal and codirectional currents in the two wires, *i.e.*, the line acts as a single conductor with the loop as a top load. If the spacing of the wires is

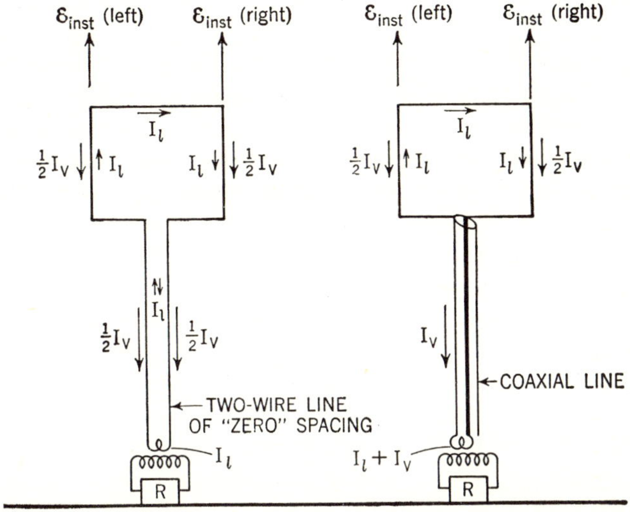

(*a*) (*b*)

FIG. 47.1.—(*a*) Symmetrical loop receiver. The vertical-antenna current I_v maintains no potential difference across the receiver R. (*b*) Unsymmetrical loop receiver. The vertical-antenna currents I_v maintain a potential difference across the receiver R because a part of I_v continues through the coupling coil. (The dimensions of the loop are assumed to be small compared with the wavelength.)

not so small compared with the wavelength that it may be taken to be zero, the line acts as an extension of the loop.

The circulating current around the loop is the desired one. It produces a true transmission-line current in the connecting line, and this maintains the required potential difference across the load. The codirectional currents, on the other hand, contribute nothing useful but cannot be avoided entirely. If, however, as considered here, they are *equal* currents in the *same direction* in the two sides of a symmetrical loop and line, they can alternately charge and discharge the ends of the line, but they cannot maintain a potential difference across the load, *provided that this is symmetrical* with respect to its terminals. If the load or the line or the loop is unsymmetrical, or if the line is attached to the side of the loop instead of at the bottom, currents of this second type in general can maintain a very significant potential difference across the load. If this is true, the sharp nulls of current in the loop when turned parallel to the magnetic field are obscured by a uniform signal that does not depend upon the purely circulating currents in the loop. Consequently, unless a loop or frame antenna with all its accessory circuits is symmetrical with respect to a plane perpendicular to the loop passing midway between its terminals, the characteristics obtained are a combination of the characteristics of a loop and a vertical antenna.

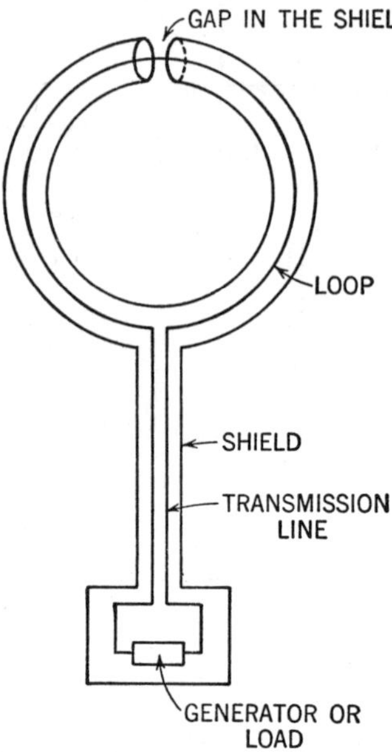

Fig. 47.2.—Shielded-loop antenna.

It is possible to achieve complete structural symmetry for a transmitting or receiving system using a loop by enclosing it entirely in a metal shield that has a gap at the top, Fig. 47.2, (or in some instances at the bottom). In the transmitting loop, essentially only currents that are on the outside of the metal shield are effective in setting up a significant distant field. But this outside surface is excited uniquely by the field maintained in the immediate vicinity of the gap by a potential difference established across it by a large current on the inner surface of the shield. Consequently, complete driving symmetry obtains, and only a circulating current around the outer surface of the circular shield can be excited; there is no codirectional up-and-down current, and a true loop field is set up. Beginning with the generator end, the transmission system consists of a length of shielded pair which connects to two coaxial lines. The load, consisting of the radiating loop, is connected in series with the two coaxial lines. It is interesting to note that the current on the shield is in effect a current on the outside surface of a coaxial line, a current that is most undesirable in coaxial feeders for linear radiators because it makes the line a part of the radiating system, but is precisely the desired current in the shielded loop. In the one case, the coaxial line must not radiate; in the present case, it must be the entire radiator.

In a receiving antenna, a load is connected to the feeder line, Fig. 47.2. A potential difference is maintained across the gap, almost wholly as a result of currents around the outside of the shield. In general, codirectional up-and-down currents also are excited, but they contribute nothing to the potential difference across the gap because they charge the edges equally with charges of the *same sign*. Under these conditions, currents on the inside of the coaxial line, excited by the electric field across the gap, are due only to true circulating currents on the outside of the shield. Therefore, the voltage maintained across the load is due only to these currents, and the ideal loop characteristics obtain.[1]

48. Radiation from Transmission Lines.—In discussing electric currents and the electromagnetic field in Sec. 1 transmission lines were used to illustrate how equal and opposite currents that are parallel and very close together (in comparison with the wavelength) set up electromagnetic fields in the far zone that practically cancel. Actually, complete cancellation is achieved for lines with equal and opposite currents only when the conductors are virtually coincident, or for a coaxial line when both ends are completely closed by metal disks.[2] When this is not the case, the field at distant points does not vanish. In the open-wire line, the field is only partly canceled because the equal and opposite currents in the wires are never infinitely close together, and because there are no equal and opposite currents to cancel the fields due to the often widely separated currents in the load and generator.

The small nonvanishing field when the ends of a coaxial line, are not closed by metal disks is due primarily to the absence of the radial currents in these disks. With the disks removed and the line left open or connected to a short straight wire, the currents along the interior of the line readjust themselves in magnitude and phase so that the field due to them again vanishes in the metal of the conducting walls. This distribution involves a slight unbalance which is compensated by a small but significant current on the *outside surface* of the outer conductor. The line currents inside but near the open or bridged end together with the currents on the

[1] The operation of the shielded loop is explained popularly by first stating that the desired loop current is due to the magnetic field, the undesired up-and-down current to the electric field, and then maintaining that the metal shield cannot be penetrated by the electric field but can be penetrated by the magnetic field. All these statements are incorrect in the light of fundamental electromagnetic principles.

[2] Strictly, all conductors would have to be perfect.

outside surface and in the terminating wire (if there is one) maintain a small uncanceled field at distant points. The contribution to the distant field due to the currents in a line driving an antenna may be obtained quite accurately if the line is assumed to be terminated in a straight wire with a concentrated load equal to the impedance of the antenna at the center of the wire. Since the amplitude and phase relations of the currents on the line depend on this impedance, the contributions to the distant field are not the same with different loads. If a coaxial line is used to center-drive a symmetrical antenna, the actual load includes not only the antenna proper but also the outside surface of the coaxial line, as discussed in Sec. 27.

The power radiated from a closed circuit depends on the distribution of current and charge around the *entire* circuit configuration. It is usually not possible to allot portions of the radiated power to specific parts of a circuit, because of the intricate intercancellation of the distant fields due to currents in different parts of a circuit. For example, the radiation resistance referred to maximum current for a resonant section of two-wire line of length $\lambda/2$ with both ends open is $30\beta^2b^2$ ohms assuming a sinusoidal distribution of current. The radiation resistance of a resonant two-wire line with open ends and of length two, three, or four half wavelengths is not two, three, or four times $30\beta^2b^2$. It is precisely $30\beta^2b^2$ so long as a sinusoidal distribution of current is a reasonable approximation. Clearly, it is not possible to state what fraction of $30\beta^2b^2I^2_{\max}$ is radiated from each half wavelength of the line. If the same two-wire line is bridged at each end by straight wires and is adjusted in length to resonance (which occurs approximately at lengths that are an integral multiple of $\lambda/2$ minus the spacing b), the radiation resistance of line and terminating bridges referred to maximum current is also $30\beta^2b^2$ if the line is adjusted in length to be resonant near an *even* integral multiple of $\lambda/2$. On the other hand, if it is resonant near a length that is an odd multiple of $\lambda/2$, more complete cancellation of the distant field due to currents in line and terminating bridges occurs, so that the radiation resistance is proportional to the very much smaller quantity β^4b^4, instead of to β^2b^2. The radiation resistance of a coaxial line with two open ends and a sinusoidally distributed current is proportional to $\beta^4(b^2 - a^2)^2$ with b the inner radius of the outer conductor and a the radius of the inner conductor. If the coaxial line is bridged at each end by straight pieces of wire of length $b - a$, the rotational symmetry of the system is destroyed at the ends, and the radiation due to the entire circuit becomes comparable with that of a two-wire line with spacing equal to $b - a$.

The radiation resistance of a four-wire line with open ends or with *symmetrically* bridged ends, Fig. 48.1, is proportional to $\beta^4 b^4$.

An exact analysis has not been carried out for a completely symmetrical circuit consisting of a transmission line of small but nonvanishing spacing, driven at one end by a generator and terminated at the other end by a center-driven antenna of arbitrary length and radius. However, it seems reasonable to conclude that the difference between the power radiated from such a complete system and that radiated from the antenna with generator connected to it directly without transmission line does not exceed in order of

FIG. 48.1.—Resonant four-wire line with symmetrically bridged ends.

magnitude $KI^2_{max}\beta^2 b^2$ for the two-wire line and $KI^2_{max}\beta^4 b^4$ for the four-wire line with symmetrically connected generator and load. (The K's are numerical factors that may be assumed less than 50.) *These differences are in almost all cases so small compared with the total power radiated from the antenna as to be negligible.* Possible exceptions are loops and short straight antennas of such small dimensions compared with the wavelength that the radiation resistance is so minute as to be comparable with that of a two-wire line.

The radiation from a coaxial line used to drive an antenna is not estimated easily, because the outside surface of the line almost inevitably forms part of the antenna system, unless the line is actually buried so that the surface of the earth takes its place as part of the antenna. The contribution to the resultant electric field and to the total radiated power by currents on the outer surface varies with the magnitude of the current, and this in turn depends on many factors, including especially the symmetry of structure of the system and the dimensions of the outer surface and of all conductors attached to it. If the current on the outer surface can be kept sufficiently small, its contribution to the far-zone field and to the radiated power can be made negligible, but not otherwise.

If transmission lines are used to connect a receiving antenna to its load, somewhat different considerations are involved. Although it follows directly from the reciprocal theorem that the receiving qualities of a line are like its transmitting qualities, so that a line that sets up a small field when driven will pick up or receive only a small signal when connected to a receiver, it may not be concluded therefrom that a negligible contribution to radiated power in the driven case automatically means a negligible pick-up in reception. This would be true if only the electric field due to a single distant transmitter were involved. But this is seldom if ever the case. There are always electric fields maintained at a receiving system due not only to the distant transmitter the signal from which is to be received, but to countless other moving charges in other antennas, in the atmosphere, or elsewhere both near and far. The receiving qualities of the line for such interfering fields may be sufficient to maintain across the load voltage differences that are undesired. This may be true even though all conditions of dissymmetry (as discussed in conjunction with the shielded loop antenna) are avoided. It follows that a line that has as small a radiation resistance (hence effective length) as possible and that lends itself to complete symmetry of installation is a prime requisite to good reception. The four-wire line fills these specifications better than either the two-wire or the coaxial line except in those cases involving base-loaded antennas erected on a conducting earth in which an entire coaxial line may be buried.

49. Rhombic Antenna.—The input impedance of a section of transmission line terminated in its characteristic impedance is equal to this characteristic impedance, and the amplitude of the current decreases exponentially. Since the terminating impedance is for practical purposes a pure resistance that can be constructed to be approximately independent of frequency over a fairly wide range, the input impedance of the line is correspondingly independent of frequency. If a section of line about 12 wavelengths long is pulled apart at the center to form a rhombus, Fig. 49.1, with the shorter diagonal approximately equal to the leg length (six wavelengths in this case) the fundamental property of the transmission line, *viz.*, equal and opposite currents close together, ceases to obtain, and the widely spaced currents in the rhombus exert large forces on charges in distant conductors, *i.e.*, by changing the parallel wires with a uniform and small spacing into a rhombus with a spacing at the center of the order of wavelengths, the two-wire *transmission line* is transformed into a rhombic *antenna*. The distribution of current and the input impedance of the rhombic antenna are *not* the same as those of the two-wire line, because the configuration of conductors is different and the load due to the coupled distant universe has been added. However, although no longer independent of frequency, the input impedance varies only relatively little with frequency. Indeed, by readjusting the terminating resistance to an optimum value, it is possible to obtain an input impedance that fluctuates as the frequency is varied but that in a typical case has resistive and reactive components neither of which deviates by more than 20 per cent from a mean value of $800-j2{,}650$ ohms for a frequency range from 4 to 20 mcps. It is clear, therefore, that, for a broad-band operation in this frequency range, the rhombic antenna has extremely good characteristics when compared

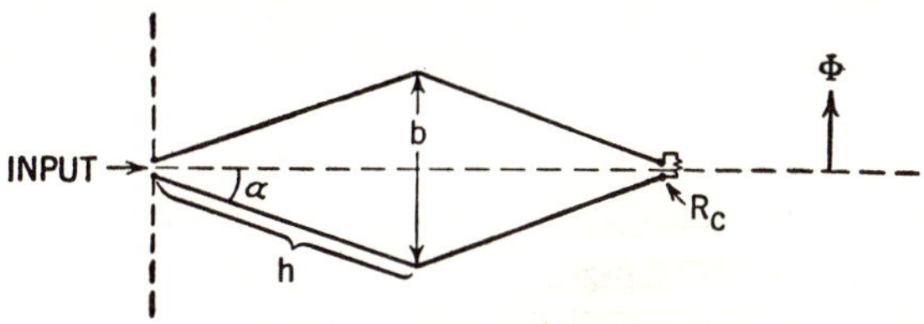

FIG. 49.1.—Rhombic antenna.

with the open-circuit type of antenna. On the other hand, in addition to a relatively small amount of power dissipated in heating the wire, 50 per cent of the power supplied to the rhombic antenna is dissipated in the terminating resistance. (It will be recalled that, for cylindrical antennas made of copper, a *negligible fraction* of power is dissipated in heat.) To offset this disadvantage, the directivity of the rhombic antenna is high. With optimum design and a leg length of 6λ, a gain of about 16 db referred to a $\lambda/2$ dipole is possible. Furthermore, although the direction of maximum field changes and the directivity decreases as the frequency is lowered, the variations are not rapid. As a result, while optimum performance is achieved at one frequency only, reasonably good performance is possible with no change in the array over a wide range of frequencies. Typical computed field characteristics of a rhombic antenna placed horizontally over a perfectly conducting plane are shown in Fig. 49.2. A large number of minor lobes or ears occur in the horizontal pattern.

Rhombic antennas are particularly useful in continuous long-distance point-to-point transmission because they permit the use of any one of a range of frequencies as required by ionospheric conditions at different times of day. Furthermore, variation in the vertical angle with frequency, Fig. 49.2, is such that it agrees very well with the mean vertical angle of arrival or departure of long transmission paths such as are encountered, for example, in transatlantic transmission.

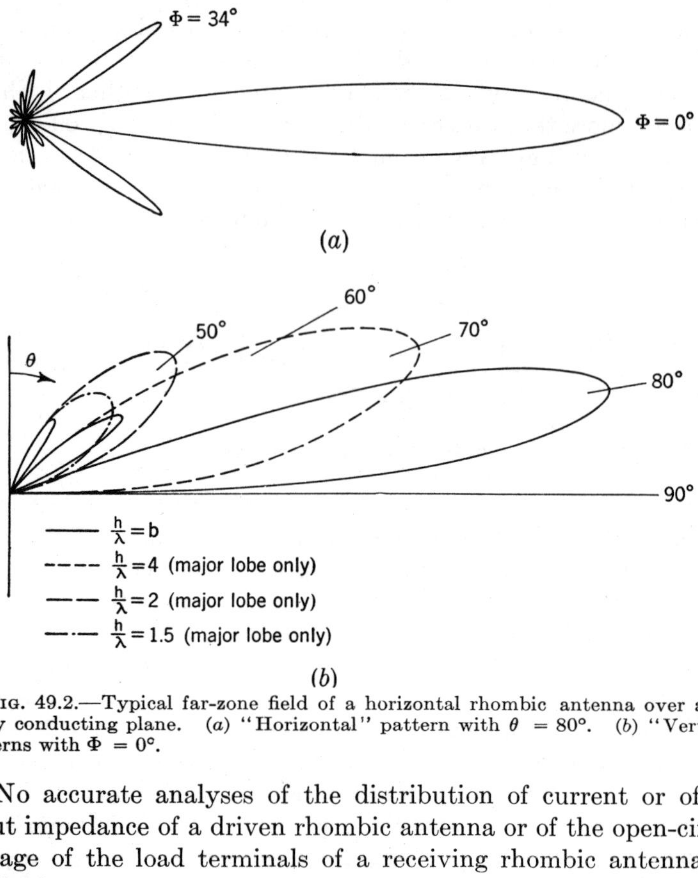

FIG. 49.2.—Typical far-zone field of a horizontal rhombic antenna over a perfectly conducting plane. (*a*) "Horizontal" pattern with $\theta = 80°$. (*b*) "Vertical" patterns with $\Phi = 0°$.

No accurate analyses of the distribution of current or of the input impedance of a driven rhombic antenna or of the open-circuit voltage of the load terminals of a receiving rhombic antenna are available. In effect, the rhombic antenna is a special case of an unrestricted loop. Computations of the distant electromagnetic field that have been made are based on the assumption that the current is the same around the rhombus as along a nonresonant line. It is to be expected that this distribution is about as good an approximation for the rhombic antenna as the sinusoidal distribution is for the center-driven cylindrical antenna, *i.e.*, it may be assumed to be adequate for computing distant fields but not satisfactory in determining input impedance. An expression for the radiation resistance computed using a distribution of current like that in a dissipationless nonresonant line is

$$R_e = 240[\ln (\beta b \sin \alpha) + 0.577] \text{ ohms} \tag{49.1}$$

where b is the short diagonal, Fig. 49.1.

Supplementary Reading

R. M. FOSTER: Directive Diagrams of Antenna Arrays, *Bell System Tech. J.*, Vol. 5, pp. 292–307, April, 1926.

G. C. SOUTHWORTH: Factors Affecting the Gain of Directive Antennas, *Proc. I.R.E.*, Vol. 18, p. 1502, September, 1930.

E. J. STERBA: Directional Transmitting Systems, *Proc. I.R.E.*, Vol. 19, p. 1184, July, 1931.

R. S. GLASGOW: "Principles of Radio Engineering," pp. 413–463, McGraw-Hill Book Company, Inc., 1936.

H. E. HOLLMAN: "Physik u. Technik der ultrakurzen Wellen," Vol. 2, pp. 58–87, Verlag Julius Springer, 1936.

G. H. BROWN: Directional Antennas, *Proc. I.R.E.*, Vol. 25, p. 78, January, 1937.

F. E. TERMAN: "Radio Engineering," 2d ed., Chap. XV, McGraw-Hill Book Company, Inc., 1937.

F. VILBIG: "Lehrbuch der Hochfrequenztechnik," Chap. V, Akademische Verlagsgesellschaft m.b.H., 1937.

E. HALLÉN: Theoretical Investigations into the Transmitting and Receiving Qualities of Antennas, *Nova Acta*, Series IV, Vol. 11, No. 4, Uppsala, 1938.

G. GRAMMER and B. GOODMAN: "The ARRL Antenna Book," American Radio Relay League, Inc., West Hartford, Conn., 1939.

H. BRÜCKMANN: "Antennen," S. Hirzel, 1939.

J. A. STRATTON: "Electromagnetic Theory," Chap. VIII, McGraw-Hill Book Company, Inc., 1941.

A. E. HARPER: "Rhombic Antenna Design," D. Van Nostrand Company, Inc., 1941.

S. A. SCHELKUNOFF: Theory of Antennas of Arbitrary Size and Shape, *Proc. I.R.E.*, Vol. 29, p. 493, September, 1941.

R. KING and F. G. BLAKE: The Self-impedance of a Symmetrical Antenna, *Proc. I.R.E.*, Vol. 30, pp. 335–349, July, 1942.

R. KING and C. W. HARRISON, JR.: Distribution of Current in Symmetrical Antenna, *Proc. I.R.E.*, Vol. 31, pp. 548–566, October, 1943.

R. KING: Coupled Antennas and Transmission Lines, *Proc. I.R.E.*, Vol. 31, pp. 626–640, November, 1943.

C. W. HARRISON, JR. and R. KING: Radiation Field of Symmetrical Center-driven Antenna of Finite Cross-section, *Proc. I.R.E.*, Vol. 31, pp. 693–698, December, 1943.

S. A. SCHELKUNOFF: "Electromagnetic Waves," D. Van Nostrand Company, Inc., 1943.

S. A. SCHELKUNOFF: Mathematical Theory of Linear Arrays, *Bell System Tech. J.*, Vol. 22, p. 80, 1943.

F. E. TERMAN: "Radio Engineers' Handbook," pp. 770–871, McGraw-Hill Book Company, Inc., 1943.

S. A. SCHELKUNOFF: Antenna Theory and Experiment, *J. Appl. Phys.*, Vol. 15, pp. 54–60, January, 1944.

M. C. GRAY: A Modification of Hallén's Solution of the Antenna Problem, *J. Appl. Phys.*, Vol. 15, pp. 61–65, January, 1944.

R. KING and C. W. HARRISON, JR.: The Receiving Antenna, *Proc. I.R.E.*, Vol. 32, pp. 18–35, January, 1944.

C. W. HARRISON, JR. and R. KING: Receiving Antenna in a Plane-polarized Field of Arbitrary Orientation, *Proc. I.R.E.*, Vol. 32, pp. 35–49, January, 1944.

R. KING and C. W. HARRISON, JR.: Impedance of Long, Short, and Capacitively Loaded Antennas with a Critical Discussion of the Antenna Problem, *J. Appl. Phys.*, 15, pp. 170–185, February, 1944.

R. KING and C. W. HARRISON JR.: Mutual- and Self-impedance of Coupled Antennas, *J. Appl. Phys.*, Vol. 15, pp. 481–495, June, 1944.

J. AHARONI: "Antennae," Clarendon Press, Oxford, 1946.

S. SILVER: "Microwave Antenna Theory and Design." McGraw-Hill Book Company, Inc., 1949.

E. C. JORDAN: "Electromagnetic Waves and Radiating Systems," Prentice-Hall, Inc., 1950.

J. D. KRAUS: "Antennas," McGraw-Hill Book Company, Inc., 1950.

S. A. SCHELKUNOFF: "Advanced Antenna Theory," John Wiley and Sons, Inc., 1952.

S. A. SCHELKUNOFF and H. T. FRIIS: "Antenna Theory and Practice," John Wiley and Sons, Inc., 1952.

H. ZUHRT: "Elektromagnetische Strahlungsfelder," Springer-Verlag, 1953.

R. W. P. KING: "Transmission-line Theory," McGraw-Hill Book Company, Inc., 1955; Dover Publications reprint, 1964.

R. W. P. KING: "Theory of Linear Antennas," Harvard University Press, 1956. (This book contains a complete analytical and experimental treatment of the material in this chapter together with many tables and charts. It also contains a very extensive bibliography.)

R. KING: Linear Arrays: Currents, Impedances, and Fields, I, *Trans. I.R.E.*, Vol. AP-7, pp. S440-S457, December, 1959.

A. Z. FRADIN: "Microwave Antennas," Pergamon Press, 1961.

R. W. P. KING: "Fundamental Electromagnetic Theory," Dover Publications, Inc., 1963.

R. W. P. KING and S. S. SANDLER: Theory of Broadside Arrays; Theory of End-fire Arrays, *Trans. I.R.E.*, Vol. AP-12, pp. 269–280, May, 1964.

CHAPTER III

ULTRA-HIGH-FREQUENCY CIRCUITS

The general laws governing electric circuits of dimensions comparable with or greater than a wavelength were formulated qualitatively in Chap. II and applied to determine the properties of a variety of circuits called antennas. This chapter deals with no new subject. It is, rather, a sequel to the study of antennas and is concerned with the application of the *same general principles* to circuits of a different type.

I. CLASSIFICATION OF CIRCUITS

1. Circuits Not Confined to Near Zone.—The designation ultra-high-frequency circuits strictly means circuits to be used within a limited range of the frequency spectrum. However, the fundamental characteristics of circuits used in the ultra-high-frequency band are associated with the fact that *their dimensions are of the same order of magnitude as the wavelength.* The term ultra-high-frequency circuits will be used in this broader sense to include all circuits not confined to the near zone, in at least one dimension, regardless of the frequency involved. Such circuits may be open, quasi-closed, or closed, as described in Chap. II, Sec. 6.

2. Radiating and Nonradiating Circuits.—Ultra-high-frequency circuits may be divided into two groups: those which function as antennas owing to their radiating property and those which do not radiate or radiate to a negligible degree. Circuits of the first type are studied in Chap. II; those of the second type are the subject of this chapter. Border-line cases, such as open-wire lines, open-end coaxial lines, and small loop antennas were discussed in Chap. II from the point of view of radiation. It will be recalled that the coupling of a circuit to the distant universe, and hence its radiation, can be kept small either by arranging equal and opposite currents close together, as in open-wire lines, or by enclosing the entire circuit in a metal shield, as discussed in Chap. II, Sec. 1. In the latter case, currents distribute themselves automatically in such a way that the electromagnetic field reduces essentially to zero in the interior of the metal and everywhere in space outside the closed shield. Nonradiating or only slightly radiating circuits of the closely spaced or metallically enclosed transmission-line types are important at the highest frequencies for the transmission of power, as circuit elements in networks, and as complete tank circuits in oscillators.

3. Nonresonant and Resonant Circuits.—Circuits that are not confined to the near zone in one or more dimensions are extremely complicated except in the special cases of nonresonant and resonant circuits, which can be described approximately in terms of relatively simple distributions of current and charge. They are of great practical importance.

If the circuit is constructed so that electric charge oscillates along the conductors in such a way that maximum concentrations of charge do not occur at fixed locations as measured parallel to the axis of the conductor but travel continuously in one direction *along the conductor*, the circuit is nonresonant. The configurations of charge move in an axial direction and constitute the axial current; and, to a close approximation, maximum values of charge and of axial current occur *simultaneously* at the *same* cross section. *Traveling waves* are said to obtain.

If electric charge oscillates back and forth along the conductors forming a circuit that is adjusted so that maximum concentrations of charge of the largest possible amplitude occur periodically at definite and *fixed locations*, the circuit is resonant. In this case, the maxima of current density occur at instants *differing in time by a quarter period* and at locations *differing in position by the equivalent of a quarter wavelength* from those of the maximum density of charge. *Standing waves* are said to obtain.

If a circuit is neither resonant with a fixed standing-wave pattern of maximum possible amplitude nor nonresonant with traveling waves, it is strictly *not* resonant. However, all circuits that are not *nonresonant* sometimes are said loosely to be resonant. The following discussion is limited to nonresonant and resonant circuits in the strict sense.

II. TRANSMISSION CIRCUITS

4. Properties of Transmission Circuits.—Circuits that are designed primarily for transmitting power from a generator to a load may be called *transmission circuits* if, as in the present treatment, attention is focused primarily on electric charge and current while the electromagnetic field is regarded as a convenient intermediate step in a mathematical development. The circuits may be called *wave guides* if the emphasis is on the field. They include conventional open-wire and coaxial lines with cross-sectional dimensions that satisfy the condition for the near zone; they also include more general types with cross sections not so restricted. Important properties of transmission circuits (wave guides) of all types include the following:

1. Low power loss (*a*) due to heating of the conductors and dielectrics of the circuit itself, and (*b*) due to radiation.
2. Sufficient power capacity. This implies sufficient spacing of conductors and adequate dielectric strength of insulating media to prevent spark discharges; it presupposes conductors with enough surface to carry large high-frequency currents.
3. An adequate frequency range and a useful frequency response.
4. Physical availability.
5. Special features such as rotational symmetry for swivel joints, invariance in polarization, and availability of flexible construction.

5. Qualitative Survey of the Analytical Problem of Transmission Circuits.—The fundamental problem of transmission is to provide a circuit that is designed so that if power is supplied to it from a generator at one end, a power as nearly equal as possible is transferred to a load at the other end. This is always a three-dimensional problem. It involves one variable in the direction of transmission, called the axial coordinate, which in the simplest case of a straight circuit may be taken as the z axis of a coordinate system; it involves also two transverse variables that are chosen according to the shape of the cross section of the circuit.

The problem to be solved is to represent the propagation in the axial direction of configurations of charge density when each configuration is characteristic of a particular cross section. In general, this is done in terms of the surface density of charge η in coulombs per square meter and the volume density of current i in amperes per square meter.[1] At sufficiently high frequencies and in good conductors, practically all the current is in a very thin skin along the surface of the conductors so that it is possible to make use of a quasi-surface current l in amperes per meter. This is the current below a unit width of surface at right angles to the direction of flow, as illustrated in Fig. 5.1 for a section of any conducting surface that is both large and thick compared with the skin depth δ defined

[1] Note that the volume density of current i is the charge times its mean nonrandom velocity *per unit volume.*

by

$$\delta \text{ (meters)} = \sqrt{\frac{2}{\omega\mu_r\mu_0\sigma}} \tag{5.1}$$

where μ_r is the relative permeability, σ the conductivity of the metal in mhos/meter, and $\mu_0 = 4\pi \cdot 10^{-7}$ henrys/m.

If the surface current is directed only parallel to the axis of transmission, it is possible to define a *total current* in amperes $I_z = \int l_z \, ds$, and a *total charge* per unit length of each conductor in coulombs per meter $q = \int \eta \, ds$, where ds is an element of length around the periphery of each cross section drawn perpendicular to

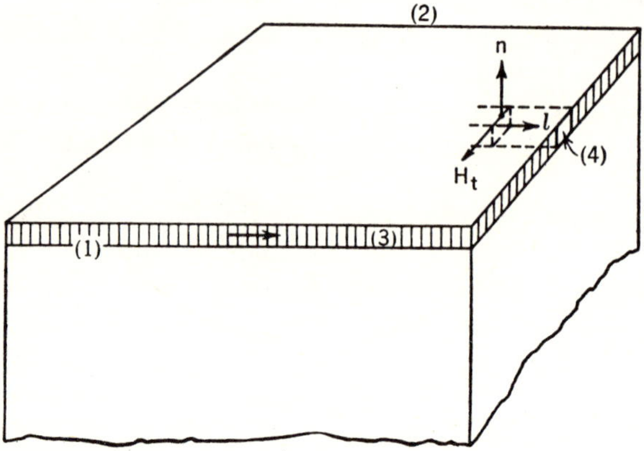

FIG. 5.1.—Surface current in a section of a large conductor. (1) Section of conductor. (2) Dielectric medium. (3) Surface layer of current, which includes all but a negligible fraction of the entire current in the conductor. (4) Surface current l in an element of unit width perpendicular to the direction of current. H_t is the component of the magnetic field tangent to the surface; n is perpendicular to the surface.

the direction of current. In the important case of a circular conductor with rotational symmetry in which η and l are each the same at all points around each circumference (as in a coaxial line), the total current and charge per unit length are obtained easily. Thus, if the subscript a refers to the radius a of the inner conductor and the subscript b refers to the inner radius b of the outer conductor, for the coaxial line of Fig. 5.2

$$q_a \text{ (coulombs/m)} = 2\pi a\eta_a \qquad q_b = 2\pi b\eta_b \tag{5.2}$$

$$I_a \text{ (amp)} = 2\pi a l_a \qquad I_b = 2\pi b l_b \tag{5.3}$$

Note that a total charge per unit length and a total current per unit cross section are adequate to describe the distributions of charge and current *only* if the surface current is wholly axial. This is true of all conventional circuits and also of cylindrical antennas. It is not true in more general types of transmission circuit to be considered later.

In order to determine the surface densities η and l it is in general necessary to determine the components of the electromagnetic field in the dielectric bounding the conducting surfaces, and then to

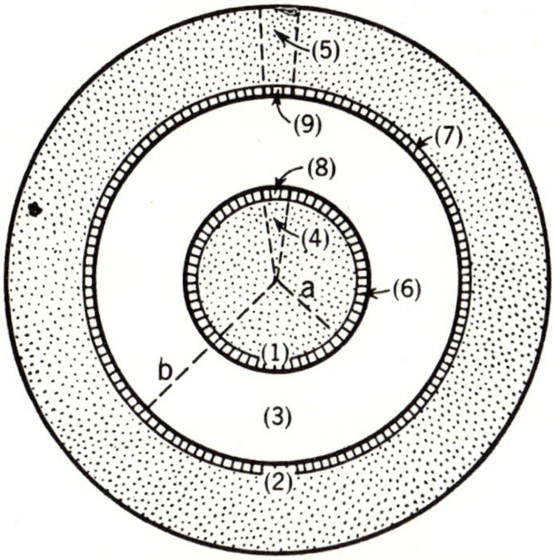

FIG. 5.2.—Cross section of a conventional coaxial line ($b <<< \lambda$). (1) Inner conductor (radius a). (2) Outer conductor (inner radius b). (3) Dielectric medium. (4) Element of inner conductor with unit width at $r = a$. (5) Element of outer conductor with unit width at $r = b$. (6) Surface layer of upward current (or positive charge) which includes all but a negligible fraction of the entire current (or charge per unit length) in the inner conductor. $I_a = 2\pi a l_a$; ($q_a = 2\pi a \eta_a$). (7) Surface layer of downward current (or negative charge) which includes all but a negligible fraction of the entire current (or charge per unit length) in the outer conductor $I_b = 2\pi b l_b$; ($q_b = 2\pi a \eta_b$). (8) Surface current l_a (or charge η_a) in an element of unit width perpendicular to the direction of current. (9) Surface current l_b (or charge η_b) in an element of unit width perpendicular to the direction of current.

compute the densities from the field. Once the electromagnetic field is known, it is easy to obtain η and l from the following simple boundary conditions, valid for good conductors bounded by good dielectrics:

$$\eta \text{ (coulombs/m}^2) = \epsilon_0 \text{ (farads/m) } \mathcal{E}_n \text{ (volts/m)} \tag{5.4}$$
$$l \text{ (amp/m)} = \mathcal{H}_t \text{ (amp/m)} \tag{5.5}$$

where η and l are the surface density of charge and the quasi-surface current, $\mathcal{E}_n$ is the component of the electric field perpendicular to the element of surface bearing the charge η per unit area, and $\mathcal{H}_t$ is the component of the magnetic field tangent to the surface element in which l flows and perpendicular to its direction of flow.

This is illustrated in Fig. 5.2. The quantities l, n, $\mathcal{H}_t$ form a right-hand system if n is a normal pointing from the surface of the conductor (below which is the surface current l) into the dielectric where $\mathcal{H}_t$ is defined.

In transmission circuits of uniform cross section, it is possible to separate the problem of determining $\mathcal{E}$ and $\mathcal{H}$ into two parts, one longitudinal in terms of the axial coordinate only, the other cross-sectional involving only the two transverse coordinates. This is possible because, in nonradiating circuits constructed of good conductors and good dielectrics, the longitudinal variation of the electromagnetic field depends upon a one-dimensional wave equation that is the same for all types of cross section; the transverse solution depends on a two-dimensional equation with different form for each shape and size of cross section. Every cross section is a plane characterized by a definite pattern or distribution of amplitude and phase for each component of the electromagnetic field and for the associated charge and current densities on the surfaces of the conductors. The distribution obtaining at a given instant travels axially with a characteristic phase velocity in nonresonant circuits.

6. Longitudinal Problem for Infinitely Long Transmission Circuit (or Its Equivalent).—If a transmission circuit of unspecified cross section is infinitely long or is terminated in such a way that the termination is equivalent to infinite length, the longitudinal solution has the simple form associated with traveling waves along conventional nonresonant transmission lines. Each component of the electromagnetic field and its associated surface current and charge have the complex form, written for $\mathcal{E}_x$, to illustrate, as

$$(\mathcal{E}_x)_z = (\mathcal{E}_x)_{z=0}\epsilon^{-(\alpha_g+j\beta_g)z} \tag{6.1}$$

where $\mathcal{E}_x$ is in general a function of the transverse coordinates x and y but not of the axial variable z; α_g is an attenuation constant in nepers per meter that can be divided into two independent additive parts (good conductors and good dielectrics are assumed), of which one depends on the high conductivity of the conductor and the other on the low conductivity of the dielectric; and β_g is the phase constant in radians per meter. Both α_g and β_g vary with the shape and size of the cross section and thus depend on the solution of the transverse problem. β_g always can be written in the form

$$\beta_g = \frac{\omega}{v_{pg}} = \frac{2\pi}{\lambda_g} \tag{6.2}$$

where v_{pg} is the axial phase velocity of propagation of each surface of constant (but not necessarily the same) phase for the components of the field in the dielectric medium or for the densities of charge or of surface current on the conductors, and λ_g is the wavelength or distance along the circuit between two consecutive surfaces that have the same transverse distribution of phase at the same instant. The subscript g stands for guide. It is used to designate quantities along a transmission circuit, because, from the point of view of traveling surfaces of constant phase, the so-called "waves" of the electromagnetic field, such a circuit may be called a wave guide.

7. Cross-sectional Problem for Transmission Circuit.—The solution of the longitudinal problem is the same for all components of the field vectors and for the densities of current and charge in all transmission circuits. The solution of the transverse problem differs greatly with the size, shape, and nature of the cross section. Two major classes of cross section are conveniently treated separately. These are cross sections that satisfy the condition for the near zone and those which are unrestricted. They will be discussed in turn. Before proceeding to do so, it is well to recall that one fundamental characteristic of all transmission circuits, in both classes, is negligible loss of power by radiation.

8. Transverse Solution—Near-zone Cross Section.—Transmission circuits with cross sections that satisfy the condition for the near zone include primarily conventional open-wire and coaxial transmission lines. Because the transverse separation of the conductors is small and because not less than two conductors with exclusively axial surface currents are involved, it is always possible to define total current for each conductor, also potential difference between conductors and between the necessarily closely spaced input and output terminals. It is possible, therefore, to define both input impedance and load impedance for a line in terms of the complex ratio of potential difference across the terminals to total current entering (or leaving) the terminals just as in near-zone circuit analysis. In fact, if the very small radiation is ignored, conventional lines can be treated in terms of near-zone theory insofar as the electromagnetic field outside the conductor and the *longitudinal* distribution of surface current and charge in the conductor are concerned. (The transverse distribution in the interior of the conductors is not a near-zone problem, as is pointed out in Sec. 3 of Chap. II. Its solution includes the determination of the skin-effect impedance per unit length of conductor.) In conventional lines, it is not necessary to determine the electromagnetic field as an intermediate step in solving for distribution of current and charge (or potential difference). Because of its close relationship to the solution of more general circuits, the electromagnetic field distri-

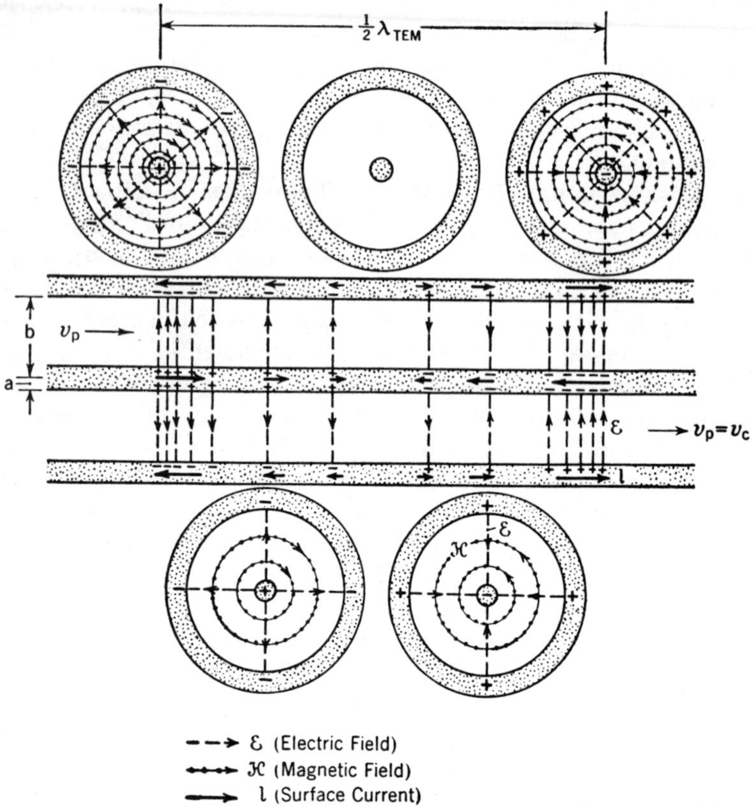

FIG. 8.1.—*TEM* mode in infinitely long (or equivalent) coaxial line with sufficiently small cross section ($b << \lambda_{TEM}/2\pi$).

bution in a conventional coaxial line that is infinitely long (or terminated to be equivalent) is shown in Fig. 8.1 under the assumption that the conductors and the dielectric are perfect. The electric field is radial, the magnetic field circular. This field distribution (which is the only one possible subject to the condition for near-zone spacing, but one of many for larger spacings) is called *transverse electromagnetic*, abbreviated *TEM*. The corresponding wavelength is $\lambda_{TEM} = (v_p)_{TEM}/f$. For highly conducting, air-filled, coaxial lines, $(v_p)_{TEM}$ is practically $v_c = 3 \cdot 10^8$ m/sec; if filled with a good dielectric of relative permeability μ_r and relative dielectric constant ϵ_r, $(v_p)_{TEM} = v_c/\sqrt{\mu_r\epsilon_r}$.

9. Transverse Solution for Circuit with Unrestricted Cross Section—Open-wire and Coaxial Line.—If the cross-sectional dimensions of a circuit become so large that they approach or exceed the wavelength, while the circuit as a whole is required to have negligible radiating properties, open-wire lines cannot be used. Although two or more parallel wires spaced an appreciable fraction of a wavelength, or a long single wire (either near or completely removed from the earth) may transmit power from one point to another, they do not satisfy the condition for low power loss; they are effective antennas and hence good radiators. This is merely stating the obvious, *viz.*, that it is not possible to have equal and opposite currents close together if the conductors carrying them are far apart. Consequently, there remains only the second type of nonradiating circuit, *viz.*, a circuit completely enclosed in metal. A coaxial line is a circuit of this type even if not restricted in size of cross section. It remains to examine what new distributions of surface current and charge are possible in a coaxial line when the cross section is increased in size. Two distinct groups of distributions or modes may be distinguished, of which the first, called the transverse-magnetic or *TM* type, is the simpler. It is characterized by surface currents that are *parallel* to the axis so that the magnetic field is always transverse. The second type, called transverse-electric or *TE*, involves more complicated distributions of surface current and charge.

10. The *TM* Type of Distribution in Coaxial and Hollow Cylindrical Conductors.—The cross section of a coaxial line may be increased in two somewhat different ways. While increasing the inner radius b of the outer conductor, it is possible to increase the radius a of the inner conductor so as to maintain the thickness of the dielectric $b - a$ so small that

$$(b - a) << \frac{\lambda}{2\pi} \tag{10.1}$$

with no restriction on b. If precautions are taken to exclude the *TE* type, described in the next section, there is no change in the configurations of surface current and charge and hence of the electromagnetic field as b and a are increased indefinitely while $(b - a)$

is kept very small compared with the wavelength, as required in (10.1). The *TEM* mode alone continues to prevail, Fig. 10.1*a*.

If $b - a$ is increased with λ fixed, or λ is decreased with $b - a$ fixed, so that (10.1) is not satisfied, in particular if $b - a$ approaches $\lambda/2$ in magnitude, a new and entirely different distribution of

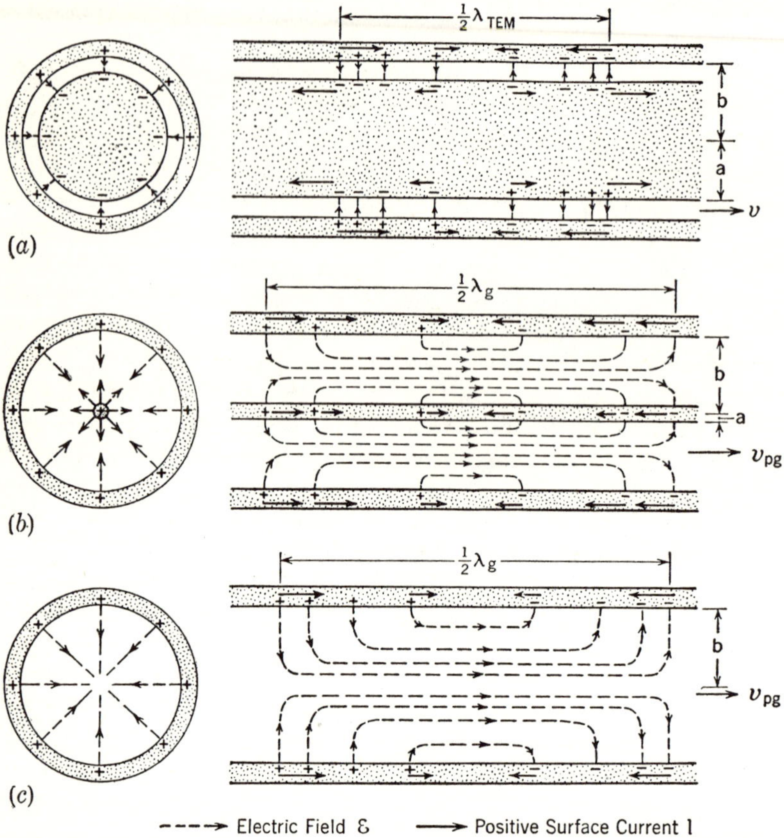

FIG. 10.1.—Distributions of surface charge, surface current, and electric field for (*a*) *TEM* mode in a coaxial pipe; *b* unrestricted; $(b - a) << \lambda/2\pi$ (*b*) $TM_{0,1}$ mode in coaxial pipe with $b > \lambda_{TEM}/2.61$; $a << b$; (*c*) $TM_{0,1}$ mode in hollow cylinder with $b > \lambda_{TEM}/2.61$. All pipes are infinitely long or the equivalent.

surface current and charge and hence of the electromagnetic field in the dielectric is possible, Fig. 10.1*b*, *in addition to the TEM mode.* (This figure and those that follow have been estimated, not computed.) It is called the $TM_{0,1}$ mode. Instead of being characterized, as is the *TEM* type, by *equal* and *opposite* total currents and charges per unit length at opposite points of the two conductors, the $TM_{0,1}$ distribution consists of currents in the *same* direction and charges of the *same* sign on opposite surfaces of the conductors. Moreover, the total current and charge per unit length on the outer conductor are greater than on the inner conductor. The $TM_{0,1}$ distribution is possible because $(b - a)$ is so large that time delays comparable with the half period are involved in determining the forces acting on the charges in one conductor due to charges in the other. As a result, the positive charge at a given cross section of one conductor at a particular instant does not experience the effects of the positive charge on the opposite surface of the other conductor at the same instant; instead, it experiences the effect of the negative charge that was there at the appropriate earlier instant. In order that the currents and electromagnetic field may reduce to vanishingly small values in the interior of both conductors, the axial distances between cross sections traversed by currents in the same phase is increased in the $TM_{0,1}$ mode as compared with the corresponding distance in the *TEM* type. In other words, the $TM_{0,1}$ mode involves a *smaller* phase constant β_g and hence a *greater wavelength* λ_g than the *TEM* type. This means that the phase velocity v_{pg} is greater in the $TM_{0,1}$ mode than in the *TEM* type and, therefore, greater than $v_c = 3 \cdot 10^8$ m/sec.

The current and charge on the inner conductor in the $TM_{0,1}$ mode not only are smaller than those on the outer conductor, but they also decrease continuously as the radius a is reduced. The inner conductor thus plays a less and less significant part in the transmission as the inner conductor is made smaller, and there is no reason why it should not be reduced indefinitely or completely removed. If this is done, a hollow conductor or pipe remains with a distribution of surface current and charge that is essentially the same as when the inner conductor is present. The phase constant β_g increases and the wavelength λ_g decreases as a is reduced, but λ_g is always greater than λ_{TEM}. The distribution of surface current and charge and the associated electric-field characteristic of the $TM_{0,1}$ mode in a pipe are shown in Fig. 10.1c. This mode is possible specifically only when

$$b > \frac{\lambda_{TEM}}{2.61} \tag{10.2}$$

If the inner conductor is present, b must be still larger than is required by (10.2). Specific values are given in Fig. 13.1.

If the radius b is increased so that $b > \lambda_{TEM}/1.63$, another and quite different TM distribution in the transverse plane is possible. It is denoted by $TM_{1,1}$, and its charge and field distributions are shown in Fig. 10.2. With $b > \lambda_{TEM}/1.2$, yet another distribution

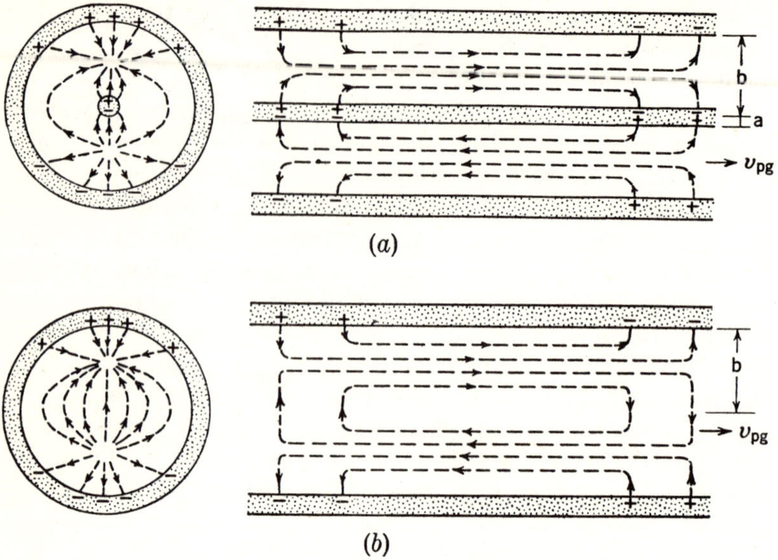

FIG. 10.2.—Surface charge and electric-field distributions for (a) $TM_{1,1}$ mode in a coaxial line with $b > \lambda_{TEM}/1.63$; $a << b$. (b) $TM_{1,1}$ mode in a hollow cylinder with $b > \lambda_{TEM}/1.63$. All surface currents are parallel to the axis; the line and the pipe are assumed to be infinitely long (or equivalent) so that cross-sectional surfaces of constant phase travel along the pipe with the appropriate phase velocity v_{pg}. Lines which appear to end in space actually go vertically down into or come up from the paper.

may be excited. By increasing b more and more, innumerable distributions or modes are possible of which a few are shown in Fig. 10.3.[1]

[1] A large variety of TM modes is possible in sufficiently large pipes. In the axially nonresonant case, these modes can be distinguished with the aid of two subscripts n and m, as $T_{n,m}$. (In axially resonant systems, described later, a third subscript p must be used as $TM_{n,m,p}$.)

The subscript n gives the *order* of the Bessel function $J_n(kb)$, where k is a constant which occurs in the mathematical solution. It characterizes the symmetry of the distribution of surface charge and current or of the electromagnetic field with respect to θ. Thus $n = 1$, Fig. 10.2, means a single cycle in the distribution around the circumference; $n = 2$, Fig. 10.3a, b, means two cycles in the distribution around the circumference; etc.

The subscript m is numerically equal to the number of the root of the equation $J_n(kb) = 0$, beginning with 1, *i.e.*, the first value of kb that satisfies the equation is root number 1 with $m = 1$. It characterizes the radial periodic-

All the modes of the *TM* type of distribution of oscillating surface charge on both coaxial and cylindrical transmission circuits

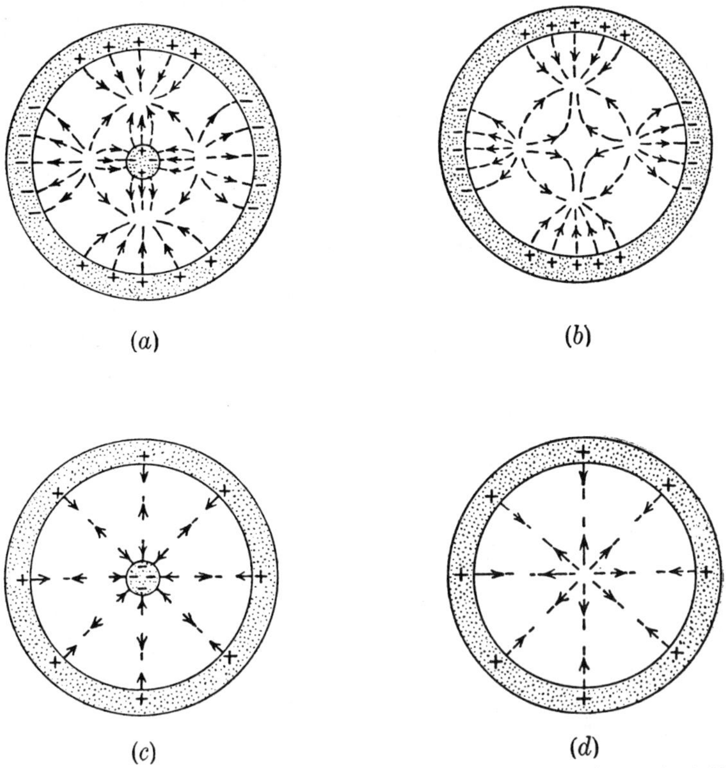

FIG. 10.3.—Cross-sectional distribution of electric charge and electric field for (a) coaxial $TM_{2,1}$ with $b > \lambda_{TEM}/1.2$; (b) cylinder $TM_{2,1}$ with $b > \lambda_{TEM}/1.2$; (c) coaxial $TM_{0,2}$ with $b > \lambda_{TEM}/1.14$; (d) cylinder $TM_{0,2}$ with $b > \lambda_{TEM}/1.14$. Note that the $TM_{0,2}$ modes are rotationally symmetrical—all $TM_{0,m}$ modes are. Electric lines which appear to end in space actually bend down into or up from the paper.

of infinite length (or equivalent) are characterized by currents *parallel* to the axis. The modes of type $TM_{0,m}$, with m any integer,

---

ity in the electromagnetic field pattern. If only a single pattern exists in the radial direction from the center out, $m = 1$, however complicated the pattern may be in the θ-direction. If a complete pattern of any one of the many belonging to the type $m = 1$ is contained within a circle smaller than the enclosing pipe and a second complete pattern exists between this circle and the metal wall, $m = 2$. This is illustrated in Fig. 10.3c, d, where the patterns of Fig. 10.1b, c are contained within a small circle, and essentially the same radial pattern is repeated outside it with directions reversed.

are rotationally symmetrical, with the density of surface current uniform and in the same axial direction at all points around a given circumference. The other modes are not rotationally symmetrical, since the direction and magnitude of the density of surface current and of surface charge while fixed in a definite phase sequence yet

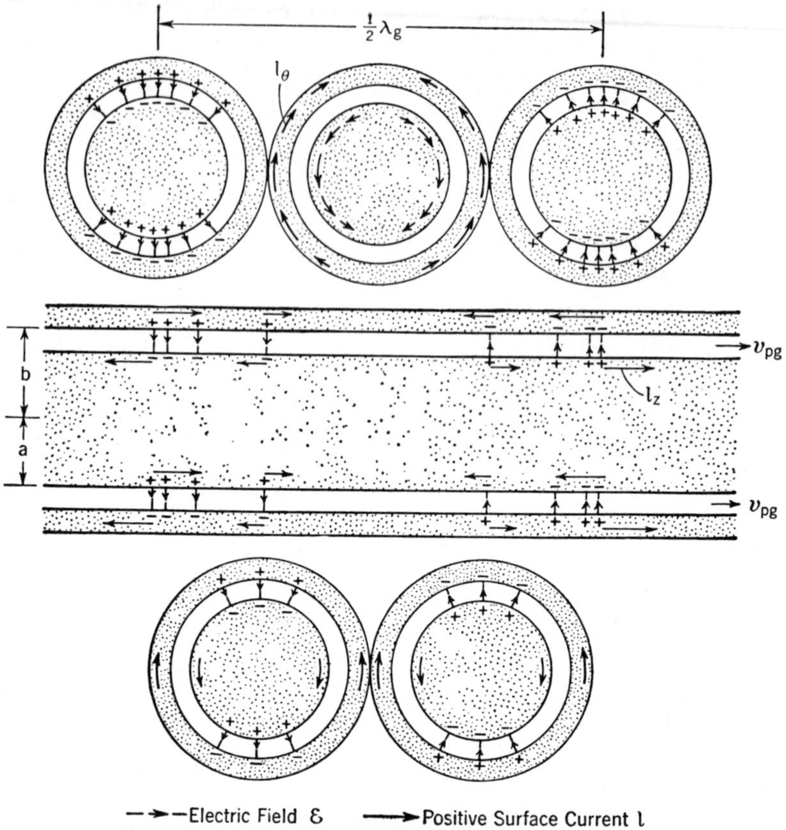

FIG. 11.1.—$TE_{1,1}$ mode in a coaxial pipe subject to $(b - a) << \lambda_{TEM}/2\pi$; $b > \lambda_{TEM}/2\pi$. The circles represent cross sections in the pipe directly above or below. $\lambda_g > \lambda_{TEM}$; $v_{pg} > v_c$.

have different magnitudes and directions at different points around a given circumference at a particular instant. But there is no motion of charge around the circumference.

11. *TE* Distribution in Coaxial and Hollow Cylindrical Conductors.—Even if the condition $\beta(b - a) << 1$ is maintained so that the two conductors of a coaxial line differ only slightly in radius, a *TE* mode is possible if

$$b > \frac{\lambda}{2\pi} \tag{11.1}$$

i.e., as soon as the circumference $2\pi b$ exceeds the wavelength. This is the $TE_{1,1}$ mode. If the inner conductor is reduced in radius, b must have progressively larger values if this mode is to remain

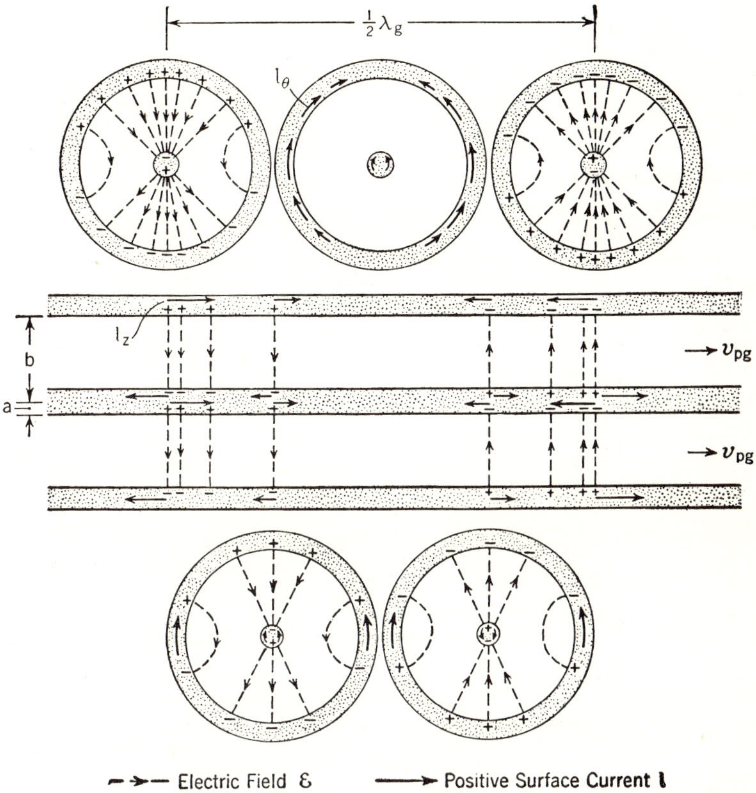

FIG. 11.2.—$TE_{1,1}$ mode in a coaxial pipe subject to $a << b$; $b > \lambda_{TEM}/3.41$. $\lambda_g > \lambda_{TEM}$; $v_{pg} > v_c$.

possible. However, it is *always* possible for any radius of inner conductor *including zero*, *i.e.*, for a hollow cylinder, if the radius b satisfies

$$b > \frac{\lambda}{3.41} \tag{11.2}$$

In a pipe the $TE_{1,1}$ mode is possible at a *lower* wavelength than *any TM mode* or any other *TE* mode. It is called the *dominant mode.*

The oscillations of surface charge in the *TE* type in an infinitely long (or equivalent) circuit are more intricate than in the *TM* type because they involve a *standing-wave* pattern of oscillating

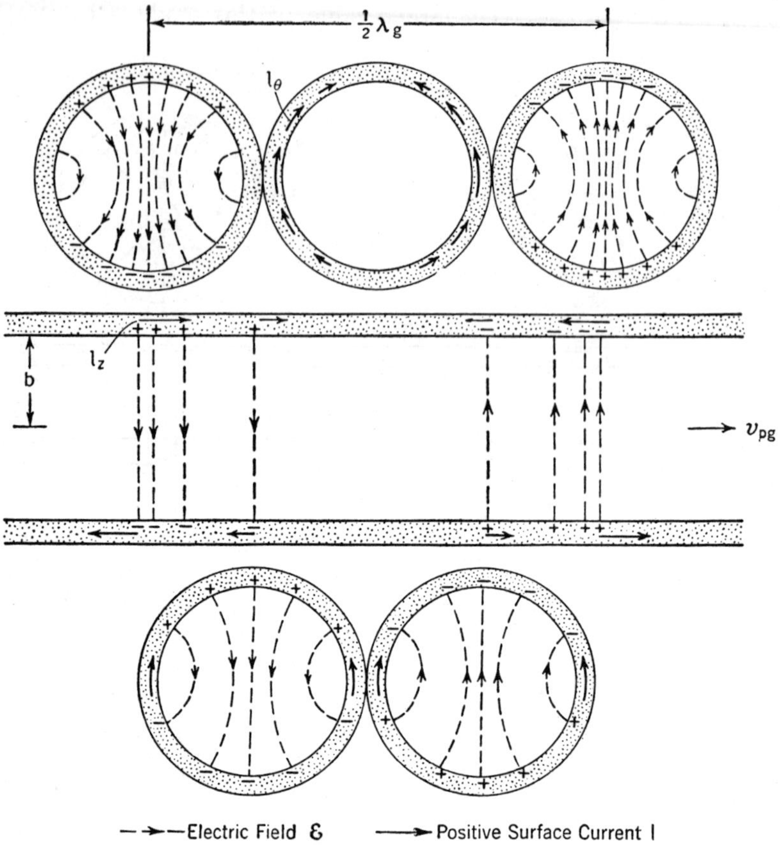

FIG. 11.3.—$TE_{1,1}$ mode in a cylindrical pipe subject to $b > \lambda_{TEM}/3.41$. $\lambda_g > \lambda_{TEM}$ $v_{pg} > v_c$.

charge *in the transverse plane* and a *traveling-wave* pattern of oscillating charge *in the axial direction.* This is best described in terms of three closely related distributions: in Fig. 11.1 for a coaxial pipe with cross-sectional dimensions such that $(b - a) << \lambda_{TEM}/2\pi$ and $b > \lambda/2\pi$; in Fig. 11.2 for a coaxial pipe such that $a << b$, $b > \lambda_{TEM}/3.41$; in Fig. 11.3 for a hollow cylinder with $b > \lambda_{TEM}/3.41$. In each, a section through the axis of an infinitely long (or equivalent) pipe is shown with surface distribution of charge, surface current, and electric field for a half wavelength assuming perfectly conducting walls. The magnetic field is not represented because of its intricate form. Cross sections through the pipe are shown with centers opposite the corresponding part of the axial section. Thus, the circular section on the upper right gives the surface charge, surface current, and electric-field distribution in the particular transverse plane where at a given instant the charge density, the axial component l_z of the surface current, and the transverse electric field are all maximum with negative charge at the top. The axial current (in this longitudinally nonresonant case) is simply the concentration of positive charge moving to the right with the phase velocity v_{pg}. (A concentration of negative charge moving to the right is equivalent to an equal concentration of positive charge moving to the left.)

In an infinitely long, perfectly conducting pipe (or its equivalent) the distribution of surface current and surface charge at any one cross section at a particular instant is propagated *without change* along the pipe with a definite phase velocity characteristic of the distribution. Accordingly, the cross-sectional distributions that pass a fixed cross section of the pipe have exactly the same sequence in time as the distribution back along the pipe at a fixed instant.

In Fig. 11.3, for example, the concurrent cross-sectional distributions at intervals of one-eighth wavelength back along the pipe beginning on the right side of the half wavelength shown must pass this cross section on the right at intervals of one-eighth period. From the point of view of a fixed observer at this cross section, the sequence is as follows: at $t = 0$, maximum concentration of negative charge is at the top; maximum concentration of positive charge is at the bottom; and both move to the right with the velocity v_{pg}. Accordingly, a positive surface current parallel to the axis and directed to the left at the top and to the right at the bottom is observed.

At $t = \frac{1}{8}$ period later than $t = 0$ and at the same location in the pipe, the concentrations of negative charge at the top and positive charge at the bottom have decreased in density for two reasons. First, maximum concentrations of negative charge have moved to the right in the form of a negative axial current to the right (equivalent to a positive current to the left) along the top surface and a positive axial current to the right along the bottom surface. Second, the negative charge at the top has started downward around the circumference; the positive charge has started upward around the circumference, *i.e.*, there is a positive current upward around both sides of the circumference.

At $t = \frac{1}{4}$ period later than $t = 0$, again at the same cross section in the pipe, there is no concentration of charge anywhere. There is no charge passing in the axial direction, and hence there is no axially directed surface current. On the other hand, the currents that started upward around each side of the circumference have reached their maximum values in a standing-wave sequence.

At $t = \frac{3}{8}$ period later than $t = 0$ (at the same cross section), the upward currents around the sides have diminished; positive charge has accumulated at the top, negative charge at the bottom. This charge concentration is moving to the right with phase velocity v_{pg} so that there is a positive current to the right along the top and to the left along the bottom.

At $t = \frac{1}{2}$ period later than $t = 0$, and at the same cross section, the currents around the circumference are everywhere zero; maximum concentrations of positive charge at the top and negative charge at the bottom obtain, so that *axial currents* are at a maximum.

The cycle continues in a similar way. It is seen to be made up of a standing-wave distribution around the circumference (in a standing-wave sequence that has maximum charge concentration at top and bottom, and maximum current at the sides a quarter period earlier and later) superimposed on a traveling-wave distribution axially. The successive time phases of the standing-wave distribution are spread out along the pipe, all traveling to the right with velocity v_{pg}. The distribution of current on the inside of the pipe of Fig. 11.3 is shown in Fig. 11.4 with the pipe represented as though cut lengthwise along the "top of cylinder" and rolled out flat. The locations of maximum current at the instant represented occur near the arrowheads. The current is instantaneously zero at the points from which or toward which the lines of flow are directed.

It is important to note that, if it is desired to insert a probe (small antenna) in the pipe or move one along it in a slit to determine the standing-wave ratio of the axial current, this can be done with a minimum effect on the distribution of current if the slit is cut *along* the lines of flow, *not across them.* For the distribution of Figs. 11.1 to 11.3, the slit should be along the center of the top or bottom.

The distribution of current on the inner surface of a coaxial conductor is similar to that in a hollow conductor if the same mode obtains. The distribution along the outer surface of the inner conductor in the coaxial pipe resembles that on the outer conductor with directions reversed. If $(b - a) << \lambda_{TEM}/2\pi$, the currents

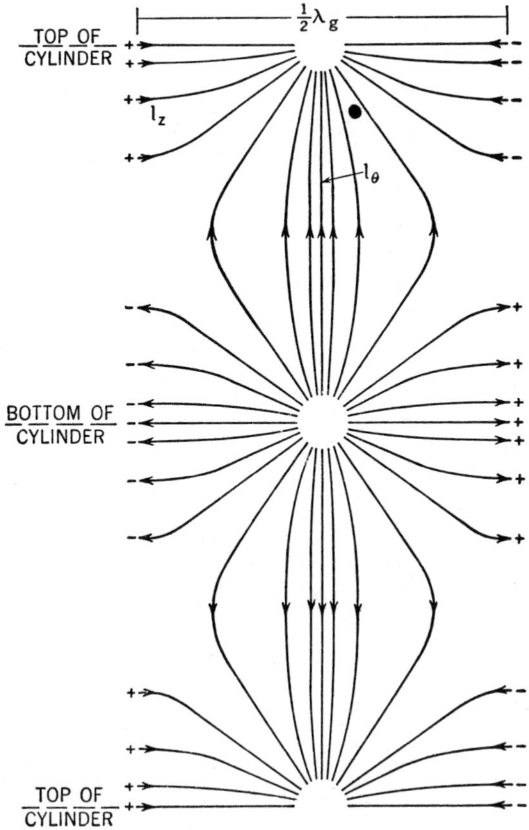

FIG. 11.4.—Surface current of the $TE_{1,1}$ mode on the inside of the metal cylinder of Fig. 11.3. The cylinder is represented as cut along the top center and laid flat.

crossing the same transverse plane in the two conductors are equal and opposite. If the radius a is appreciably smaller than the radius b, the current on the inner conductor is smaller; it approaches zero as a vanishes.

Distributions of charge density and electric field in a transverse plane of a nonresonant, hollow, circular cylinder are shown in Fig. 11.5a, c, d for three of the possible higher modes. A fourth, the circular electric mode, is shown in Fig. 11.5*b* and differs considerably from the other modes in that the surface density of charge is always zero everywhere. There is no axial current, only circulating currents in the transverse plane which periodically reverse their direc-

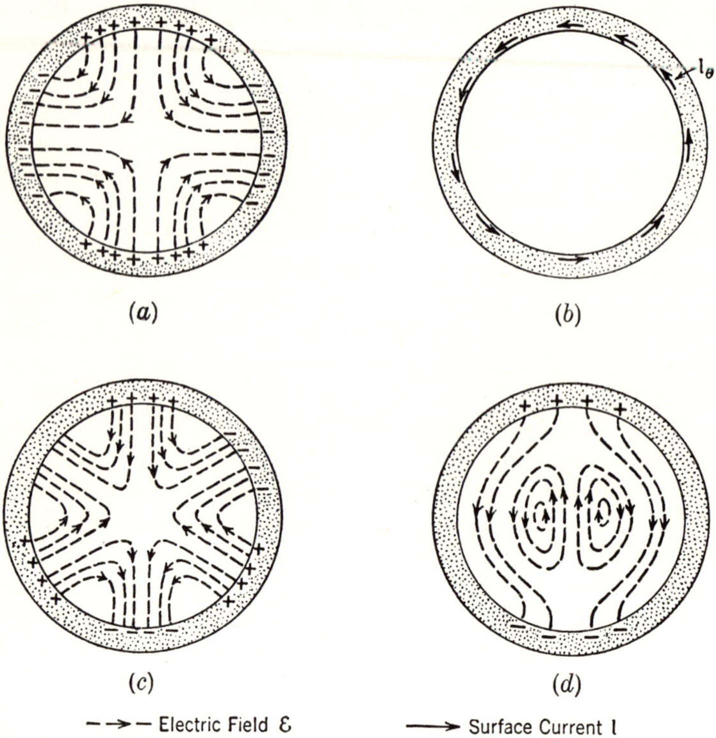

FIG. 11.5.—Higher *TE* modes in an axially nonresonant cylinder. (*a*) $TE_{2,1}$; ($b > \lambda_{TEM}/2.06$) at instant of maximum charge density. (*b*) $TE_{0,1}$; ($b > \lambda_{TEM}/1.64$) at instant of maximum surface current density (charge density always zero). ℰ in the dielectric is also circular but with maximum a quarter period later. (*c*) $TE_{3,1}$; ($b > \lambda_{TEM}/1.50$) at instant of maximum charge density. (*d*) $TE_{1,2}$; ($b > \lambda_{TEM}/1.18$) at instant of maximum charge density.

tion, while a given amplitude travels unchanged axially along the pipe with a definite phase velocity. This mode is very difficult to excite but has the interesting property of an attenuation constant that decreases with frequency.[1]

[1] The subscripts n, m in $TE_{n,m}$ have a meaning similar to those for the *TM* mode. As before, n is the order of the Bessel function; it characterizes the rotational periodicity of a pattern. Thus, in Fig. 11.3, $n = 1$; in Fig. 11.5*a*, $n = 2$; in Fig. 11.5*c*, $n = 3$. m is the number of the nonvanishing root of the equation $J_n'(kb) = 0$; it characterizes the radial periodicity. Thus, in Fig. 11.3, $m = 1$; in Fig. 11.5*d*, $m = 2$.

Except for the *TEM* mode in the coaxial line, no mention has been made of the magnetic field. This does not mean that there is no magnetic field or that it is less important than the electric field that has been represented. As is pointed out in Sec. 1 of Chap. II, the electromagnetic field in the present treatment is considered to be merely a mathematically useful intermediate step in calculating distributions of current and charge and the circuit properties associated with them. In general transmission circuits, all distributions of oscillating surface charge and current are determined from a previously determined electromagnetic field, using (5.4) and (5.5). Since a representation of the electric and magnetic field distributions is primarily useful in determining the distribution of current and charge, this representation in terms of the field is not necessary if the problem has been carried to its conclusion and the current and charge are represented. It is a simple matter to obtain the field distributions from surface current and charge distributions, using (5.4) and (5.5) and the fact that electric and magnetic lines cross at right angles, that electric lines become perpendicular and magnetic lines parallel to a perfectly conducting boundary surface as this is approached more and more closely. In order to show this, the electric lines have been drawn in the several patterns for the current and charge distributions. It is left as a simple exercise in the *TM* type and as a more complicated one in the *TE* type to sketch sectional patterns of the magnetic field.

12. Hollow Conductors of Rectangular Cross Section.—The use of pipes as transmission circuits or wave guides is not restricted to those of circular cross section. Both *TM* and *TE* types, each with a large number of different modes, are possible in pipes of any cross-sectional shape, including rectangular, triangular, elliptical, and others. The distributions of surface current and charge correspond in ways appropriate for the particular shape to analogous ones in the pipe of circular section. The longest wavelength or dominant mode in a rectangular pipe of sides a and b is the $TE_{1,0}$ (which corresponds to the $TE_{1,1}$ in a circular pipe). The $TE_{1,0}$ and $TE_{0,1}$ modes can be excited if

$$a > \frac{\lambda_{TEM}}{2} \text{ for the } TE_{1,0} \text{ mode}; \qquad b > \frac{\lambda_{TEM}}{2} \text{ for the } TE_{0,1} \text{ mode} \qquad (12.1)$$

These two modes differ only in having the transverse (x and y) axes interchanged. The distributions of surface current and charge

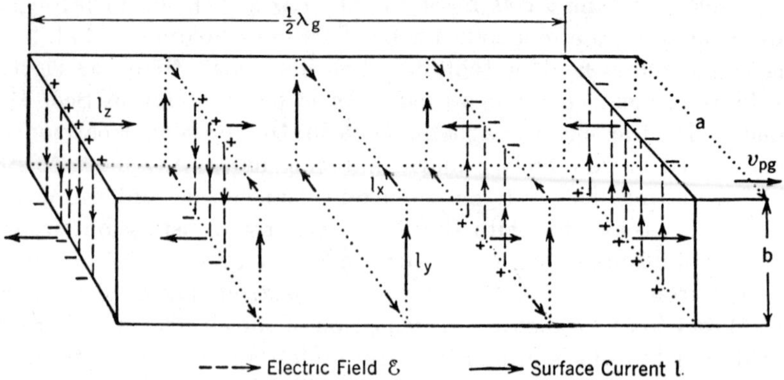

FIG. 12.1.—$TE_{1,0}$ mode in a rectangular pipe; $a > \frac{1}{2}\lambda_{TEM}$; $b < \frac{1}{2}\lambda_{TEM}$. Arrows indicate the direction of positive surface current; *all* charge concentrations move to the right.

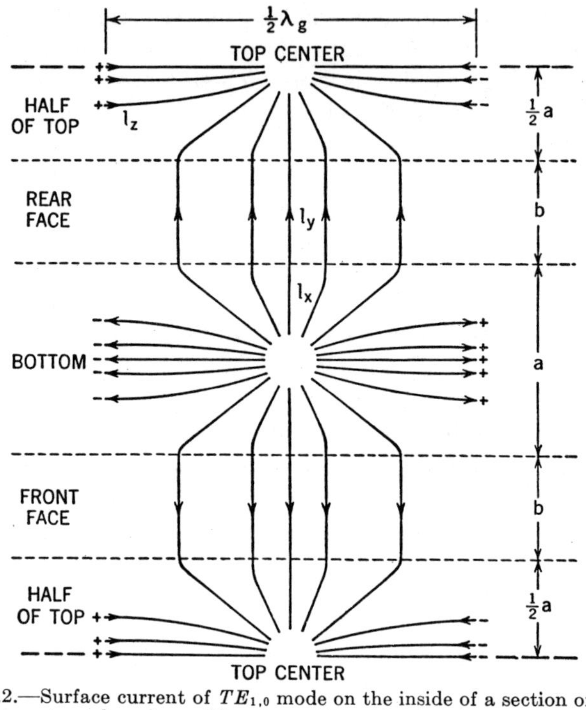

FIG. 12.2.—Surface current of $TE_{1,0}$ mode on the inside of a section of an axially nonresonant rectangular pipe. The pipe of Fig. 12.1 is represented as cut lengthwise along the center of the top and laid flat. Note that arrows indicate the direction of positive surface current; *all* charge concentrations move to the right.

and of the electric field for the $TE_{1,0}$ mode are shown in Fig. 12.1. As in all TE modes, there is a standing-wave pattern around the periphery of each cross section. The direction of flow of surface current on the inner surfaces is shown in Fig. 12.2 with the pipe represented as though cut lengthwise along the center of the top and folded out flat. The magnitude of the current is greatest near

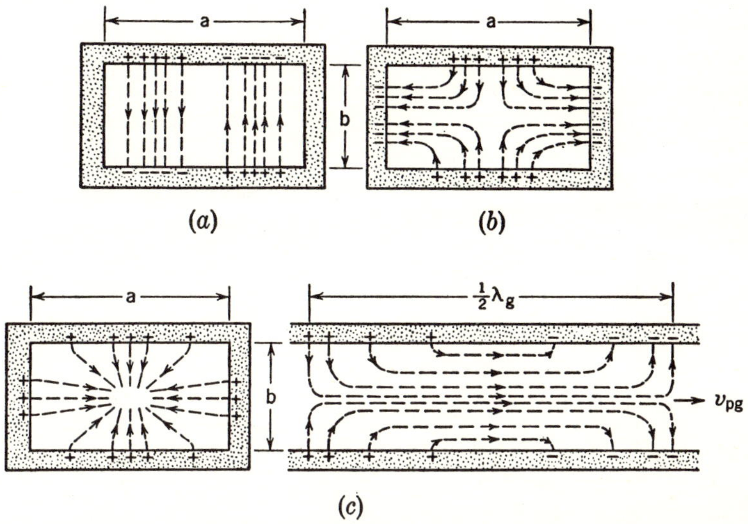

FIG. 12.3.—Higher modes in a rectangular pipe. (a) $TE_{2,0}$ mode; $b > \lambda_{TEM}$. (b) $TE_{1,1}$ mode; $\frac{2ab}{\sqrt{a^2 + b^2}} > \lambda_{TEM}$. (c) $TM_{1,1}$ mode; $\frac{2ab}{\sqrt{a^2 + b^2}} > \lambda_{TEM}$.

or at the arrows, and reduces to zero at the points toward which the lines of flow converge or from which they diverge. The distribution is similar to that of Fig. 11.4 for a pipe of circular cross section. Note that *on the side surfaces the current is transverse only.* If a single mode is to obtain, it is necessary that the longer side a satisfy the inequality

$$\lambda_{TEM} > a > \tfrac{1}{2}\lambda_{TEM} \tag{12.2}$$

with the shorter side b such that

$$b < \tfrac{1}{2}\lambda_{TEM} \tag{12.3}$$

It is interesting that there is no lower limit for b. It can be made as small as desired (at the expense of increasing attenuation).

Cross sections showing distributions of charge and electric field in several higher modes[1] are given in Fig. 12.3.

13. Parameters of Transmission Circuits.—Nonresonant transmission circuits of all the types described are characterized by two fundamental parameters, the phase constant β_g and the attenuation constant α_g. These appear in the solutions of the general form (6.1), which characterize the axial behavior of a traveling surface of constant phase for density of charge, for each component of surface current, and for individual components of the electromagnetic field. The attenuation constant must be small in every

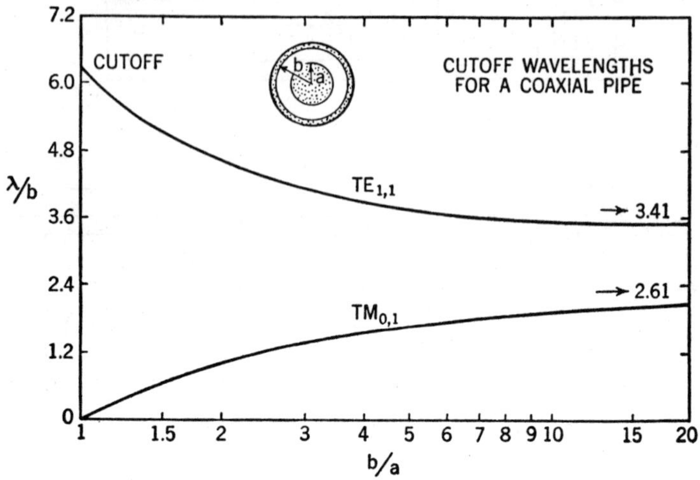

FIG. 13.1.—Cutoff wavelengths for two modes in a coaxial pipe.

practically important transmission circuit, at least over a useful range of frequencies. The band of frequencies for which the attenuation constant is small is called the low-attenuation band or pass band; the band of frequencies for which the attenuation constant is very large is called the high-attenuation band or stop band. In the ideal case of perfect conductors and perfect dielectrics, the attenuation constant is zero in the pass band and rises abruptly and steeply at the beginning of the stop band. The frequency at which this happens is called the cutoff frequency, f_{cutoff}. The

[1] The several modes in a rectangular pipe may be distinguished by the integral subscripts m and n in the axially nonresonant case (a third subscript is required in the axially resonant case). These specify the number of half-wavelength distributions of surface charge or current along the x (width a) or y (width b) directions. The transverse distribution of charge is always sinusoidal with zero density at the corners.

corresponding wavelength as measured using the TEM mode is called the cutoff wavelength. It should be denoted by $(\lambda_{TEM})_{\text{cutoff}}$. Since cutoff wavelengths will always be expressed in terms of the TEM wavelength, the subscript TEM will usually be omitted and λ_{cutoff} written. Where convenient, the number subscripts m and n may be used to distinguish between cutoff values for different modes.

All coaxial-line and hollow-conductor transmission circuits are high-pass filters in that the pass band lies above a certain cutoff frequency (or below a certain cutoff wavelength). The cutoff

TABLE 13.1.—WAVE-GUIDE PARAMETERS AND CONSTANTS

Parameter \ Mode	TEM: coaxial	$TM_{0,1}$: circular	$TE_{1,1}$: circular	$TE_{1,0}$: rectang.
b	Inner radius of outer conductor	Inner radius of cylinder		Shorter side
a	Radius of inner conductor	Not defined		Longer side
λ_{cutoff}	∞	$2.61b$	$3.41b$	$2a$
Wavelength range for single mode $(\lambda = \lambda_{TEM})$	$2\pi b < \lambda < \infty$; $(b - a) < \frac{\lambda}{2\pi}$	$2.06b < \lambda < 2.61b$	$2.61b < \lambda < 3.41b$ ($TM_{0,1}$ also possible)	$a < \lambda < 2a$; $b < \lambda/2$
$N \equiv \frac{\lambda_{TEM}}{\lambda_{\text{cutoff}}}$	0	$\lambda_{TEM}/2.61b$	$\lambda_{TEM}/3.41b$	$\lambda_{TEM}/2a$
α_g	$\frac{R^s}{\mathcal{R}}\left[\frac{\frac{1}{a}+\frac{1}{b}}{2\ln\frac{b}{a}}\right]$	$\frac{R^s}{\mathcal{R}b}\frac{1}{\sqrt{1-N^2}}$	$\frac{R^s}{\mathcal{R}b}\left[\frac{0.436+N^2}{\sqrt{1-N^2}}\right]$	$\frac{R^s}{\mathcal{R}}\left[\frac{\frac{1}{b}+\frac{2}{a}N^2}{\sqrt{1-N^2}}\right]$
β_g	$\beta\sqrt{1-N^2}$			
λ_g	$\lambda/\sqrt{1-N^2}$			
v_{pg}	$v_p/\sqrt{1-N^2}$			
v_{gg}	$v_p\sqrt{1-N^2}$			

Surface resistance, $R^s = \sqrt{\frac{\omega\mu_r\mu_0}{2\sigma}} = 2\pi\sqrt{\mu_r f\rho}\cdot 10^{-7}$ ohms
(σ is conductivity in mhos/m; $\rho = 1/\sigma$ is resistivity in ohm-m; for copper $\sigma = 5.8\cdot 10^7$ mhos/m

Characteristic resistance $\begin{cases}\text{in free space: } \mathcal{R}_c = \sqrt{\frac{\mu_0}{\epsilon_0}} = 376.7 \text{ ohms} \\ \text{in a dielectric: } \mathcal{R} = \sqrt{\frac{\mu_r}{\epsilon_r}}\,\mathcal{R}_c\end{cases}$

Characteristic phase velocity $\begin{cases}\text{in free space: } v_c = \frac{1}{\sqrt{\mu_0\epsilon_0}} = 3\cdot 10^8 \text{ m/sec} \\ \text{in a dielectric: } v_p = v_c/\sqrt{\mu_r\epsilon_r}\end{cases}$

Phase constant $\begin{cases}\text{in free space: } \beta = \omega/v_c \\ \text{in a dielectric: } \beta = \omega/v_p\end{cases}$

Wave length $\begin{cases}\text{in free space: } \lambda = v_c/f = 2\pi/\beta \\ \text{in a dielectric: } \lambda = v_p/f\end{cases}$

wavelength in a coaxial line depends upon both the inner radius b of the outer conductor and the radius a of the inner conductor. Moreover, the formula is intricate. Curves showing the ratio $(\lambda/b)_{\text{cutoff}}$ as a function of b/a are given in Fig. 13.1 for the two most important modes, the $TM_{0,1}$ and the $TE_{1,1}$. It follows from this figure that, for the $TM_{0,1}$ mode, $(\lambda/b)_{\text{cutoff}}$ decreases to zero as $(b - a) \to 0$. On the other hand, $(\lambda/b)_{\text{cutoff}}$ increases to 2.61 as $a \to 0$. This is the value for the circular pipe *without inner conductor.* $(\lambda/b)_{\text{cutoff}}$ for the $TE_{1,1}$ mode behaves differently. It approaches 2π (circumference equal to one wavelength) as $(b - a) \to 0$; it diminishes to $(\lambda/b)_{\text{cutoff}} = 3.41$ (circumference only slightly more than one-half wavelength) for the pipe ($a = 0$). Cutoff wavelengths for several pipes are summarized in Table 13.1, where formulas are given for phase and attenuation constants for the wavelength λ_g, for the phase velocity v_{pg}, and the group velocity v_{gg}. As in the electromagnetic field near an antenna, Chap. II, Sec. 36, the phase velocity exceeds the group velocity, the relationship being

$$v_{pg}v_{gg} = v_c^2 \tag{13.1}$$

if air is the dielectric. If any other dielectric is used, $v = v_c/\sqrt{\mu_r\epsilon_r}$ must be written for v_c, as explained in Chap. II, Sec. 2.

14. Comparison and Summary of the Properties of Transmission Circuits.—Since it is possible to select from a variety of transmission circuits, it is well to summarize the properties of each type in the light of the basic criteria listed in Sec. 4.

1. *a. Power loss in heating conductors and dielectrics that are part of the line.*—For pipes of *comparable surface area,* the attenuation due to imperfect conductors does not differ greatly for any of the four modes for which formulas are given in Table 13.1. Since coaxial lines may be and usually are much smaller than corresponding pipes, the losses in heating are usually greater in the coaxial lines. Because insulating supports are not required in the pipe, the attenuation due to them is eliminated. Losses in all types of transmission circuits are increased if more modes than the one mode actually used for power transmission are excited.

b. Power loss by radiation.—Radiation from metal pipes of all types including the coaxial is practically nonexistent if the ends are closed. If the ends are open, significant radiation occurs because of the uncanceled forces exerted by moving charges inside but near the open ends and because of currents that may be on the outside surface of a coaxial or other pipe if it has an open end or slits or apertures of any kind anywhere along its length. An open end on a sufficiently large pipe is essentially equivalent to an electromagnetic horn, as discussed in Chap. II.

2. *Power Capacity: a. Dielectric breakdown.*—Hollow conductors are superior to coaxial lines in their ability to handle larger concentrations of charge, because these concentrations can be kept farther apart than in a conventional coaxial line so that the electric field is less intense.

b. Current capacity.—In practice, a pipe is likely to have a much greater conducting surface than a coaxial line, and therefore its current-carrying capacity is greater.

3. *Frequency Range and Frequency Response.*—Efficient operation of any transmission circuit requires that only a single mode be excited, *viz.*, the one actually used. Although it is possible and necessary for some purposes to eliminate undesired higher or lower modes by appropriate methods of driving, discussed at a later point, the simplest and best method wherever possible is to make use of the longest wavelength and the lowest frequency mode, the so-called "dominant mode," and to design the size of the pipe so that the operating frequency is in the pass band of this mode and in the stop band of all higher frequency modes. In effect, this means that the useful operating range of a given transmission circuit lies between a lower frequency limit defined by the cutoff of the dominant mode and an upper frequency limit defined by the cutoff of the next higher mode. These ranges are given in Table 13.1.

Because the lower cutoff of the TEM mode in a coaxial line is at zero frequency, the frequency range of this dominant mode in the coaxial line exceeds the range of any other mode. The upper-frequency limit of the $TM_{0,1}$ mode from the point of view of single-mode operation is the cutoff for the $TE_{2,1}$ mode. The wavelength ranges for single-mode operation are also specified in Table 13.1.

4. *Physical Availability.*—The practical and economical possibility of constructing coaxial and hollow conductors for use as transmission systems depends on the dimensions required to assure single-mode operation at a specified frequency. Since the TEM mode in a coaxial line has no lower cutoff frequency, coaxial lines (and open-wire lines) may be used at frequencies as low as desired and may be constructed of any convenient size permitted by considerations of power capacity. On the other hand, at higher and

higher frequencies, the coaxial line must be made smaller and smaller if single-mode operation is to be assured. Ultimately a point is reached at which adequate power capacity is impossible. This usually occurs at wavelengths in the centimeter range. Pipe transmission is not available at longer wavelengths because of the large size of pipe required. Where possible, pipes using only the dominant $TE_{1,1}$ mode in a circular pipe, or the $TE_{1,0}$ mode in a rectangular pipe, are preferred.

5. *Special Features.*—For connecting to rotating directional antennas, a transmission circuit that permits the use of a swivel joint is often convenient. Requirements for this purpose include not only a circular pipe but also complete rotational symmetry of the distributions of oscillating charge and current. Both these conditions obtain in the TEM and $TM_{0,1}$ modes in a coaxial system and in the $TM_{0,1}$ mode in circular pipes. Since the TEM mode is dominant, it is normally used with a coaxial line. On the other hand, with the pipe the $TE_{1,1}$ mode has a lower (longer wavelength) cutoff frequency than the rotationally symmetrical $TM_{0,1}$ mode. If a pipe with a swivel joint is designed for the pass band of the $TM_{0,1}$ mode, it is also large enough for the pass band of the $TE_{1,1}$ mode. The $TE_{1,1}$ mode must be avoided either by using a driving device that does not excite it, or it must be suppressed by suitably designed wire grids, for example.

In any transmission system, it is essential that the cross-sectional distribution of current and charge at the load end be known. In rotationally symmetrical systems, there is no difficulty. On the other hand, in the $TE_{1,1}$ mode in a coaxial or in a circular pipe, the transverse standing-wave pattern may spiral down a pipe that is not absolutely uniform. This makes it difficult to operate a load efficiently, since it must be designed for a particular orientation of the transverse standing-wave pattern or of the polarization of the associated electric field. The rectangular pipe using the $TE_{1,0}$ mode has the great advantage of a fixed transverse standing-wave pattern, with charge concentrations always on the wider side and an electric field parallel to the narrower side.

For some purposes, flexible transmission cable is required. Such cable can be constructed for both coaxial and hollow transmission circuits. Attenuation is usually greater than in rigid circuits, and difficulties may be encountered because of small changes in the cross-sectional shape or size as the cable is bent.

III. TRANSMITTING AND RECEIVING SYSTEMS USING NONRESONANT CIRCUITS

15. Methods of Driving.—A variety of possible distributions of oscillating charge on the interior conducting surfaces of axially nonresonant coaxial and hollow transmission circuits have been described. But nothing has been said as to the connections to the device driving the circuit, *i.e.*, to a generator or a receiving antenna, or to the load terminating the circuit. These are described in this and the following sections.

In conventional coaxial (and open-wire) lines, the two terminals of a symmetrical generator always can be connected or coupled directly to the two conductors (or groups of conductors) of the line because these carry equal and opposite currents and charges. Most ultra-high-frequency generators designed for use with coaxial or with single pipes have a coaxial-line output, although the dimensions of the line may be such that the $TM_{0,1}$ rather than the TEM mode is excited. In either case, the currents on the inner surface of the outer conductor and the outer surface of the inner conductor are axially directed and rotationally symmetrical so that total currents can be defined easily for both the TEM mode (in which they are opposite and equal) and the $TM_{0,1}$ mode (in which they are codirectional, with the inner current smaller). The transfer from one coaxial line to another is made easily if both lines use the TEM mode or the $TM_{0,1}$ mode, or if one uses the TEM and the other the $TM_{0,1}$ mode. The simplest connection for the TEM mode is illustrated in Fig. 15.1*a*. In the arrangement shown, the small coaxial line from the generator is terminated in its characteristic impedance by means of a double-stub matching section. The movable single-stub section of Fig. 15.1*b* may be used instead of the double-stub tuner. Instead of changing abruptly from the line of smaller to that of larger size, as shown in the figure, it is desirable to insert a tapered section between the two lines or pipes. If this is made a wavelength or more long, the matching stubs are usually not necessary. In practice, such a tapered length may be less convenient than a matching section.

It is a simple matter to transfer from a coaxial-pipe TEM or $TM_{0,1}$ mode to a single-pipe mode such as the cylindrical $TM_{0,1}$ or the rectangular $TM_{1,1}$. These are characterized by exclusively axial surface currents that obviously may be excited using the arrangement of Fig. 15.1*b*, where the inner conductor of the coaxial line is extended a short distance into the pipe. The arrangement is now similar to that of an antenna erected perpendicular to a conducting plane of infinite extent. In this case, the infinite plane is folded over into an infinitely long (or equivalent) pipe. A standing-wave distribution of oscillating current and charge is maintained on the projecting inner conductor or antenna, and a rotationally

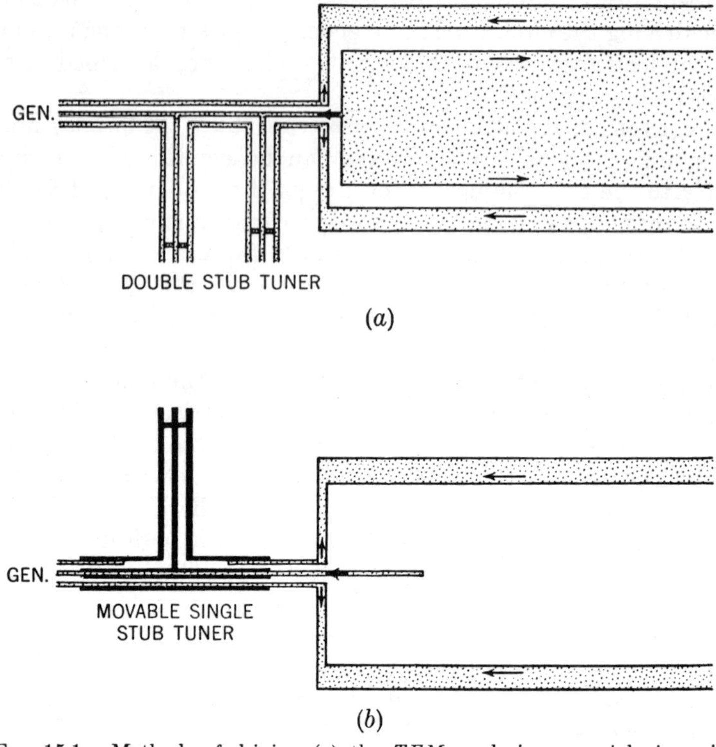

FIG. 15.1.—Methods of driving (a) the *TEM* mode in a coaxial pipe with b unrestricted and $(b - a) << \lambda_{TEM}/2\pi$; (b) the $TM_{0,1}$ mode in a hollow cylinder with $b > \lambda_{TEM}/2.61$ using a conventional coaxial line. Arrows show the location of maximum currents at the appropriate instant.

symmetrical traveling-wave distribution is maintained on the inner surface of the pipe. Because the electromagnetic field of an antenna decreases rapidly outward *along the axis*, Chap. II, Fig. 35.1, the field and hence the current and charge distributions on the pipe even a short distance beyond the end of the antenna are not affected *directly* by the periodically varying charge in the antenna. The distribution of current and charge is that of the cylindrical $TM_{0,1}$ mode or rectangular $TM_{1,1}$ mode.

In order to excite the TE modes, in particular the dominant $TE_{1,1}$ in a circular pipe or the $TE_{1,0}$ in a rectangular pipe, a standing-wave pattern must be excited in the transverse plane. This may be accomplished by the arrangement in Fig. 15.2. The inner

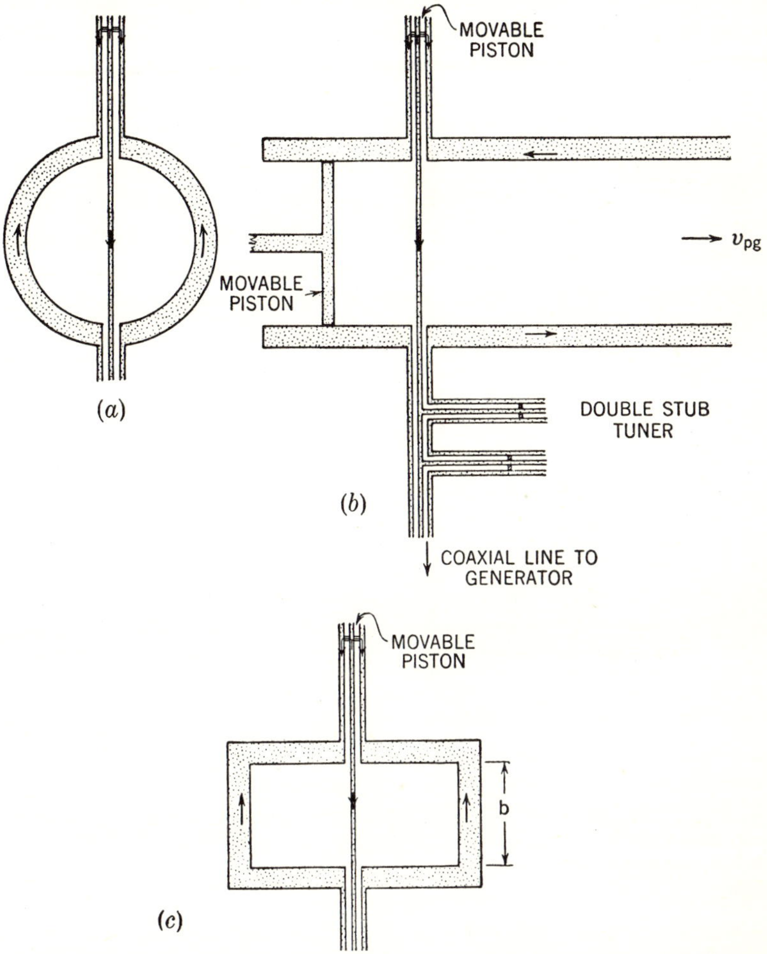

FIG. 15.2.—Method of driving (a) and (b) $TE_{1,1}$ mode in a hollow cylinder; and (c) and (b) $TE_{1,0}$ mode in a rectangular pipe. Arrows denote maximum current.

conductor of the resonant end of a coaxial line is arranged as an antenna to extend across the center of and perpendicular to the axis of the pipe. By means of the adjustable stub at the top, the maximum current may be fixed at the center of the antenna. A standing-wave distribution is excited in the transverse plane, Fig.

15.2*a*. By suitably adjusting the large movable piston in the pipe, the electromagnetic field due to currents up and down on the inner face of the piston and the field due to currents in the antenna itself may be adjusted to be essentially in phase at points to the right of the antenna much as in an antenna with parabolic reflector. The resulting distributions of oscillating charge and of the electromagnetic field at a distance of a wavelength or more from the

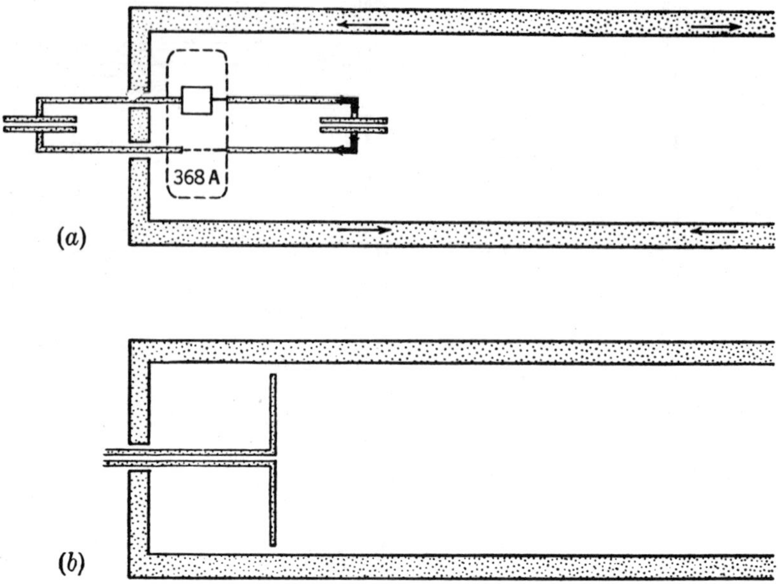

FIG. 15.3.—Methods of exciting the *TE* modes. (*a*) Vacuum tube and part of two-wire line tank circuit of UHF generator built into hollow pipe. Currents in the line exert forces on top and bottom to set currents in motion there as required for *TE* mode. (*b*) Two-wire feeder and antenna placed at optimum distance from the closed end to excite *TE* mode.

antenna are characteristic of the $TE_{1,1}$ mode alone if the size of the pipe is appropriately chosen. Near the antenna a complicated superposition of many modes is to be expected. The corresponding arrangement for a rectangular pipe is shown in Fig. 15.2*c*. A double-stub tuner (or a movable single-stub tuner) must be provided at the end of the coaxial line if this is to be nonresonant. If the line is short and resonant, a telescoping section of coaxial line may be used instead of the tuner. If the ultra-high-frequency generator has a tank circuit consisting of a section of two-wire line, or if a two-wire line is to be used as a feeder, the arrangements of Fig. 15.3 may be used.

16. Matching Circuits for Pipes.—In describing the distributions of current and of charge in the conducting surfaces of a pipe (or the distribution of the electromagnetic field in its interior) it has been assumed that the circuit is axially nonresonant. The mathematical form for a traveling-wave distribution can always be

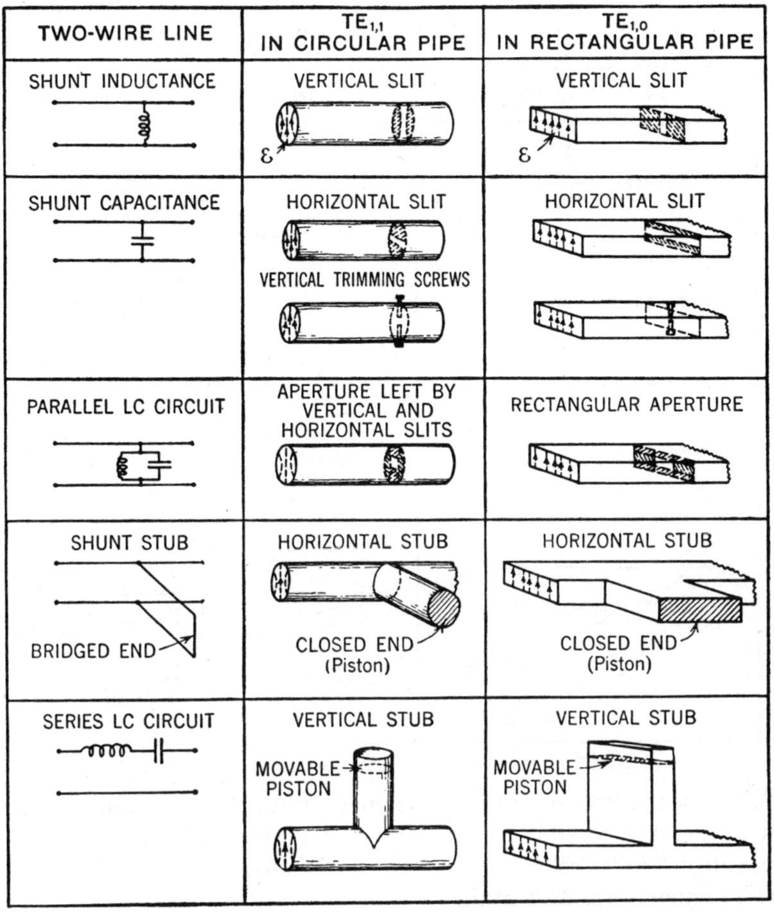

FIG. 16.1.—Roughly comparable circuit elements.

obtained by allowing the circuit to become infinitely long. In practice, nonresonant transmission circuits may be approximated closely by utilizing suitably designed terminating devices.

In conventional line theory, in which impedance may be defined by the ratio of potential difference to total current, a formula for the input impedance of an infinitely long line is derived readily.

By definition it is the characteristic impedance Z_c and is largely resistive. Obviously any termination for a line that equals Z_c is the equivalent of an infinitely long line. In pipes (or coaxials operated in other than the *TEM* mode), it is not possible to define in the conventional way an input impedance for an infinitely long pipe. A generalized impedance concept, nevertheless, is useful in a similar but more restricted sense in pipes as discussed in Sec. 26.

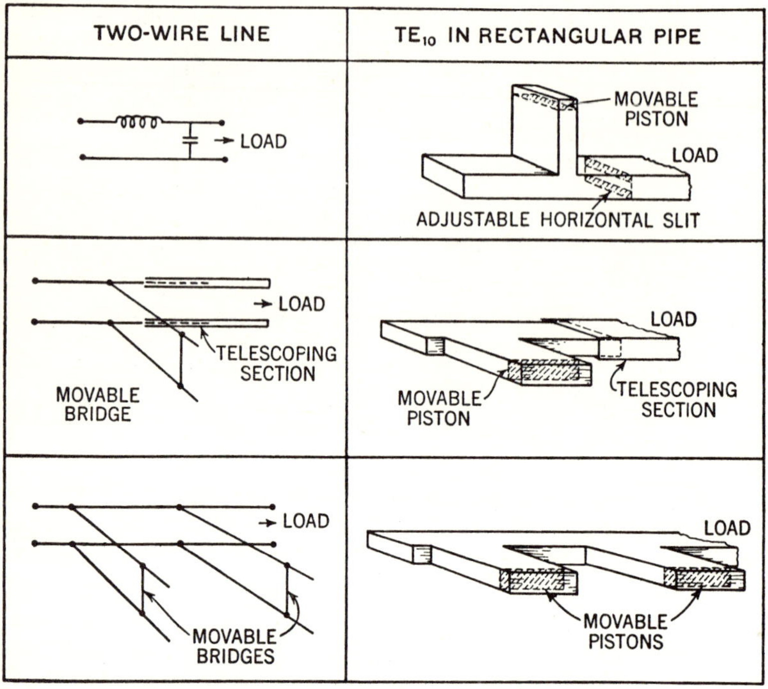

FIG. 16.2.—Roughly comparable matching sections.

For present purposes, it is sufficient to state that, for a given mode in a pipe, it is possible to construct circuit elements (in a generalized meaning) that are equivalent in their effect to shunt and series coils, capacitors, and stub sections in conventional lines. A few roughly equivalent circuit elements for a two-wire open line and the dominant modes in circular and rectangular pipes are shown in Fig. 16.1. A combination of two suitably chosen circuit elements may be arranged in a matching section that can be inserted between the long pipe and the load in ways that are comparable to the various special cases obtained in more conventional circuits using a T section of reactive elements or the general section of a resonant line with adjustable taps, Chap. II, Sec. 28. Several of many possible combinations are shown in Fig. 16.2. They include an arrangement of inductor and capacitor, a movable single-stub tuner and a fixed double-stub tuner.

17. Load Circuits for Pipes.—In general terms, two types of load may be described for pipe transmission circuits. The first type consists of loads that are designed especially for a particular mode in a given pipe and therefore have no input terminals in the sense used in ordinary networks confined to the near zone. Loads of the second type are conventional. They have two closely spaced input terminals for which an input impedance can be defined in terms of potential difference and total current.

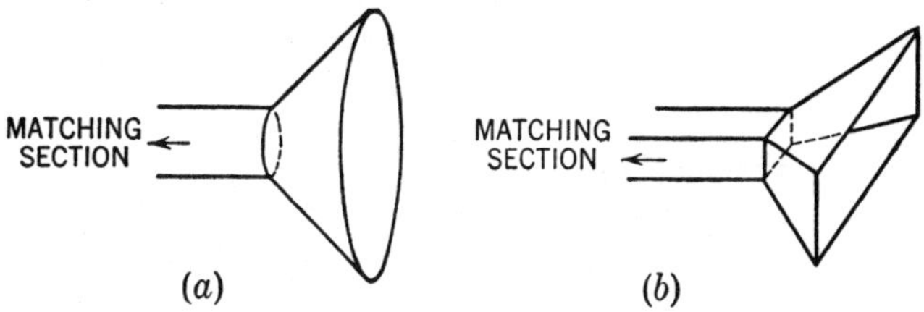

FIG. 17.1.—(a) Horn for $TE_{1,1}$ mode in a circular pipe. (b) Horn for $TE_{1,0}$ mode in a rectangular pipe.

Loads that are designed especially for a single mode in a particular pipe may be connected directly to the output end of the pipe following the matching section. Important examples are antennas of the electromagnetic horn type, Fig. 17.1. If the shape of the cross section of a horn designed for the TE modes is the same as that of the pipe, but flared gradually, each cross-sectional plane continues to have essentially the same standing-wave distribution of current and charge as along the axially nonresonant pipe. However, there is no traveling-wave distribution in the axial direction along the horn and along a part of the matching section. These are resonant in the axial direction as well as in the transverse plane. A detailed discussion of axially resonant sections of pipe is reserved for Secs. 19 to 27. The lines of flow of sheets of current on the inner surfaces of a horn of rectangular cross section driven in the $TE_{1,0}$ mode are sketched in Fig. 17.2 at an instant when the current is a maximum and the charge density is everywhere zero in the standing-wave distribution. The electromagnetic field at distant points is due primarily to these sheets of current, as discussed in Chap. II, Sec. 42. Field patterns are not the same if the pipe is excited in different modes, so that a variety of directional properties is available.

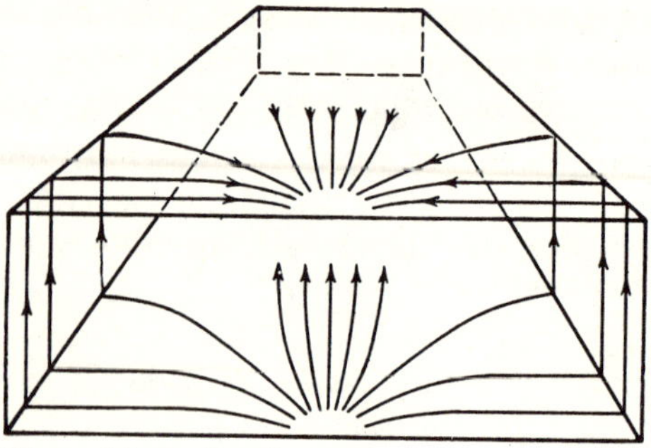

FIG. 17.2.—Lines of flow of surface current in an electromagnetic horn.

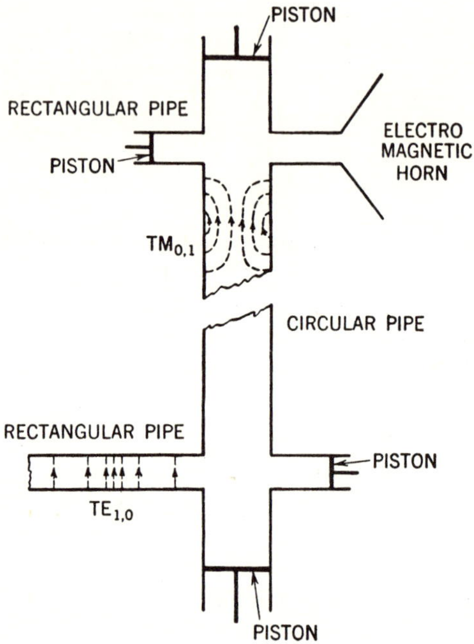

FIG. 17.3.—Cross section showing the transfer from the $TE_{1,0}$ mode in a rectangular pipe to the rotationally symmetrical $TM_{0,1}$ mode in a circular cylinder and back to a rectangular pipe. Pistons for matching are also shown. Dotted lines show the approximate distribution of electric field for a short distance in each pipe at a particular instant.

In addition to electromagnetic horns of various types, the load for a particular mode in a pipe may be another pipe of different size or of different cross-sectional shape. For example, if it is desired to make use of a swivel joint in a pipe transmission system using the dominant $TE_{1,0}$ mode in a pipe of rectangular section, it is necessary to transfer to a rotationally symmetrical mode, such as the $TM_{0,1}$ in a circular cylinder. A simple arrangement for accomplishing this is shown in Fig. 17.3. Since the $TE_{1,0}$ mode in the lower rectangular pipe has exclusively an up-and-down electric field, the $TE_{1,1}$ mode should not be excited in the vertical circular pipe, even though the radius is necessarily well above the cutoff for this mode. By appropriate adjustment of the two pairs of pistons, the rectangular as well as the circular pipe may be made

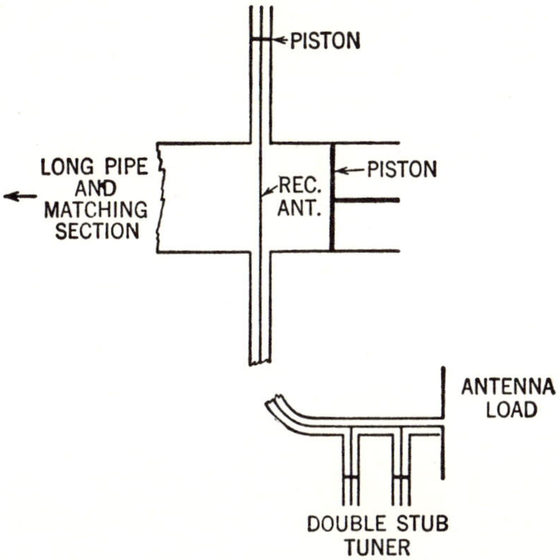

FIG. 17.4.—Transfer section for connecting a *TE* pipe mode to a conventional coaxial line with a two-terminal load.

nonresonant. As a second example, a rectangular pipe of long side a may be connected to a pipe with long side $2a$, $3a$, or na, where n is any integer, preferably odd, so that a symmetrical connection is possible. The wider pipe is then excited in the $TE_{n,0}$ mode, although lower modes also may be excited simultaneously.

In order to drive conventional two-terminal antennas or other loads from a pipe transmission circuit, it is usually necessary to make use of a closely spaced coaxial line or a two-wire line to connect directly to the load. The problem then is reduced to driving a coaxial line or a two-wire line from a pipe. This is accomplished readily if it is recalled that, according to the reciprocal theorem (Sec. 43), the relative directional properties of a transmitting array with one driven antenna are the same as the directional properties of the same array used as a receiver. Accordingly, the effective and simple arrangements of Fig. 15.1*b* for driving a pipe from a coaxial line using an antenna may be inverted for use with a receiver, Fig. 17.4. It is necessary to change the position of the matching section, since it is now the long pipe that is loaded and the coaxial line that is driven. This means that a section of the type shown in Fig. 16.2 should be inserted between the long pipe and the receiving antenna. Since the properties of the receiving array (including the antenna and the piston beyond it) as a load can be modified by

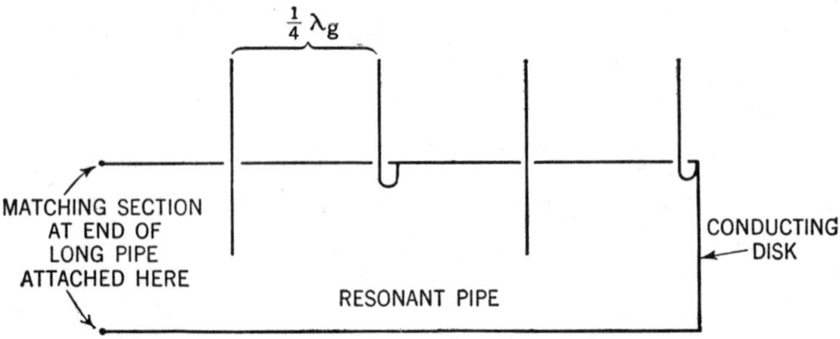

FIG. 17.5.—Plane view of an axial section of a resonant load for a pipe driven in a dominant *TE* mode. The array of four antennas is driven with progressive phase differences of a quarter period. The loops are at points of maximum surface current and maximum magnetic field perpendicular to the loops; the straight antennas at points of maximum surface charge and maximum vertical electric field in the standing-wave pattern.

moving the piston in the pipe as well as the piston in the coaxial stub above the antenna, it is often possible to obtain a satisfactory match by adjusting these alone without using a matching section between the receiving antenna and the long pipe. If an antenna is to be driven from a two-wire line rather than from a coaxial line, an arrangement like Fig. 15.3*b* may be used at the load end and provided with a suitable matching section in the pipe.

A single antenna or the several units of an entire array may be driven directly from a pipe in a way resembling Fig. 41.15 of Chap. II for a coaxial line. Since as much as a quarter wavelength of an antenna can be inserted, a greater variety of connections is possible. Two methods of coupling are shown in Fig. 17.5 for a *TE* mode. Each unit is erected vertically on the pipe with a part extending through a hole into the pipe. If it merely projects into the pipe, it must be located at a point of maximum charge concentration or maximum transverse electric field in the axial standing-wave pattern. If it is bent into a loop and connected to the inner surface of the pipe, it must be placed at a point of maximum axial current or maximum transverse magnetic field. Note that distances between maxima of current or between maxima of charge are spaced in terms of λ_g, the wavelength in the pipe, not in terms of λ_{TEM}.

Since λ_g is always greater than the wavelength λ_{TEM} in space, arrays may not have the desired spacing in terms of λ_{TEM}. It is possible, however, to reduce λ_g by increasing the diameter of the pipe, by inserting dielectric disks, or by inserting an inner conductor of such size that the *TEM* mode is alone possible. In this case, $\lambda_g = \lambda_{TEM}$, and the problem is essentially the same as that discussed under end-driven arrays in Chap. II, Sec. 42. In the present case, however, the coaxial line may have as large an outside diameter as the pipe, if the inner conductor is correspondingly enlarged so that $(b - a)$ is sufficiently small.

In all cases in which antennas project into a pipe or into a coaxial conductor through holes or slits, surface currents pass around the edges of the hole to the outside surface of the pipe, where they are distributed in more or less complicated patterns. Their amplitudes depend on the size and number of the holes and whether all or a part of the outside surface of the entire pipe is resonant in one of many possible modes. Currents on the outside surface necessarily contribute directly to the distant electromagnetic field, and the array therefore consists of the projecting antennas and the outer surface of the pipe. Even if there are no projecting antennas, the presence of holes in a pipe involves currents on the outside surface, and a nonvanishing field due to them as well as partly uncanceled fields due to currents on the inside. The magnitudes and directions of such fields depend on the size, the number, and the location of the holes, as well as on the dimensions of the pipe and the mode that is excited. A section of an imperfectly enclosed pipe (sometimes called a leaky wave guide) thus is seen to be an antenna.

18. Receiving Systems.—A receiving system differs from a transmitter only in the nature of the load and the method of driving. The load may consist of a detector or mixer specially designed for a particular mode in a pipe. It may be built into a section of the pipe and connected to it directly after a suitable matching section. If the load is of a more conventional two-terminal type with coaxial or two-wire line connections, a suitable transfer section for passing from the pipe to a coaxial or a two-wire line must be used with appropriately designed and placed matching sections.

The reciprocal theorem indicates that any of the transmitting arrays appropriate for use with pipes may be inserted and used as receiving arrays with comparable directional properties. The electromagnetic horn is an example. For maximum transfer of power to a pipe transmission circuit or wave guide from a receiving array, a suitably designed matching section between the array and the long pipe is required. A second matching section between pipe and load permits adjustment to make the pipe nonresonant, as described for conventional lines in Chap. II, Sec. 28. If the pipe is short, it may be used as a resonant section with one of the matching sections omitted.

IV. RESONANT CIRCUITS

19. Introduction and Notation.—In the analysis of conventional lines, Chap. I, emphasis was placed primarily on the transmission of power from a generator attached to one end of the line, called the sending end, to a load at the other end, called the receiving end. The symbols Z_S and Z_R were introduced, with subscripts S and R, to designate the sending- and receiving-end impedances.

At ultra-high-frequencies, conventional lines as well as more general transmission circuits are used for many purposes besides the transmission of power. A transmission circuit may be used for making measurements such as the determination of wavelength, impedance, and phase; or a resonant section may form part of a tank circuit in an ultra-high-frequency generator. In some of these applications, it is convenient to couple the generator and the load to the circuit at other points than at the ends. In such cases, the names sending end and receiving end no longer apply; it becomes necessary to designate points along a transmission circuit in terms of an axis of coordinates. For example, the z coordinate may be measured along the axis of the circuit with origin at one end. Then $z = 0$ locates one end, $z = s$ the other end. The terminating impedance at $z = 0$ of a conventional line is then conveniently denoted by Z_0, that at $z = s$ by Z_s. The subscripts 0 and s merely locate the impedance on the z axis.

In more general types of transmission circuit for which it is not possible to define terminating impedances in the conventional sense because the circuit does not end in a pair of terminals that are in the near zone with respect to each other, factors that measure the over-all attenuation and phase shift at a termination are introduced to characterize the properties of the termination. They are defined in such general terms that they are as useful for conventional lines as for general transmission circuits. Factors characterizing the termination at $z = 0$ have the subscript 0; those for the termination at $z = s$ have the subscript s.

20. Resonant Sections of Transmission Circuits.—If a transmission circuit of the coaxial or the hollow type is closed at both ends by a metal piston, *axially* resonant distributions of oscillating current and charge may be excited. Depending upon the method of driving and the size of the pipe, these distributions may be of the *TM* or the *TE* type. If the matching section at the end of a

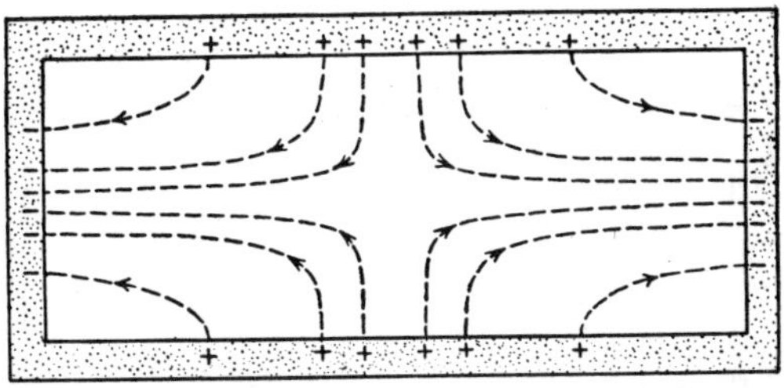

FIG. 20.1.—Charge and electric-field distribution at the instant when these are maximum in the standing-wave $TM_{0,1,1}$ mode in a closed circular cylinder of inner radius b and length s. Rotational symmetry about the axis obtains. A quarter period later, charge and electric field are everywhere zero; sheets of maximum current are from + to − along the walls; a circular magnetic field obtains with oppositely directed maxima at the two circular end walls and zero at the center.

$$TM\lambda_{0,1,1} = \frac{2}{\sqrt{\frac{1}{s^2} + \frac{0.58}{b^2}}}$$

nonresonant transmission circuit excited in *one particular mode* is removed and replaced by a metal piston, the circuit may be tuned to resonance by moving the piston until the amplitudes of the current and charge (as well as of the electric and magnetic fields) are maximized. The resulting axial standing-wave distributions differ from the traveling-wave case, as is discussed in general terms in Sec. 3. The transverse distribution characteristic of the particular mode is essentially the same as when the circuit is nonresonant, but it is now the *transverse* pattern of a standing-wave instead of a traveling-wave distribution in the axial direction. Distributions of maximum charge (and of maximum transverse electric field) are axially fixed at intervals of $\lambda_g/2$ instead of traveling with a characteristic phase velocity. Extreme amplitudes of axially directed current (and of maximum transverse magnetic field) occur a quarter period after the maxima of charge and midway between them. This is illustrated in Fig. 20.1 for the $TM_{0,1,1}$ mode and in Fig. 20.2 for the $TE_{1,1,1}$ mode in a circular pipe.

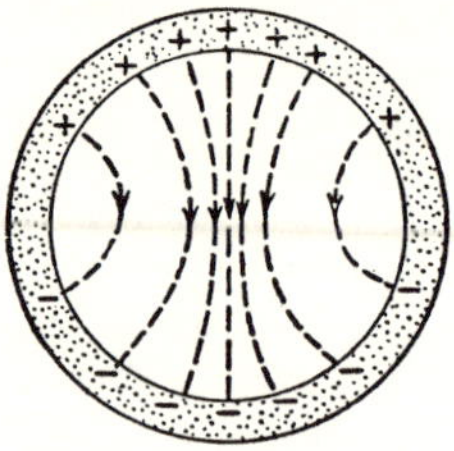

(a)

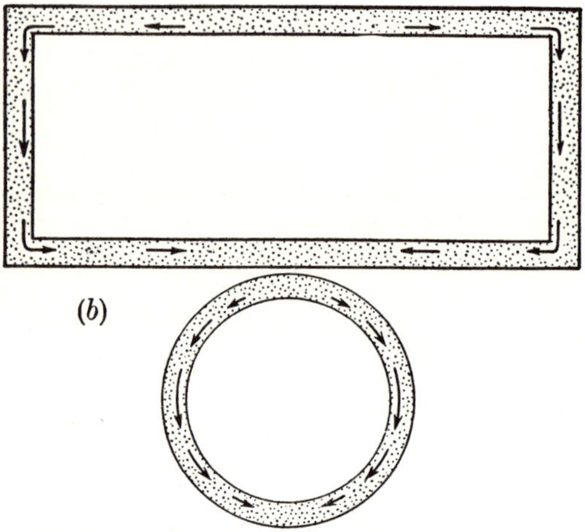

(b)

FIG. 20.2.—(a) Distribution of charge and electric field at the instant of maximum in the standing-wave $TE_{1,1,1}$ mode in a closed circular cylinder of inner radius b and length s. The distribution is the same for each cross section, but the density of the charge and the intensity of the electric field are largest at the center and diminish cosinusoidally to zero at the end walls; (b) distribution of surface current a quarter period after (a). There is a standing-wave pattern in the transverse and in the axial planes. Current is from top center to bottom center around side and end walls. The distributions are not rotationally symmetrical.

$$TE\lambda_{1,1,1} = \frac{2}{\sqrt{\frac{1}{s^2} + \frac{0.34}{b^2}}}$$

An additional resonant mode in a circular cylinder, the $TM_{0,1,0}$, has no nonresonant counterpart, Fig. 20.3. This mode is independent of the length of the cylinder. The $TM_{0,1,p}$ mode is like Fig. 20.1, the $TE_{1,1,p}$ mode like Fig. 20.2 but with p half wavelengths contained between the end surfaces. Similar resonant modes can be excited in hollow metal parallelepipeds of length s, longer side a, and shorter side b. The $TE_{1,0,1}$ mode has a resonant wavelength[1]

$$TE\lambda_{1,0,1} = \frac{2}{\sqrt{\left(\frac{1}{a^2} + \frac{1}{s^2}\right)}}$$

If a transmission circuit of the hollow or the coaxial type is closed at both ends by metal pistons, the condition for resonance

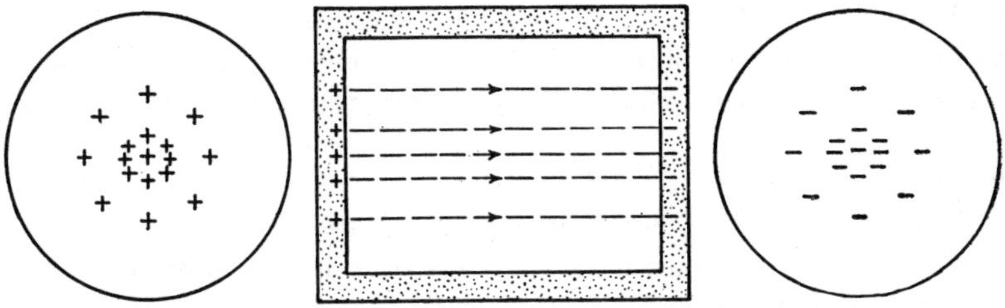

FIG. 20.3.—Distribution of charge and electric field at the instant of maximum in the standing-wave $TM_{0,1,0}$ mode in a closed circular cylinder of inner radius b. A quarter period later radial currents are on the inner surface of each end face and uniform currents parallel to the axis are on the cylindrical surface. The magnetic field is circular and the same in magnitude and direction in every cross section. $TM\lambda_{0,1,0} = 2.61b$

for any particular mode is

$$\beta_g s_p = p\pi \qquad p = 1, 2, \cdots \tag{20.1}$$

where β_g is the phase constant for the particular mode; and s_p is the pth resonant length of the circuit, or distance between the inner surfaces of the pistons for an integral value of p. The shortest

[1] General formulas for resonant wavelengths for closed cylinders of circular and rectangular cross section are given below. *Rectangular cylinder* of sides a, b, s in the x, y, z directions,

$$\lambda_{m,n,p} = \frac{1}{\sqrt{(m/a)^2 + (n/b)^2 + (p/s)^2}}$$

Circular cylinder of inner radius b and length s,

$$TM\lambda_{n\,m\,p} = \frac{2}{\sqrt{(p/s)^2 + (k_{n,m}/\pi b)^2}}; \qquad TE\lambda_{n,m,p} = \frac{2}{\sqrt{(p/s)^2 + (k_{n,m}'/\pi b)^2}}$$

where $k_{n,m}$ is the mth root of $J_n(x) = 0$, and $k_{n,m}'$ is the mth root of $J_n(x)' = 0$. A few values are

$k_{0,1} = 2.40$	$k_{0,2} = 5.52$	$k_{0,1}' = 3.83$	$k_{0,2}' = 7.02$
$k_{1,1} = 3.83$	$k_{1,2} = 7.02$	$k_{1,1}' = 1.84$	$k_{1,2}' = 5.33$
$k_{2,1} = 5.14$		$k_{2,1}' = 3.05$	

length for resonance is defined by $p = 1$. By measuring s_1 experimentally, β_g can be determined theoretically directly from (20.1) with $p = 1$. In practice, it is not possible to excite the resonant section and to determine resonance without inserting small driving and receiving antennas at appropriate points that depend upon the mode. Since these driving and detecting circuits, however small and loosely coupled they may be, inevitably affect the condition of resonance, this method is not satisfactory.

21. Generalized Condition for Resonance; Phase Factors of Terminations; Measurement of Wavelength.—In Sec. 16, it is pointed out that generalized circuit elements can be devised for transmission circuits that are not restricted to the near zone in the transverse plane. Elements corresponding in their effect on the circuit to variable inductances, capacitances, tuned circuits, and stub sections of conventional lines are described specifically for the dominant modes in pipes of circular and rectangular section. By analogy with reactive terminations for conventional lines, they are called reactive elements, since energy losses in heating them are negligible. When such a reactive element terminates a transmission circuit or wave guide, its principal effect is to change the condition for resonance. Thus, if a part of a transmission circuit extending from $z = 0$ to $z = s$ is terminated at $z = 0$ by one reactive element and at $z = s$ by another, the condition for resonance may be written in the form

$$\beta_g s_p + \Phi_0 + \Phi_s = (p + 1)\pi; \qquad p = 0, 1, 2, \cdots \tag{21.1}$$

where Φ_0 is a phase factor characteristic of the termination at $z = 0$ and Φ_s is a corresponding factor for the termination at $z = s$. These factors are defined by (21.1) subject to the arbitrary convention that the value of Φ for a metal piston is chosen to be $\pi/2$ if β_g is in radians per meter, or 90° if β_g is in degrees per meter. (Zero might be chosen equally well; for some purposes, a factor $\Phi' = \Phi - \pi/2$ is defined with $\Phi' = 0$ for a piston.) If both terminations are pistons, (21.1) reduces to (20.1). If driving and receiving antennas are arranged near the end $z = s$ of a transmission circuit, a phase factor Φ_s may be defined for the entire section containing them. By tuning the circuit to resonance by moving a piston ($\Phi_0 = \pi/2$) at $z = 0$ to locate two successive resonance peaks that occur when the line has the lengths s_p and s_{p+1}, from (21.1)

$$\beta_g(s_{p+1} - s_p) = \pi \tag{21.2}$$

This expression is true for all types of transmission circuits including open-wire lines using reactive terminations of any description. If s_{p+1} and s_p are both determined, preferably with $p = 1$, the phase constant β_g of the line may be computed directly from (21.2). The wavelength λ_g of a particular mode in a transmission circuit is *defined* by

$$\lambda_g = \frac{2\pi}{\beta_g} \tag{21.3}$$

Hence, from (21.2),

$$\tfrac{1}{2}\lambda_g = s_{p+1} - s_p \tag{21.4}$$

This is a standard formula for determining λ_g in a resonant transmission circuit from the experimentally determined distance $s_{p+1} - s_p$ between successive resonance peaks.

The phase factor Φ_0 for any termination at $z = 0$ may be determined experimentally by using a piston at $z = s$ and moving this until the circuit is tuned to resonance, preferably at $p = 1$, when

$$\Phi_0 = \frac{3\pi}{2} - \beta_g s_1 \tag{21.5}$$

If β_g has been calculated using Table 13.1 or determined experimentally, as explained in Sec. 20, and s_1 is measured directly, Φ_0 may be computed from (21.5). The same formula is correct for open-wire lines. (The value of Φ for a straight piece of wire of the same size as the line, when used to terminate a two-wire line, is given very nearly by $\Phi \doteq \pi/2 + \beta_g D/2$, where D is the separation of the wires.)

If the termination is not purely reactive so that energy is dissipated in heating it or a circuit coupled to it, a phase factor may still be defined and measured experimentally in the same way, provided that the circuit is not so nearly nonresonant that standing waves of appreciable amplitude are impossible. In order to obtain comparable values of Φ for slits, stub sections, etc., as reactive elements, it is usually desirable to connect a closed-end section of pipe exactly $\lambda_g/4$ in length beyond them so that the pipe is as completely closed as possible when measurements are made. The $\lambda_g/4$ section is the analogue of a high impedance or an insulating stub on conventional lines. Experimental curves of Φ for horizontal and vertical slits in a rectangular transmission circuit as determined with such a $\lambda_g/4$ section are shown in Fig. 21.1. If it is desired to determine Φ for a particular termination, such as an open end or a horn, it should be measured exactly as used. In all cases, Φ is determined with respect to a particular cross section such as the cross section at $z = 0$ or $z = s$. A value of Φ so determined may be referred to any other cross section, *e.g.*, at a distance s' from $z = s$, by writing $\Phi_s = \Phi_s + \beta_g s$.

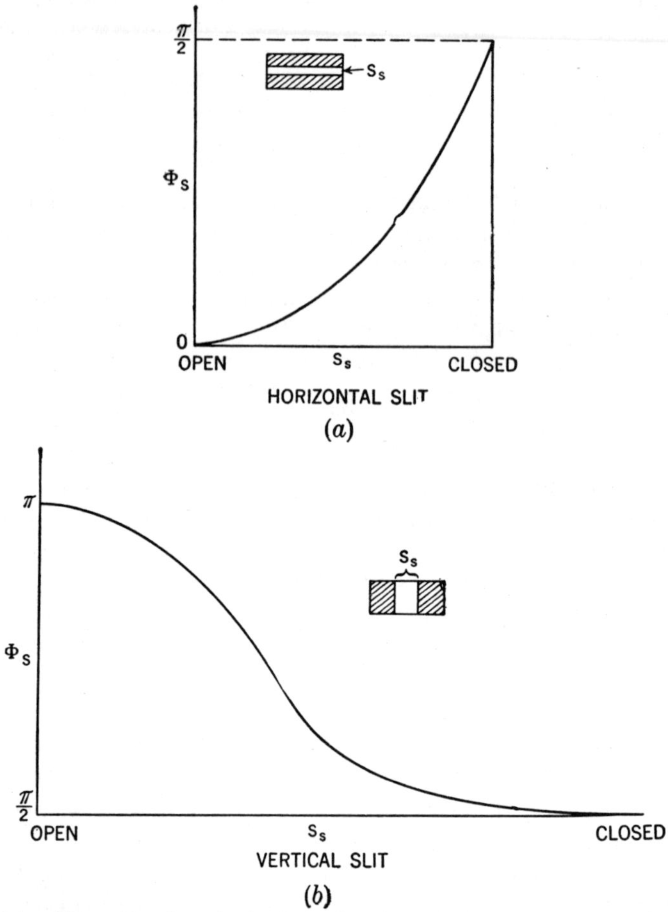

FIG. 21.1.—Phase functions for horizontal and vertical slits in a rectangular transmission circuit using the $TE_{1,0}$ mode.

22. Attenuation Factors of Terminations.—The amplitude of a standing-wave distribution in a transmission circuit terminated in a metal piston at each end and tuned exactly to resonance with the pistons separated a distance s_p is given by

$$\hat{K}_p = \frac{\hat{K}}{\sinh\,(\alpha_g s_p)} \tag{22.1}$$

This equation can be applied to any property related to the amplitude of the standing wave, such as charge density, current density, magnetic or electric field, where $\hat{K}_p$ stands for the resonant amplitude of the chosen quantity, α_g is the attenuation constant of the transmission circuit operated in a given mode (as defined in

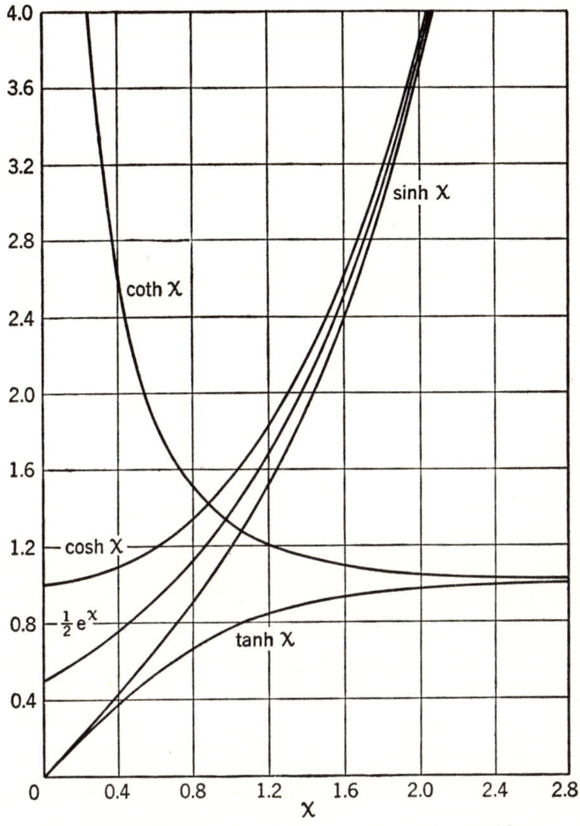

FIG. 22.1.—Hyperbolic functions of a real variable χ.

Table 13.1), $\hat{K}$ is a constant that depends upon the excitation. The length s_p is defined by (20.1). The quantity sinh $\alpha_g s_p$ may be obtained from the curve marked sinhχ, Fig. 22.1, by letting $\chi = \alpha_g s_p$. When $\alpha_g s_p$ is small, an approximation of (22.1) is

$$\hat{K}_p \doteq \frac{\hat{K}}{\alpha_g s_p} \tag{22.2}$$

If the two terminations are *nondissipative* and are characterized by phase functions Φ_0 and Φ_s, the same formulas apply with s_p obtained from (21.1).

If the terminations are *dissipative*, it is possible to define attenuation factors A_0 and A_s for them to take account of their share in the attenuation. The attenuation factors are so defined that the amplitude at resonance is

$$\hat{K}_p = \frac{\hat{K}}{\sinh\,(\alpha_g s_p + A_0 + A_s)} \tag{22.3}$$

When $\alpha_g s_p + A_0 + A_s$ is small,

$$\sinh\,(\alpha_g s_p + A_0 + A_s) \doteq \alpha_g s_p + A_0 + A_s.$$

For a circuit that is not nonresonant or nearly so,[1] the values $0.707\hat{K}_p$ occur at $s_p \pm \frac{1}{2}\,\Delta s_p$ where Δs_p is the width of the resonance curve that has its maximum at s_p. This width satisfies the equation

$$\frac{\beta_g\,\Delta s_p}{2} = \alpha_g s_p + A_0 + A_s \tag{22.4}$$

This relation is useful in the experimental determination of the attenuation constant α_g for a transmission circuit and the attenuation factors A_0 and A_s of the terminations. It is correct for conventional lines as well as for pipes. Since a termination at $z = s$ that makes a transmission circuit nonresonant is equivalent to an infinitely long line, it follows from (22.3) that A_s is infinite for such a termination.

The attenuation constant α_g of a transmission circuit may be determined experimentally by measuring the full widths Δs_p and Δs_{p+1} of two successive resonance peaks at 0.707 of maximum amplitude. The terminations of the line preferably should be metal pistons of which one is movable. A small receiving antenna must be inserted near the fixed piston and connected by a coaxial line to a sensitive and accurately calibrated detector. A small driven antenna may be arranged to project into the pipe at a convenient point. It must be so small that the generator is so loosely coupled to the circuit that no significant reaction on the generator is observed when the line is tuned. The locations s_p and s_{p+1} of

[1] Specifically if

$$\frac{\alpha_g^2}{\beta_g^2}\sinh^2\,(\alpha_g s_p + A_0 + A_s) << 1 \tag{22.3a}$$

and

$$(\tfrac{1}{2}\,\Delta s_p)^2 << s_p^2 \tag{22.3b}$$

the movable piston for two successive resonance adjustments, and the locations $s_p \pm \frac{1}{2}\,\Delta s_p$ and $s_{p+1} \pm \frac{1}{2}\,\Delta s_{p+1}$ for the 0.707 amplitudes in each case may be determined experimentally. By subtracting (22.4) from the same formula with $p + 1$ written for p,

$$\tfrac{1}{2}\beta_g(\Delta s_{p+1} - \Delta s_p) = \alpha_g(s_{p+1} - s_p) \tag{22.5}$$

With s_{p+1}, s_p, Δs_{p+1}, Δs_p, and β_g known, α_g may be calculated directly. It is to be noted that the attenuation due to the detector is subtracted out.

The attenuation factor A_s of a terminating device, in particular of the receiving antenna with the attached circuit and reflecting piston, may be computed directly from (22.4), using the values of s_p and Δs_p measured in determining α_g. A_0 for a metal piston is negligibly small, so that

$$A_s = \tfrac{1}{2}\beta_g\Delta s_p - \alpha_g s_p \tag{22.6}$$

The attenuation factor of any other termination may be measured by using it in place of the movable piston. (If necessary for experimental reasons, the receiving antenna with its reflecting piston and the driven antenna may be made movable in a slot while the termination to be measured is fixed. If this is done, antennas and piston must be moved equally to maintain their relative positions.) If new values of s_p and Δs_p are determined for one resonance peak, the formula (22.4) may be solved directly for A_0, using previously determined values of β_g, α_g, and A_s.

23. Standing-wave Ratio.—The standing-wave ratio ρ is defined as the ratio $\hat{K}_{max}/\hat{K}_{min}$, where $\hat{K}$ is the amplitude of the current, or of the charge density, or of the electric or the magnetic field, along a transmission circuit. ρ is given by the general formula

$$\rho = \frac{\hat{K}_{max}}{\hat{K}_{min}} = \frac{\cosh\,(\alpha_g w_{max} + A_s)}{\sinh\,(\alpha_g w_{min} + A_s)} \tag{23.1}$$

where w_{max} is the distance from the termination at $z = s$ to the first cross section where $\hat{K} = \hat{K}_{max}$; w_{min} is the distance from the termination of the circuit to the first cross section where $\hat{K} = \hat{K}_{min}$; w_{min} differs from w_{max} by $\lambda_g/4$, and A_s is the attenuation factor of the termination. Since w_{max} and w_{min} always may be kept less than $\lambda_g/2$, it follows that for a transmission circuit with low attenuation terminated in a power-absorbing load, it can usually be assumed that

$$\alpha_g w_{min} << A_s; \qquad \alpha_g w_{max} << A_s \tag{23.2}$$

whence

$$\rho = \coth A_s \tag{23.3}$$

The quantities $\cosh(\alpha_g w_{\max} + A_s)$ and $\sinh(\alpha_g w_{\min} + A_s)$ in (23.1) and $\coth A_s$ in (23.3) may be read directly from the curves, Fig. 22.1. It follows from (23.3) that it is possible to determine A_s from an experimental determination of the standing-wave ratio instead of from the width of a resonance curve at 0.707 of maximum amplitude. If the resonance curve is blunt so that the condition $(\Delta s_p/2)^2 << s_p^2$ is not well satisfied and the ratio $\hat{K}_{\max}/\hat{K}_{\min}$ is small, a determination of A_s from the standing-wave ratio is to be preferred. On the other hand, if Δs_p easily satisfies this condition, the determination from the width of the resonance curve is usually much more accurate. This is due to the fact that in determining the standing-wave ratio experimentally, it is necessary to move a receiving antenna, a coupling loop, or a probe along the line. Since the detuning effect of the receiving antenna and the power absorption in its load vary with the position of the antenna along the standing-wave pattern, it is difficult to determine accurately the standing-wave ratio due to the effect of the termination alone. It is no less difficult to obtain an accurate measure of the distorting effect on the standing-wave ratio due to the detector, as its antenna or coupling loop is moved relative to the standing-wave pattern from a point of maximum effect to one of minimum effect, or vice versa. In the resonance-curve method, on the other hand, it is possible to take accurate account of the attenuation due to the detector.

It is clear from (23.3) that a standing-wave ratio of unity corresponds to $A_s = \infty$, the value for a termination that makes a line nonresonant.

24. *Q* of a Resonant Transmission Circuit.—The condition for resonance in a transmission circuit expressed by (21.1) may be disturbed by varying either the length s or the frequency in $\beta_g = 2\pi f/v_{pg}$. If the change in frequency required to reduce the resonant amplitude $\hat{K}_p$ to 0.707 of its maximum value is sufficiently small so that Φ_0, Φ_s, α_g (which are, in general, functions of frequency) change by a negligible amount, the ratio of the resonant frequency to the band width Δf is the same as the ratio of resonant length to the width of the resonance curve. Hence, using (22.4),

$$Q = \frac{f_r}{\Delta f} = \frac{s_p}{\Delta s_p} = \frac{\beta_g s_p}{2(\alpha_g s_p + A_0 + A_s)} \tag{24.1}$$

If the attenuation functions of the terminations are zero, (24.1), reduces to the simple form

$$Q = \frac{\beta_g}{2\alpha_g} \tag{24.2}$$

for the Q of a resonant section of transmission circuit with dissipationless terminations. These formulas apply to conventional open-wire lines as well as to pipes.

25. Efficiency of Transmission.—The ratio of the power P_s transferred to a terminating load characterized by an attenuation function A_s to the power P_0 supplied to a transmission circuit of length s and attenuation constant α_g is

$$\frac{P_s}{P_0} = \frac{\sinh 2A_s}{\sinh 2(\alpha_g s + A_s)} \tag{25.1}$$

This ratio is maximized when A_s is infinite and the load matched, when

$$\frac{P_s}{P_0} = e^{-2\alpha_g s} \tag{25.2}$$

The decibel loss is given by

$$\text{Loss (db)} = 10 \log_{10} \frac{P_0}{P_s} = 10 \log_{10} \left(\frac{\sinh 2(\alpha_g s + A_s)}{\sinh 2A_s} \right) \tag{25.3}$$

The formulas apply to conventional lines and to pipes.

26. Definition of Generalized Impedance.—Impedance is defined for conventional transmission lines by the ratio of potential difference between two closely spaced terminals to the current entering and leaving these terminals. When impedance is determined be measurements on a transmission line, calculations are based on experimental determinations of the locations and widths of resonance curves or on the standing-wave ratio. Thus the phase factor Φ_s and the attenuation factor A_s for an impedance Z_s must be determined explicitly or implicitly. The input impedance

$$Z_{\text{in}} = R_{\text{in}} + jX_{\text{in}}$$

of a section of conventional transmission line of characteristic resistance R_c and length s, terminated in an impedance Z_s characterized by the phase and attenuation factors Φ_s and A_s, is given in convenient ratio form as $z_{1\text{in}} = Z_{\text{in}}/R_c = r_{1\text{in}} + jx_{1\text{in}}$, where

$$r_{1\text{in}} = \frac{\sinh 2(\alpha s + A_s)}{\cosh 2(\alpha s + A_s) - \cos 2(\beta s + \Phi_s)} \tag{26.1}$$

$$x_{1\text{in}} = \frac{-\sin 2(\beta s + \Phi_s)}{\cosh 2(\alpha s + A_0) - \cos 2(\beta s + \Phi_s)} \tag{26.2}$$

The terminal impedance $Z_s = R_s + jX_s$ is obtained in ratio form by writing $s = 0$ in (26.1) and (26.2).

$$r_{1s} = \frac{\sinh 2A_s}{\cosh 2A_s - \cos 2\Phi_s} \tag{26.3}$$

$$x_{1s} = \frac{-\sin 2\Phi_s}{\cosh 2A_s - \cos 2\Phi_s} \tag{26.4}$$

The relations (26.3) and (26.4) may be looked upon as generalized definitions of the resistance and reactance ratios, $r_{1s} = R_s/R_c$, $x_{1s} = X_s/R_c$, of a termination for a transmission circuit of any type on which A_s and Φ_s can be measured.

The formulas (26.1) to (26.4) are true if $g_1 = R_cG$ replaces r_1, $b_1 = R_cB$ replaces x_1, and $\Phi_s' = \Phi_s - \pi/2$ replaces Φ_s. G and B are defined by

$$Y = G + jB = \frac{1}{Z} \tag{26.5}$$

r_1 and x_1 (or g_1 and $+b_1$) may be determined directly from the circle diagram of Fig. 26.1 on which are drawn circles of constant $(\alpha s + A)$ in nepers, and circles (scaled in degrees from 0 to 180°) of constant $(\beta s + \Phi)$ for use with r_1 and x_1, or of constant $(\beta s + \Phi')$ for use with g_1 and $+b_1$.

The characteristic resistance R_c in the ratios is defined in the conventional way for two-wire and coaxial lines with near-zone cross sections. In the general analysis for unrestricted cross sections, generalized potential functions and stream functions with no direct physical significance appear instead of voltage difference and total current. The formula for R_c for these functions is simple. For all TM modes in pipes of any shape,

$$R_{cTM} = \mathcal{R}\sqrt{1 - \frac{\lambda_{TEM}^2}{\lambda_{\text{cutoff}}^2}} \tag{26.6}$$

For all TE modes,

$$R_{cTE} = \frac{\mathcal{R}}{\sqrt{1 - \frac{\lambda_{TEM}^2}{\lambda_{\text{cutoff}}^2}}} \tag{26.7}$$

where $\mathcal{R} = 376.7\sqrt{\mu_r/\epsilon_r}$ ohms. These formulas may be used for

R_c if resistances and reactances are desired rather than ratio factors *so long as exclusively a single transmission mode is involved in a transmission circuit of given cross section.* If two sections of transmission circuits with different cross sections or excited in different

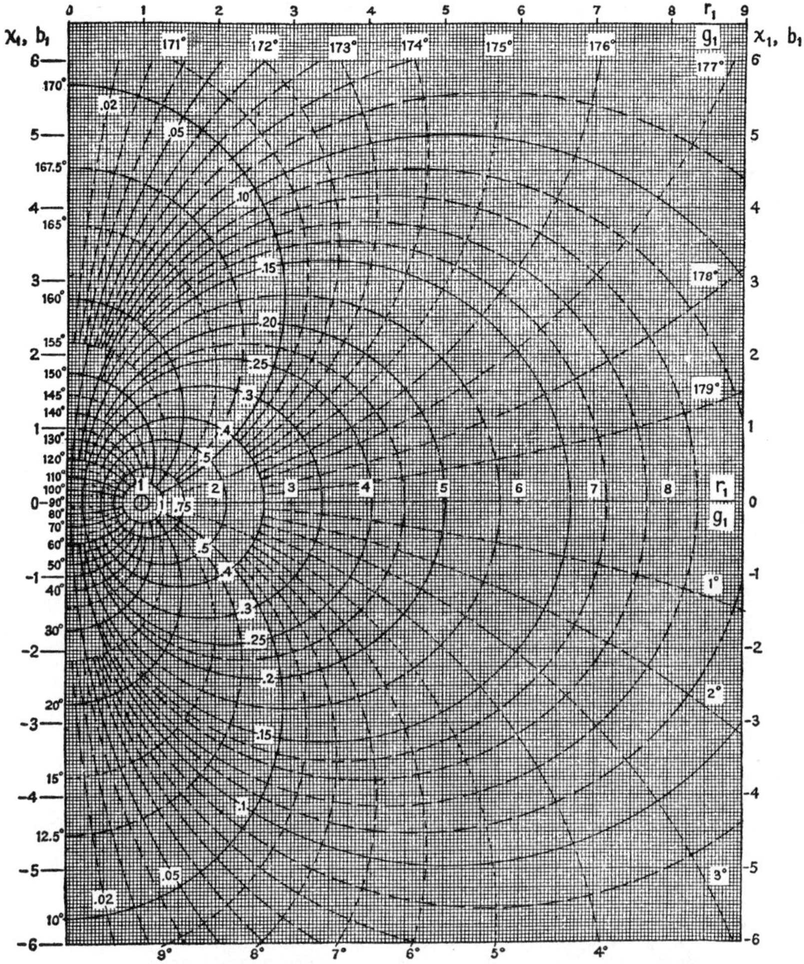

FIG. 26.1.—Circle dragram showing circles of constant attennation (circles marked in nepers from 0.02 to 1) and constant phase (circle marked indegrees). With r, x, as coordinate, the circles of constant phase give $(\beta s + \Phi)$ with g_1 and b_1 as coordenates, the circles of constant phase give $(\beta s + \Phi')$.

modes are involved, the different values of R_{cTM} or R_{cTE} may not be treated like different values of R_c for conventional lines, *i.e.*, if a nonresonant transmission circuit with a given R_{cTM} or R_{cTE} is connected to a transmission circuit with the same numerical value of R_{cTM} or R_{cTE} but for a different mode or shape of cross section, the second circuit is not therefore also nonresonant.

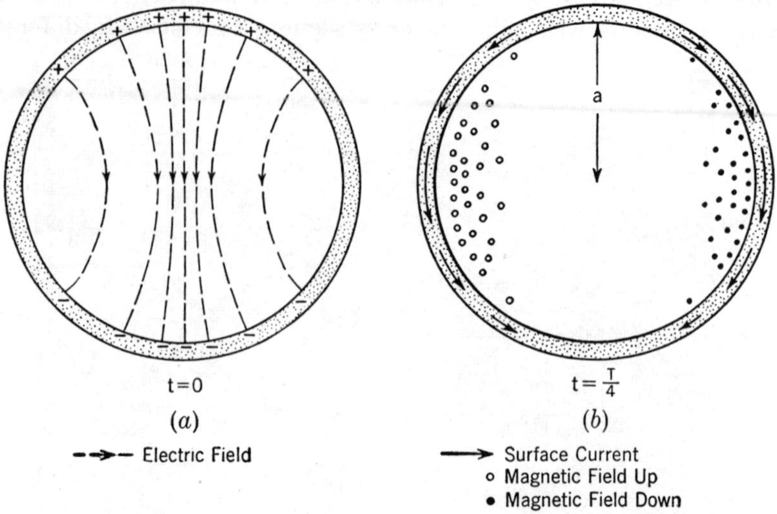

FIG. 27.1.—Dominant *TM* mode in a sphere; (*a*) and (*b*) show the same meridian plane. $\lambda_r = 2.28a$.

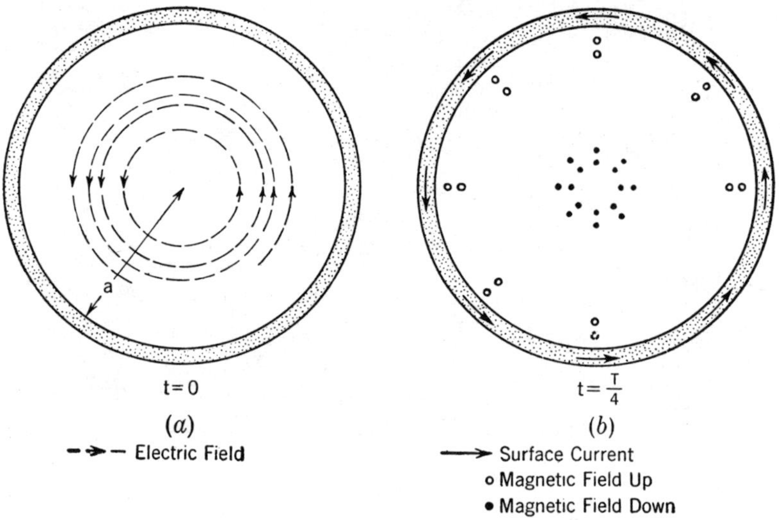

FIG. 27.2.—Longest wavelength *TE* mode in a sphere, $\lambda_r = 1.40a$.

Although it is possible to generalize the impedance concept formally to define input impedances and terminal impedances for pipes, the functions so obtained are not so generally useful as ordinary impedances in conventional lines. Actually a knowledge of the attenuation and phase functions A and Φ for a section of a circuit or a termination is often all that is required for solving problems in transmission circuits of all types. These functions can be deter-

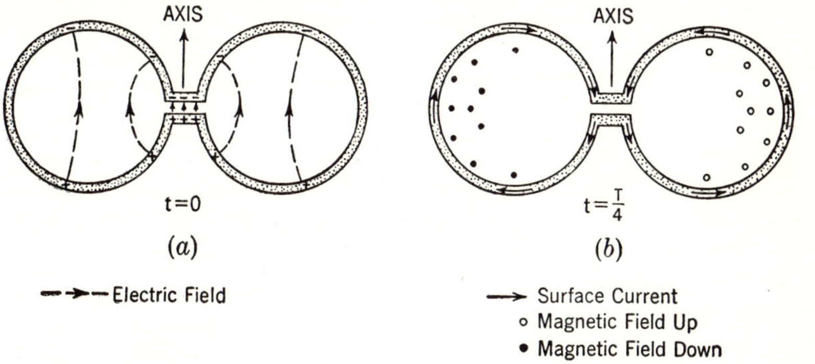

FIG. 27.3.—Dominant *TM* mode in a toroidal resonator. The resonator is generated by rotating each of the above figures about the axis shown.

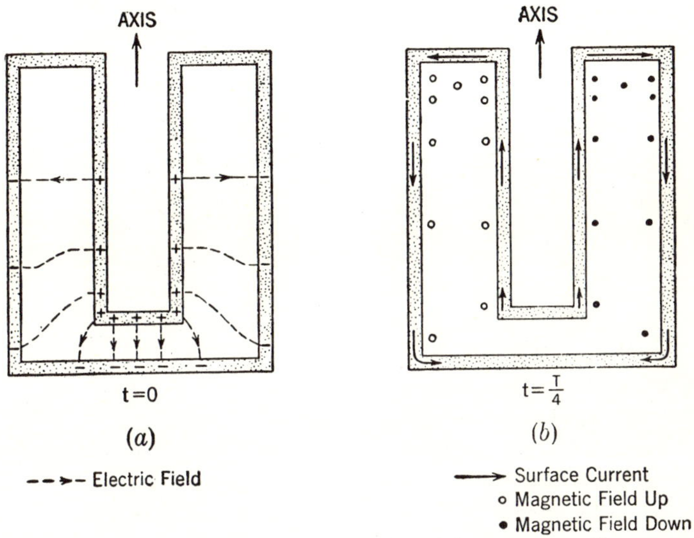

FIG. 27.4.—Dominant *TM* mode in a quasi-coaxial resonator. The resonator is generated by rotating each of the above figures about the axis shown.

mined from simple formulas expressed in terms of experimentally measurable magnitudes.

27. Cavity Resonators.—The essential property of resonant sections of ordinary or of coaxial pipes is that they are enclosed in metal walls and excited by a source within. Every closed, highly conducting shell can be excited in an infinite sequence of resonant modes, each characterized by a particular standing-wave distribution of surface current. The frequencies at which resonance occurs depend upon the shape and size of the enclosed cavity. The resonant mode excited by the lowest possible frequency is called the fundamental mode or the dominant mode. Depending upon the method of excitation, a mode of the *TM* or *TE* type exists. As indicated by the name, the *TM* type is characterized by surface currents that are so distributed as to set up a magnetic field that has components in only one plane, called the transverse plane. The surface currents have directions that are everywhere perpendicular to this plane. The magnitude of the surface current (l amp/m) at any point is equal to the magnitude of the magnetic field (H amp/m) tangent to the metal surface at that point. The *TE* type is characterized by standing-wave distributions of oscillating charge which set up an electric field that has components in only one plane, called the transverse plane. The surface charge is so distributed that the resulting electric field is perpendicular to the metal walls, with the lines of the electric field ending on negative charge. The magnitude of the electric field (E volts/m) at the surface multiplied by the universal electric constant (ϵ_0 farads/m) is equal to the magnitude of the surface density of charge (η coulombs/m^2). Distributions of surface current and of charge and of the magnetic and electric fields in the *TM* and *TE* modes of longest wavelength in the sphere are shown in Figs. 27.1 and 27.2. The ratio of wavelength to radius is given in each case. Corresponding diagrams for the *TM* modes of longest wavelength in resonators of other shapes which are used, for example, in the Klystron oscillator are given in Figs. 27.3 and 27.4.

Cavity resonators are characterized by extremely large values of Q, where Q is defined for each resonant mode by $Q = f_r/\Delta f$. Δf is the band width between the frequencies $f_r \pm \Delta f/2$, for which the amplitude of the surface current, the surface charge, the electric field, and the magnetic field is reduced to 0.707 of its value at the resonant frequency f_r. In general, Q becomes greater as the conductivity of the metal is increased, as the ratio of volume to enclosing metal surface is increased, and as the wavelength is increased. Largest possible values are in silvered spheres. For nonmagnetic metals filled with air, the Q of a sphere for the dominant *TM* mode is given by

$$Q = 0.725 \frac{a}{\delta} = 0.318 \frac{\lambda_r}{\delta} \qquad (27.1)$$

where a is the radius in meters, δ is the skin depth defined by $\delta = \sqrt{2\rho/\omega\mu_0}$ with δ in meters and ρ in ohms/meter. For a copper sphere

$$Q = 8.4 \cdot 10^4 \sqrt{\lambda_r} \tag{27.2}$$

A copper sphere of radius $a = 7$ cm is resonant in the dominant mode with $\lambda_r = 16$ cm. Q is 33,600. If the sphere is silvered inside, Q is 34,600.

A spherical shell of metal has resonant modes when excited from a source within, as just described. If the shell has no holes, the distribution of currents and charges on the inner surface is always such as to provide complete cancellation of all forces at outside points. No energy is radiated. The same spherical shell can be excited from a source outside the shell, as, for example, when it is immersed in the far-zone field of a driven antenna. In this case, the sphere also has an infinite sequence of resonant modes, but the surface currents are on the outside, and the sphere when excited in *TM* modes is like a thick radiating antenna. It is not a cavity resonator. The first resonant wavelength of the outside surface of the sphere is $\lambda_r \doteq 7.3a$ (as compared with $\lambda_r = 2.28a$ for the inside). Owing to large radiation, Q is very low.

Supplementary Reading

G. C. SOUTHWORTH: Hyper-frequency Wave Guides—General Considerations and Experimental Results, *Bell System Tech. J.*, Vol. 15, pp. 284–309, April, 1936.

W. L. BARROW: Transmission of Electromagnetic Waves in Hollow Tubes of Metal, *Proc. I.R.E.*, Vol. 24, pp. 1298–1328, October, 1936.

J. G. BRAINERD, G. KOEHLER, H. J. REICH, and L. F. WOODRUFF: "Ultra-high-frequency Techniques," D. Van Nostrand Company, Inc., 1942.

L. A. WARE and H. R. REED: "Communication Circuits," John Wiley & Sons, Inc., 1942.

S. A. SCHELKUNOFF: "Electromagnetic Waves," D. Van Nostrand Company, Inc., 1943.

F. E. TERMAN: "Radio Engineers' Handbook," McGraw-Hill Book Company, Inc., 1943.

R. KING: Transmission Line Theory and Its Application, *J. Applied Phys.*, Vol. 14, pp. 577–600, November, 1943.

D. MIDDLETON and R. KING: Transmission Line Theory Applied to Wave Guides and Cavity Resonators, *J. Applied Phys.*, Vol. 15, pp. 524–535, July, 1944.

R. W. P. KING: "Transmission-line Theory," McGraw-Hill Book Company, Inc., 1955; Dover Publications reprint, 1964.

CHAPTER IV

WAVE PROPAGATION

1. Introduction.—In ordinary experience, and especially in the science of mechanics, we have become accustomed to the idea of "action at a distance." Theoretically, the gravitational force acting upon a stone or upon a planet depends on the position of every other particle of matter in the universe. According to elementary ideas about gravitation, it is the *present* position of the earth that determines the force now pulling a baseball downward, not the position of the earth a fraction of a microsecond ago. At least no one has yet measured any actual time lag between gravitational cause and effect.

When dealing with electrical charges, one finds that disturbance of an electric charge of either sign produces a force upon distant electrons and protons. Again we are dealing with "action at a distance," but this time there is a definite measurable delay. In empty space, the time lag is directly proportional to the intervening distance and is the same for all types of disturbances. Many of these "disturbances" arise spontaneously within a single molecule or atom. Extra energy due to a collision or other cause is retained for a brief period and then suddenly released, producing a vibratory force at the distant electron or proton, the frequency of vibration depending directly upon the energy released. Radiant heat, visible light, ultraviolet light, X rays, gamma radiation,—all these represent the cumulative effect of the spontaneous release of energy stored in many different molecules or atoms, each release apparently being an individual event related only statistically to all the rest. These natural "radiations" differ only with respect to the location of the disturbance within the molecule or atom, and the consequent magnitude of the energy change involved.

In contrast to these spontaneous, random, and disassociated electronic disturbances, a radio wave originates in a systematic mass movement of many millions of electrons, all guided in the same general path and executing related motions. By the cumulative effect of such related disturbances one may hope to produce a planned force, of practical engineering magnitude, at a great distance. The simplest periodic mass movement that can be guided and controlled is the surge of electrons, to and fro, on a wire or other conductor. Best results are obtained by allowing the "in-phase motion" to extend over as much of the wire as possible, and to include as many electrons as possible (taking care that the planned disturbance is not offset unintentionally by any contrary mass movement of electrons in adjacent portions of the apparatus).

Accordingly, the high-frequency limit of the so-called "radio-frequency spectrum" is determined by our present ability to originate and control such systematic mass movements of hordes of electrons, all performing related movements simultaneously. This limit depends upon the practical engineering tools at hand. It has increased notably within the last decade and is likely to rise to still higher frequencies with further improvement in vacuum-tube technique.

The low-frequency limit is set by the enormous cost of building metal structures large enough and high enough to make effective use of energy liberated by rather sluggish oscillations. This limit occurs at about 10,000 cps, has not changed appreciably during the last 30 years, and shows no prospect of changing in the predictable future.

Within these two present limits, all intervening frequencies have been neatly classified into decades—each decade labeled with descriptive name and symbol, so that such loose terms as "microwaves" and "quasi-optical frequencies" may be discarded eventually. The recent recommendations of the Federal Communications Commission in this regard are presented in Table 1.1.

TABLE 1.1.—FREQUENCY CHART

Frequencies, kilocycles per second	Wavelengths, meters	Designations and abbreviations	
10 to 30 inclusive	30,000 to 10,000	Very low	vlf
Above 30 to 300 inclusive	10,000 to 1,000	Low	lf
Above 300 to 3,000 inclusive	1,000 to 100	Medium	mf
Above 3,000 to 30,000 inclusive	100 to 10	High	hf
Above 30,000 to 300,000 inclusive	10 to 1	Very high	vhf
Above 300,000 to 3,000,000 inclusive	1 to 0.1	Ultra-high	uhf
Above 3,000,000 to 30,000,000 inclusive	0.1 to 0.01	Super-high	shf

Historically, the idea of delayed action at a distance naturally suggested some form of wave propagation, made familiar by waves in water and by compressional waves of sound in material media. Early speculations were directed toward the formulation of an elaborate set of specifications for an invisible, imponderable medium called the "luminiferous aether" in which a transverse quasi-mechanical wave could be transmitted. This concept or model was never particularly useful, because the requisite mechanical properties of the medium were very difficult to accept. Later, the "aether" became a medium in which electric and magnetic fields were propagated. Gradually the emphasis was shifted from the medium to the fields themselves, until the "ether" has become synonymous with vacuum—an empty space through which electric and magnetic fields are propagated at the velocity of light—light being merely a special case of electromagnetic radiation. The resulting force upon the distant electron is computed correctly by means of Maxwell's equations, which involve the electric and magnetic fields. Such fields are essentially a mathematical concept, very useful in bridging the gap from transmitter to receiver, their presence always inferred from the observed behavior of currents in the two antennas. If one prefers, these traveling fields may be regarded as a mathematical artifice, particularly successful in predicting correctly the result of experiments and the service to be obtained from a given engineering construction.

Each band of frequencies, classified decimally in Table 1.1, is accompanied by a statement of corresponding "wavelength," the wavelength being the distance traveled during the period of one oscillation by a wave propagated in empty space at the speed of light, 300 m/microsecond. Thus a 1,000,000-cycle alternating current radiates a free-space wave 300 m long; a 3,000-megacycle current corresponds to a wavelength of 10 cm; etc.

2. General Discussion—Wave Propagation vs. Frequency.—In addition to the decimal classification referred to, it is also useful to classify radio frequencies according to their *behavior* in the neighborhood of the earth. In fact, the earth is almost always the main obstacle interfering with the desired transmission of intelligence. Radio waves surmount this obstacle in a variety of ways, depending upon the orientation and placement of the antenna, and especially upon the frequency or related wavelength. Figure 2.1[1] offers a convenient starting point for this type of classification. The main curve *ABEF* (omitting the side loop *BCDE*) and its ocean ground-wave variant represent roughly what was known about radio

[1] From *Bur. Standards Letter Circ.* 615.

transmission prior to about 1920. On this plot, the very low frequencies look particularly attractive. Note, however, that the comparison is made on the basis of 1 kilowatt *actually radiated,* not 1 kilowatt developed at the transmitter. To build a quarter-wave vertical antenna for a frequency of 10,000 cps, a metal structure $4\frac{1}{2}$ miles high would be required. Such construction being out of

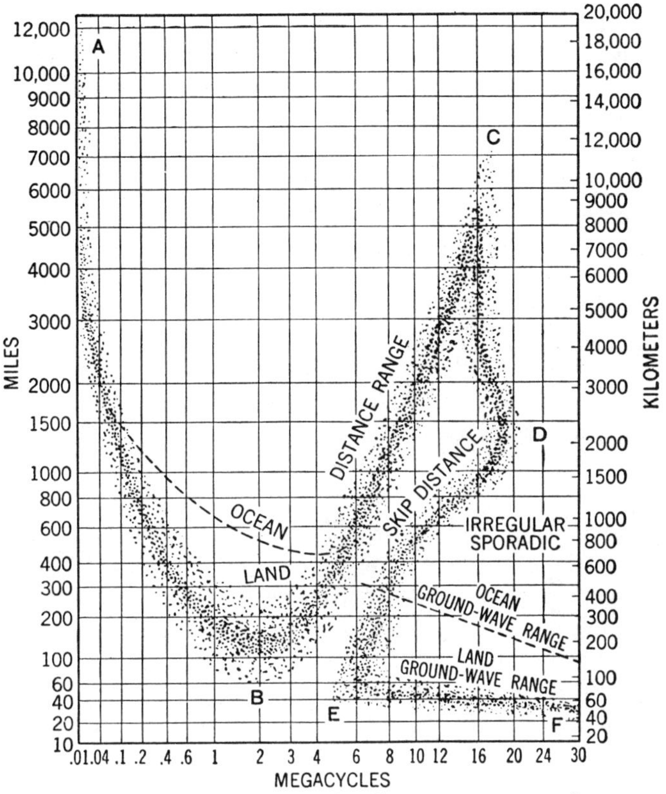

FIG. 2.1.—Distance ranges of radio waves—summer day.

the question, the antenna is built as high as economic considerations will permit, with the result that the effective height is still a minute fraction of a wavelength. The antenna is therefore very inefficient, most of the power input going into copper losses and ground losses. This is a "brute-force" method for starting the wave on its way. Furthermore, the very low frequency severely limits the total number of frequency assignments and limits each station to a narrow modulation band (say low-speed telegraphic transmission).

The very low frequencies are used by a small number of high-power shore stations, using enormous antennas, most of them built many years ago. However, these are not mere relics of a bygone age. Instead, they are valued as stand-by apparatus capable of pushing through a signal during magnetic storms when transoceanic cables and all high-frequency radio channels may be out of service. To explain this, it is necessary to consider the nature of the old-fashioned ground wave, represented by the curve *ABEF* of Fig. 2.1.

Figure 2.2 represents a comparable problem in optics. A parallel beam of visible light, projected from a searchlight, illuminates a 6-in. opaque sphere. Geometric optics, based on ray construction, would suggest the presence of a perfectly black cylindrical shadow behind the sphere, the edges of the shadow being sharply defined, with full light intensity just beyond the

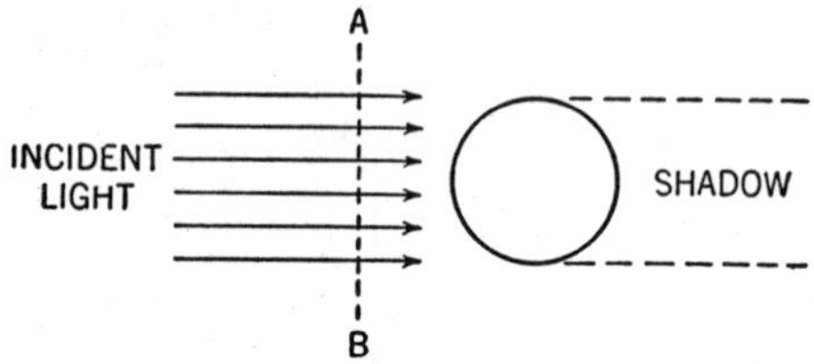

FIG. 2.2.—Experiment on diffraction of light.

boundary of the shadow. However, careful measurements made within the shadow, and especially in the boundary zone, reveal that geometric optics is merely an approximation. For a more accurate solution, one must undertake a wave study, considering each point on a wave front, such as *AB*, to be a source of wavelets spreading from these point sources. Behind the sphere, these wavelets almost cancel, but the cancellation is not complete. Some light has penetrated the "shadow" by bending or "diffraction" in much the same way that a sound wave travels around a barn. Just outside the shadow, alternate reinforcement and partial cancellation are experienced. Within the shadow, the light falls off rapidly but does not drop discontinuously. By going to lower frequencies (infrared rays or radiant heat rays), the amount of useful bending increases. On going all the way to the very low radio frequencies, bending around a sphere of the radius of the earth has become a practicable engineering expedient, not dependent upon any assistance from the upper or lower atmosphere. However, on account of the very nature of this diffraction process, the wave that reaches the receiver has been in the immediate neighborhood of the semiconducting earth over its entire path and is attenuated considerably below the level theoretically attainable by means of free-space transmission.

From the "very low" frequencies to the "low" frequencies, this attenuation increases, and the bending also becomes less effective. The ground-wave curve of Fig. 2.1 falls off steeply. In partial compensation, however, it becomes possible to build a somewhat more efficient antenna structure and even to produce some very inefficient radiation from the small antennas that can be mounted on a ship. The low frequencies have been used for reliable medium-distance communication, but their most valuable functions are aircraft guidance and ground-wave beacon service in general. Reasonable certainty that no significant sky wave is present greatly facilitates direction finding. Sky waves produce violent instrumental errors in the single-loop direction finder, necessitating its replacement by a more elaborate spaced-antenna array. Even when instrumental errors have been properly eliminated, sky waves often show an *actual* departure from the great-circle path of 90° or more. Over land, the somewhat limited range of a low-frequency beacon is not necessarily a disadvantage, since additional transmitters may be installed where necessary.

In the "medium" frequency range, used by local broadcasting stations in the United States, the dependable service area is determined by ground-wave transmission. Depending upon the power of the transmitter, the directivity of the antenna, and the local terrain, the ground-wave range is of the order of 100 or 200 miles. Toward the upper limit of this broad frequency band, the dependable coverage decreases until it barely extends over a large metropolitan district. Considering all waves below 200 m as obviously of little value, the government at one time assigned to the American amateurs all radio frequencies above 1,500 kcps. In the first few years following 1920, surprising reports were received with increasing regularity, reports of amateur low-power contacts established at ranges far removed from the ground-wave curve of Fig. 2.1. Rather vague ideas regarding sky-wave transmission had existed for many years. In 1901, both Kennelly and Heaviside had applied to the reflection of Marconi's transoceanic waves certain preexisting theories regarding conduction in the outer atmosphere. Nighttime "fading" of medium-frequency waves was later properly ascribed to sky-wave interference. However, the true explanation of the sky-wave area, Fig. 2.1, was not possible until high-frequency transmitters and sensitive short-wave receivers became commonly available in the hands of thousands of amateurs. The location and detailed shape of this area change with hour and season and with the sunspot cycle. By night, the area shifts to the left and includes the "broadcast" waves, making long-distance medium-frequency transmission possible after sunset. By choosing an operating frequency appropriate to the occasion and the route, long-distance transmission can be made extraordinarily efficient. At sufficient distances, the choice becomes rather critical, too low a frequency resulting in absorption of the wave, while too high a frequency causes the wave to penetrate the atmosphere and to be transmitted uselessly into interstellar space. These limitations determine, respectively, the left and right boundaries of the sky-wave area. On unfavorable occasions, these boundaries are much nearer together than is indicated for the average case in Fig. 2.1. In general, sky-wave transmission by day and by night are about equally good, but the choice of appropriate frequency is very different.

Before continuing with the description of sky waves from the outer atmosphere, the discussion of ground waves will be continued and extended. Even under favorable circumstances, dependable sky-wave transmission seldom extends beyond 30 mcps. Figure 2.1 indicates that the diffracted ground wave, bent around the curved earth, is also of very little value at 30 mcps and beyond; yet, in Table 1.1, a vast range of higher frequencies is listed as a useful part of the radio-frequency spectrum, and vigorous attempts are being made to push this range still higher. This apparent contradiction may be explained as follows.

The sky wave derives its value entirely from the fact that it provides an efficient detour around a great obstacle, *viz.*, the earth. Attenuation in ordinary air is slight. In its zigzag course, the wave loses some energy at each reflection from the outer atmosphere and from the ground, but accumulated losses are small in comparison with the attenuation suffered by a ground wave of similar frequency. Hence the energy obtainable at the receiver may rise toward the value theoretically available in free space at a comparable distance. The resulting compromise represents a vast improvement over the ground-wave level.

There is another way, however, in which the earth may be avoided. Between two points several hundred miles apart, a line of sight may sometimes be established, while at shorter distances this naturally becomes somewhat easier. Elevation of any sort, due to hilltops, towers, tall buildings, or aircraft, makes possible the type of transmission indicated by ray *A*, of Fig. 2.3. At first glance, this would seem to represent an ideal solution, capable of giving a signal approximating the free-space value, regardless of frequency and regardless of elevation above the horizon. This first impression is deceptive. Aside from avoidance of poor dielectrics, such as trees and houses in immediate proximity to the antenna, one secures little actual improvement by locating a medium-frequency broadcasting station on a hilltop. At medium frequencies and even at high frequencies, the "ground-reflected ray" *B* interferes destructively with the "direct ray" *A*. This

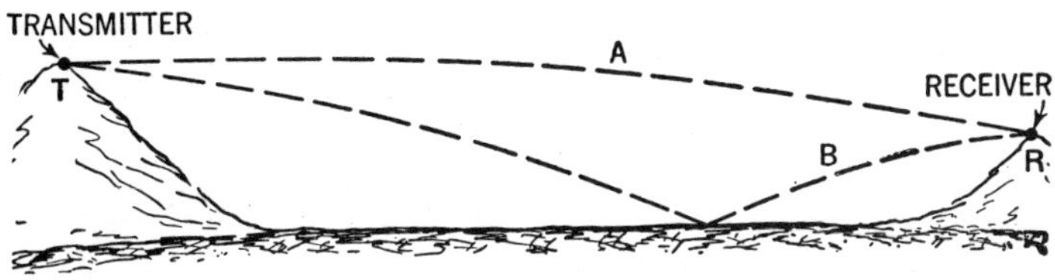

FIG. 2.3.—Direct ray and ground-reflected ray.

does not cancel completely the ground-wave signal, but it *does* cancel the major *gain* that otherwise might be expected when direct ray *A* becomes geometrically possible. The signal is little better than that obtainable over a smooth curved earth. The trouble arises from the fact that ray *B* suffers a change of phase, of nearly 180°, on reflection from the ground. The extra distance covered by ray *B* contributes a *further* shift of phase. Measured as a fraction of a relatively long *wavelength*, however, this path difference soon becomes insignificant as the distance from the transmitter increases.

Not until the "very high frequencies" are reached do attainable elevations produce much practical improvement at distances where the signal would be otherwise unserviceable. Even here the interference remains bad at low angles near the horizon. The use of "ultra-high frequencies" begins to provide a solution for this transmission problem. Because of the shorter wavelength, one can work closer and closer to the horizon before the path difference becomes negligible. Also because of the shorter wavelength, one can build reflectors of reasonable size, so sharply directive that the ground-reflected wave is not experienced appreciably until the horizon is approached. This eliminates the succession of maxima and minima otherwise to be expected in the region well above the horizon, where alternate reinforcement and cancellation would take place. Both these advantages will be greatly extended by the use of "super-high frequencies." Somewhere in this super-high-frequency range one must begin to expect excess attenuation from the air itself, from water vapor sometimes present, and from water droplets in the air. The frequencies concerned begin to approach some of the natural frequencies emitted by molecular disturbances. A corresponding absorption may be inferred. Whether such effects will terminate the practical radio spectrum or whether useful ranges of transparency exist are questions to be determined.

Very high frequencies and to a lesser extent ultra-high frequencies are sometimes bent a few degrees, or a fraction of a degree, below a distant horizon. This is particularly likely to occur under the circumstances indicated in Fig. 2.4. The lowest direct ray A

FIG. 2.4.—Diffraction of radio wave over an obstacle.

from transmitter T is somewhat too high for reception at R. However, the intermediate point D is strongly illuminated, and a part of the energy arriving at D is bent into the nominal shadow zone. The broken path TDR thus becomes available, though poor. Transmission from T to D and from D to R is efficient, but a considerable loss occurs at point D. Raising the frequency is likely to be harmful in the case of Fig. 2.4, for it makes the bending more difficult. Ground reflections occur on both sections of the broken route, further complicating the pattern. A broad flat hilltop at D would be undesirable, because the wave would be kept in close proximity to the earth over too great a distance, suffering a very rapid attenuation at such frequencies. Additional break points are admissible provided that the total change of direction does not exceed very few degrees.

A question of terminology arises. Are the direct ray and the ground-reflected ray to be included in the general designation "ground wave," or are they to be given a separate category, neither ground wave nor "sky wave"? The first alternative is officially recommended by the Institute of Radio Engineers and has achieved general acceptance. In effect, this defines a "ground wave" as "everything except a sky wave." In addition to the direct wave and ground-reflected wave, it includes diffracted waves of all sorts, whether bent over the gradual curvature of the earth (long waves) or bent over local obstacles. In practice, it also includes a vast amount of scattered radiation, reradiated and reflected from trees, buildings, overhead wires, underground pipes, and all similar obstacles near the ground-level path from sender to receiver.

3. Modification of Wave Propagation by the Lower Atmosphere. The lower atmosphere (troposphere) exerts two main influences on radio transmission, both of them important at wavelengths below 10 m. The first effect has to do with the sheer weight and compressibility of the atmosphere. Because of the average density gradient of the air, even the so-called "direct wave" *A* of Fig. 2.3 has a slight curvature, concave downward. One may visualize this in elementary terms by considering that the bottom of the

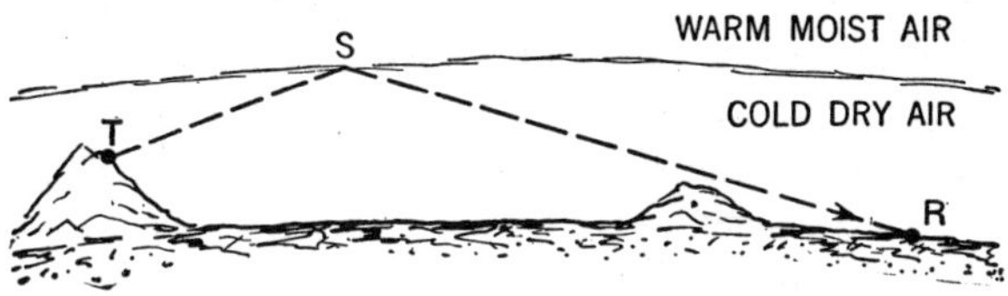

FIG. 3.1.—Reflection of radio wave from air-mass boundary.

corresponding wave front necessarily travels in a denser medium and at a slightly lower speed than the top. This curvature is beneficial. It extends the practical horizon slightly beyond that determined by a purely geometrical construction made with straight lines. The improvement of signal strength is relatively constant and at all times small. The wave is still counted as a part of the "ground wave."

The second effect varies from day to day and even from hour to hour. Occasionally a layer of warm, moist, tropical air flows over a layer of cold, dry, polar air. The resultant air-mass boundary is approximately level and may be sharply defined. Even at vertical incidence, it reflects a detectable radio echo, though the reflection coefficient may be as low as 0.001 per cent. Such direct-echo experiments provide the meteorologist with a new tool for following continuously the growth and alteration of air-mass boundaries of all sorts. Reflections from clouds, especially storm clouds, are not unusual. Air-mass boundaries commonly occur at heights between 0.5 and 3.0 km. Though the vertical-incidence reflection is so extremely small, the grazing-incidence reflection may occasionally be surprisingly good. Hence, if one modifies

Fig. 2.4 by introducing an air-mass boundary, the condition represented in Fig. 3.1 is obtained. Several times in the course of a winter season, for example, the signal at R, received by route TSR, may approach the free-space value, an enormous improvement over the usual signal represented in Fig. 2.4. Less spectacular improvements occur with greater regularity. Being undependable, the effect is rather undesirable, though it explains many of the freak ranges sometimes obtained in communication and all other applications of "microwaves."

4. Sky Waves from the Ionosphere—Basic Theory.—The wave occasionally reflected from an air-mass boundary or a cloud is properly describable as a "sky wave." However, the main "sky-wave coverage" indicated in Fig. 2.1 and employed for almost all long-distance radio transmission, is obtained in an essentially different manner. In 1902, Kennelly[1] and Heaviside[2] had assumed the outer atmosphere to be a conductor, merely because of its low pressure; but low pressure alone is not sufficient. The air must be ionized. Electrons in enormous numbers must be torn from their parent atoms and molecules, in order that they may serve as highly mobile carriers of electric charge. The term *ionosphere* is applied to that region of the outer atmosphere, lying above the stratosphere, in which it is possible to produce such great ionization. In general the energy, obtainable from the radio wave itself, is altogether insufficient to produce this ionization. Instead, the diurnal predictable changes of ionization are caused by ultraviolet light, which forms a considerable part of the sun's radiation. Fortunately, most of this ultraviolet light is absorbed by the upper levels of our atmosphere, thus preventing destruction of all plant and animal life upon the surface of the earth. The absorption is selective, certain frequencies of ultraviolet being strongly absorbed by the materials first encountered in the outermost levels. The electrons thus released form the F layer, Fig. 4.1. Above a certain height, say 300 km, the electron density decreases because the air is becoming too thin and there is insufficient material to serve as a source of electrons. Below this height, the electron density decreases because ultraviolet light of appropriate frequency is lacking, having already been absorbed at the higher levels.

At a lower level, approximately 100 km, incoming solar rays encounter molecular oxygen in large quantities for the first time.

[1] A. E. Kennelly, *Elec. World*, Vol. 39, p. 473, 1902.

[2] O. Heaviside, Encyclopaedia Britannica, 10th ed., Vol. 33.

Above this level, collisions of oxygen atoms are infrequent. Once separated, they tend to remain apart, thus providing insufficient opportunities for the formation of O_2. Hence, the ultraviolet light, of frequency appropriate to the ionization of molecular oxygen, finds no serious obstacle in its path while passing through the *F*

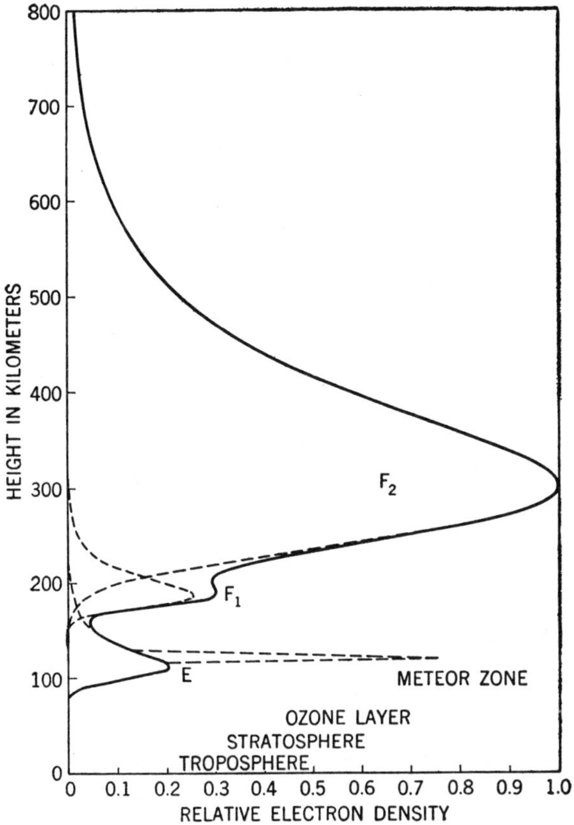

FIG. 4.1.—Layers of ionization in the atmosphere.

region. However, on arriving at the 100-km level, it is suddenly absorbed, leaving a dense concentration of electrons strewn about in the sharply defined *E* layer.

Similar selective absorption accounts for a very weak trace of ionization, called the *D* layer, at a still lower altitude where ozone (O_3) is first encountered. Sometimes a minor maximum of ionization may be observed on the lower fringe of the main *F* layer, Fig. 4.1. At other times, this is concealed within the main layer.

Other submaxima below the E region may be of importance for special transmission problems. In general, however, communication depends mainly upon the E and F_2 layers. These are about equally important, the lesser density of the E region being offset by its lesser height. In both layers, the distribution of electrons changes constantly. The main factors effecting radio transmission are

1. The levels at which density maxima occur.
2. The magnitudes of these density maxima (electrons per cubic centimeter).
3. The average collision rate of electrons with atoms and molecules (causing absorption of radio waves).

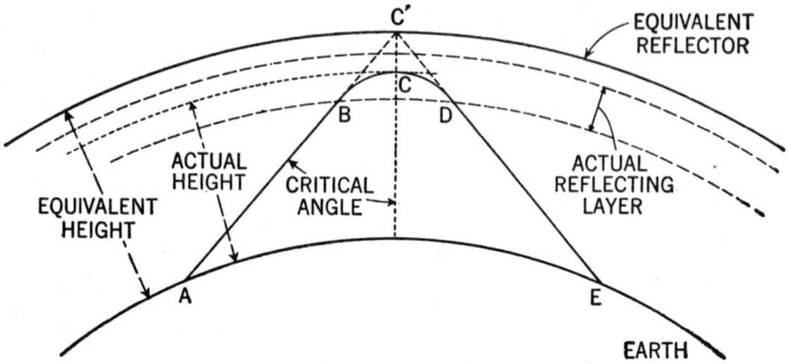

FIG. 4.2.—Actual path and path of equivalent reflected ray.

The effect upon radio transmission of a layer of free electrons may be described in several different ways. To begin with, one may consider a directed beam, or bundle of rays, entering such a layer from beneath, Fig. 4.2. As the wave progresses, a systematic mass vibration of electrons is caused, many billions of electrons swaying back and forth under the influence of the field of the incident wave. This removes energy from that wave, causing a decrease in the original components of the electric and magnetic fields. However, each moving electron reradiates, and the cumulative effect of *all* reradiation contributes *new* components of electric and magnetic field, not agreeing in direction with the incident field. As a result, the direction of each ray progressively alters until that ray may finally become horizontal and then return toward the earth.

This gradual refraction, or bending, may be considered also in terms of wave fronts. Under the conditions of Fig. 4.2, the upper part of each wave travels a little faster than the lower part, thus causing the wave front to turn over gradually as if the lower edge were dragging. Actually, at all points where free electrons are present, the "wave velocity" or phase velocity exceeds the velocity obtainable in empty space. This speeding-up process is greatest at points where the average density of free electrons is greatest. This does not mean that a pulse or telegraphic dot can be sent through the medium at a measurable velocity exceeding that of light. On the contrary, the signal velocity (group velocity) may be very much *less* than the free-space velocity. In fact, the time required to travel the actual path $ABCDE$ is the same as the time that *would* have been required to traverse the longer path $ABC'DE$, Fig. 4.2, in space not containing free electrons. Accordingly, the terms "reflection" and "refraction" are used almost interchangeably in dealing with the return of sky waves from the ionosphere. When speaking of refraction, one has in mind the *actual* curved path $ABCDE$. When speaking of reflection one introduces the concept of an *equivalent reflection* which would produce a similar down-coming wave. The equivalent height is readily measurable by means of a determination of the angle of arrival or by determining the time lag of the down-coming radiation. Computation of the actual path BCD requires detailed knowledge of the electron distribution. Equivalent heights are sufficient for most engineering purposes.

The explanation may also be phrased mathematically. Setting up simple equations for the motion of the free electrons, one obtains for the refractive index[1] n,

$$n = \sqrt{1 - \frac{4\pi N e^2}{\omega^2 m}}$$

where N = electrons per cubic centimeter.
e = electronic charge, esu.
m = electronic mass, grams.
$\omega = 2\pi f$.
f = operating frequency, cps.

Hence the refractive index of a space containing free electrons is less than unity. A radio wave, entering such a medium from the un-ionized air beneath, will be turned back by total internal reflection, provided that the angle of incidence exceeds a definite critical

[1] In this elementary summary, the effect of the earth's magnetic field upon the motion of the free electrons and also the damping effect of collisions are neglected. For an extensive summary, see The Physics of the Ionosphere, *Rev. Modern Phys.*, Vol. 9, pp. 1–43, 1937.

angle, just as with ordinary light traveling in water or glass and reflected against an air surface. For rays steeper than the critical ray, the wave penetrates the layer, Fig. 4.3. The critical angle is determined by the frequency of the transmitter and by the electron density at the most dense level of the layer. In effect, each ray acts as if it had intelligence and purpose. It bores into the layer, seeking an electron density sufficiently great to turn it back (by total internal reflection). If a sufficient density (say 200,000 electrons per cubic centimeter) exists at any height, the ray reaches its apex at that height and begins its downward travel. If such a density does not exist in the E layer at the time, the ray passes

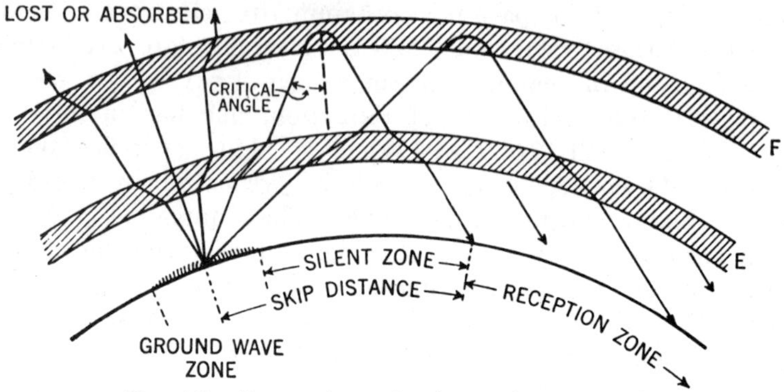

FIG. 4.3.—Penetration, reflection, and critical angle.

through to the F layer, where it repeats its search. If *again* unsuccessful, the ray passes out into interstellar space. At the critical angle, the required density is just barely attainable. All steeper rays (nearer the zenith) will be lost. Less steep rays (nearer grazing incidence) will be reflected, with a margin of electron density to spare. This results in a "skip distance." If the ground wave does not extend far enough to bridge the gap, there is also a "silent zone," Fig. 4.3. Increasing *steepness* requires a greater electron density. Increasing *frequency* has a similar effect and will eventually cause penetration even at angles approaching grazing incidence.

5. Sky Waves from the Ionosphere—Practical Applications.— In order to escape from the earth without excessive ground attenuation, a sky wave must leave the earth at an angle at least 3° above the horizontal. (Lower-angle radiation contributes to the ground wave only.) Rays slightly above this angle are often best for long-distance communication, since this "low-angle radiation" zigzags to the distant receiver with the smallest possible number of "hops," thus avoiding repeated losses due to an excessive number of reflections from sky and ground. At 3° elevation, the distance per hop is about 3,500 km (2,100 miles). Longer distances are automatically broken up into units not exceeding 3,500 km. Sometimes, plural-path transmission exists, with waves arriving by several modes, Fig. 5.1. Interference between such paths is partly responsible for "fading" of signals and for side-band distortion. Fading can be minimized by selection of the best mode at the receiver, but this requires elaborate antenna arrays and receivers, justified only at a few fixed receiving points.

In spanning distances less than 3,500 km, the beam must be elevated *above* 3°. More commonly, the transmitter radiates a

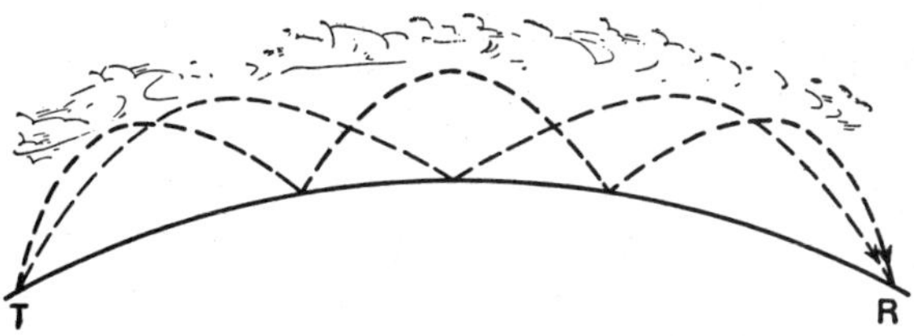

FIG. 5.1.—Plural-path transmission.

wave having a broad vertical pattern, or it radiates at *all* vertical angles. In either case, the *useful* ray becomes steeper and steeper as the receiver is brought in closer to the transmitter, Fig. 4.3, thus requiring a greater and greater electron density. Finally, at the critical distance (skip distance) and critical angle, penetration of the layer prevents any closer approach. At this particular distance, the frequency now in use is the "maximum usable frequency," abbreviated MUF, determined by the distance and the prevailing electron density at a point halfway from transmitter to receiver. Still shorter distances may be bridged by a sky wave, but the frequency must be reduced to a lower value. Thus there exists a series of values of MUF corresponding to selected distances. These distances may be reduced all the way to zero by a sufficient lowering of frequency. The lowest value of the MUF, required at zero distance, is the "critical frequency" or "penetration frequency" of the F_2 layer, Fig. 4.1. At lower frequencies, echoes are observable even in the immediate vicinity of the transmitter. At higher frequencies, rays directed toward the zenith penetrate the F_2 layer and are lost. Similar "penetration frequencies" for the E layer and the F_1 layer may be measured with precision by noting sharp breaks in a curve of echo-time lag vs. frequency, taken with a receiver adjacent to a pulsed transmitter. Curves such as the virtual-height curve of Fig. 5.2 are recorded automatically, often at 10-minute intervals, by ionospheric observatories at various points on the earth. Computations based on these observations make it possible for the Bureau of Standards to issue predictions of MUF as a function of

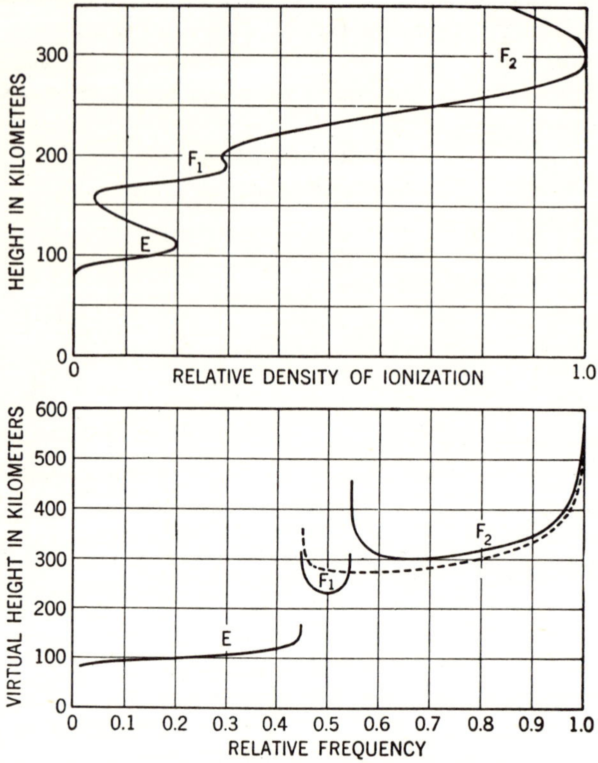

Fig. 5.2.—Virtual height of ionized layers.

geographic route, time of day, season, and solar period (the 11-year sunspot cycle).

In general, the MUF is the frequency at which strongest signals will be obtained. However, in selecting an appropriate operating frequency, it is desirable to leave some margin for variability of the ionosphere. For example, on an occasion when the MUF is predicted as 10 mcps, it would be preferable to select a frequency assignment not exceeding 8.5 mcps, thus leaving a 15 per cent margin for error. In general, excessive absorption prevents reliable long-distance communication on frequencies less than 50 per cent of MUF. However, this lower boundary also varies widely with the route and period. Hence, predictions of the "lowest useful high frequency," abbreviated LUHF, are now available. Unfortunately, on routes such as the North Atlantic route, the frequency interval between MUF and LUHF is some-

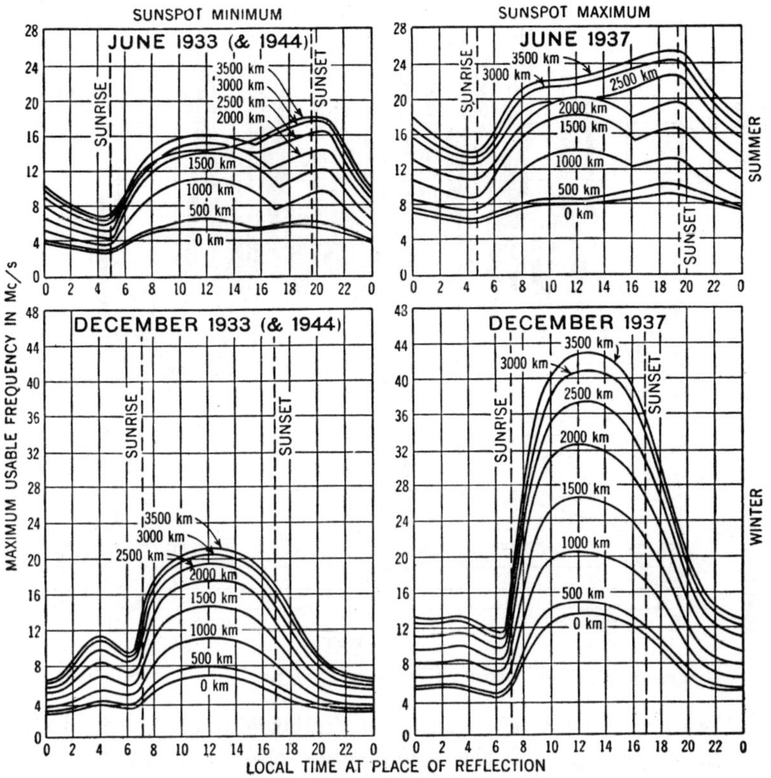

FIG. 5.3.—Typical values of maximum usable frequency.

times very limited, and it may not be possible to leave an adequate margin above and below the assigned frequency.

From the facts stated it follows that efficient operational use of all "high-frequency" radio channels can be greatly improved by a careful study of the detailed and accurate MUF and LUHF predictions now available through official channels. For security reasons, the latest methods for presenting and using these data cannot be quoted here. However, the general nature of the diurnal changes, the seasonal cycle, and the 11-year cycle are fully shown in Fig. 5.3, which is a reproduction of a chart[1] published by the Bureau of Standards in 1940, applicable to the continental United States. The sharp cusps, appearing on the June curves in the late afternoon hours at medium distances are due to transition from E-layer transmission to F-layer transmission.

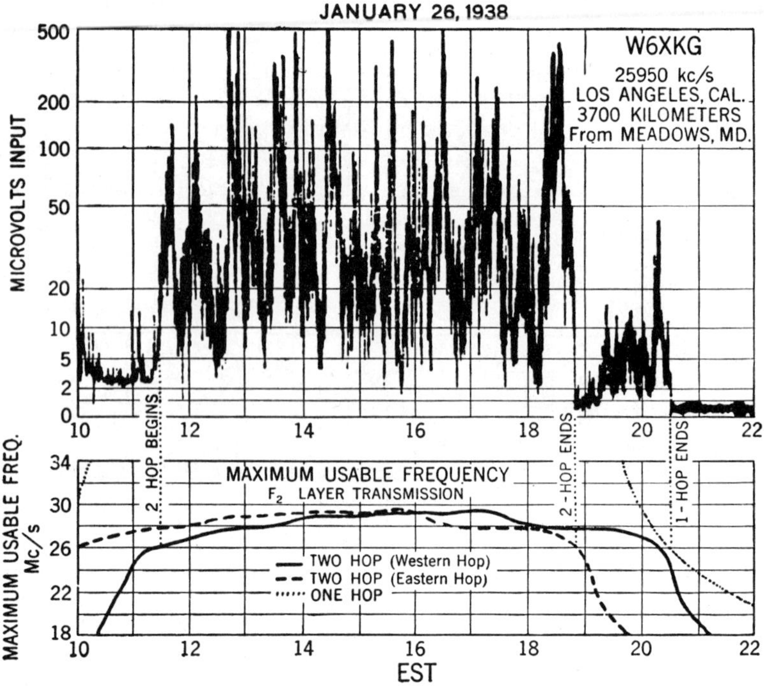

FIG. 5.4.—Field intensity record verifying predicted transmission.

Figure 5.4 is reproduced from a similar source[2] in order to show a typical example of the remarkably close agreement between predicted behavior of the ionosphere and the practical operational results that ordinarily follow. The following explanatory paragraphs accompany Fig. 5.4:

The upper part of the figure is a continuous record by the National Bureau of Standards of the field intensity of W6XKG, Los Angeles, Calif., on 25,950 kc/s, at a distance of 3,700 km. In the lower part are shown

[1] From *Bur. Standards Letter Circular* LC-614, Oct. 23, 1940.

[2] NEWBERN SMITH, SAMUEL S. KIRBY, and THEODORE GILLILAND, *Bur. Standards Research Paper*, RP 1167, January, 1939 (part of *Bur. Standards J. Res.*, Vol. 22, pp. 81–92, January, 1939).

graphs of the calculated maximum usable frequency over this distance, for both one- and two-hop transmission. This distance is slightly beyond the maximum distance for good one-hop transmission during the winter day, on the basis of the $3\frac{1}{2}°$ angle of departure. The changes between one- and two-hop transmission are well marked, and the ratio of signal strength is well over 100:1. This difference is due partly to the increased absorption over the flatter trajectory and partly to the unfavorable angle of departure. It should be noted how the times of beginning and ending of two-hop transmission agree with the times the calculated maximum usable frequency passed through about 26 Mc/s at the western and eastern hops, respectively. The two-hop transmission began as soon as the western hop permitted and ended when the eastern hop failed.

After the failure of two-hop transmission, the station still came in by one-hop. As the evening progressed the intensity increased, owing to the departure of the daytime absorption and the rise in the height of the layer, with a consequent more favorable angle of departure. The failure of single-hop transmission also is seen to agree with the fall of the calculated maximum usable frequency through 26 Mc/s.

Accurate correlations of this nature are also obtained ordinarily on signal-strength records taken at the Cruft Laboratory field station at Weston, Massachusetts.

6. Abnormalities and Interruptions of Wave Propagation.—Predicted communication conditions are occasionally interrupted by two types of disturbances: A "solar flare," an excessive emission of ultraviolet light from a small point upon the sun, can cause the sudden production of abnormal ionization at levels below the normal *E* layer. At these lower levels, electrons collide so frequently with atoms and molecules that reradiation is very inefficient. In extreme cases, all sky waves are absorbed wherever the reflection would take place on the illuminated side of the earth. Fortunately this "Dellinger fade-out" is of short duration, lasting from a few seconds to about 30 minutes in extreme cases. As the extra electrons recombine or become attached to neutral molecules, the high frequencies come back first, the lower frequencies a little later. Ten-minute fade-outs may occur several times per month in the active years of the sunspot cycle, but less frequently in quiescent years.

"Magnetic storms" interrupt communication for much longer periods. Corpuscular radiations from the sun appear to be responsible. Streams of matter shot out volcanically from the sun occasionally bombard the earth. Ions and electrons from these streams are guided toward the magnetic poles, causing visible aurora borealis and aurora australis displays. Clouds of electrons, in violent motion throughout the upper atmosphere, cause abnormal compass deviations and also cause secondary currents on the surface of the earth and in all power lines, telephone lines, and submarine cables. In extreme cases, the compact ionospheric layers are completely blown to pieces, thoroughly disrupting sky-wave communication. Such a storm may last for several days. Even after the bombardment has ceased, the layers require a considerable time to re-form. During these poststorm days, electron densities are abnormally low, requiring a downward shift in sky-wave frequencies for best results. In the past, only statistical predictions of magnetic storminess have been possible. Improved predictions now are being attempted.

The ionospheric layers are not perfectly smooth and continuous. They contain ripples and clouds. Very large clouds occasionally cover entire states, causing thousands of reports of amateur 5-m transmission over great distances by a true sky wave. Small clouds cause an erratic scattering of radio signals. Scattered radiation occasionally causes signals to be received within the normal silent zone. Time lags, detectable by the ear, give such signals an abnormal reverberent quality. Scattered radiation also causes signals to arrive from unexpected directions, greatly confusing sky-wave bearings. Until further research is concluded, such bearings should be regarded with much suspicion and should be used only where more reliable information is not obtainable. Sky-wave direction finding is necessary in locating illegal or hostile transmitters at distances beyond ground-wave range and in general radio-propagation research. It is sometimes useful in locating distress signals. For straightforward long-distance navigation, other radio aids are now available. In any event, some form of spaced antenna array is necessary when observing the direction of arrival of sky waves. Even when the sky wave accurately follows a great-circle path, a single loop antenna is subject to violent instrumental errors that make it completely unreliable. These instrumental errors are fully explainable in terms of a change of polarization of the sky wave caused by rotation of the ionospheric electrons in the magnetic field of the earth.

PROBLEMS

Chapter I

1. Calculate the resistance r, inductance l, and capacitance c, per loop meter for a transmission line of copper wire of radius $a = 0.040$ cm, separated a distance $D = 2$ cm between centers, at $\omega = 10^9$. *Ans.:* $r = 2.65$ ohms/m; $l = 1.567\mu$h/m; $c = 7.1\ \mu\mu$f/m.

2. Calculate the attenuation constant α and the phase constant β per loop meter, and the characteristic impedance Z_c, of the line of Prob. 1. What is the Q factor of this line? *Ans.:* $\alpha = 2.82 \cdot 10^{-3}$ nepers/m; $\beta = 3.33$ radians/m; $Z_c \doteq R_c = 470$ ohms; $Q = 591$.

3. The line of Probs. 1 and 2 is 5 m long and is terminated in a load equal to R_c. What is the attenuation of this line in decibels? A generator whose open-circuit voltage is 50 volts and whose internal impedance is $31 + j0$ ohms is connected to the sending end. What are the magnitude and angle of the current in and the voltage across the load? *Ans.:* 0.12 db; $I_R = 100(1 - 0.014)\underline{/-954^\circ}$ ma; $E_R = 47.0(1 - 0.014)\underline{/-954^\circ}$ volts. Use the approximation $\epsilon^{-x} = 1 - x$.

4. The line of Prob. 3 is short-circuited by a bar. Considering the line to be dissipationless, and the generator impedance to be 31 ohms as before, calculate the input current and the current in the bar. Plot the current distribution along the line, showing numerical values of the input current, and the location and values of the maximum current. *Ans.:* $Z_S = j671$ ohms; $|I_S| \doteq 74.5$ ma; $|I_R| = |I_{max}| \doteq 130$ ma.

5. Plot the voltage distribution under the same conditions as in Prob. 4, showing values of the input voltage, and the location and values of the maximum voltage. *Ans.:* $|E_S| \doteq 50$ volts; $|E_{max}| \doteq 619$ volts; $E_R = 0$.

6. What is the length in inches of a quarter-wave section of transmission line operated at a frequency of 300 mcps? *Ans.:* 9.84 in.

7. A dissipationless transmission line whose characteristic impedance is 200 ohms is connected to a load of $100 + j0$ ohms. The frequency is 300 mcps. What is the input impedance of the line if the length is (*a*) 15 cm, (*b*) 25 cm, (*c*) 50 cm, (*d*) 65 cm? What conclusion is to be drawn from the similarity of some of the answers? Solve by using the formula and also by using the circle diagram. *Ans.:* (*a*) $196 + j140$; (*b*) $400 + j0$; (*c*) $100 + j0$; (*d*) $196 + j140$; impedance repeats every half wavelength.

8. If the line of Prob. 7 is one wavelength long, and the voltage across the load is 1,000 volts, sketch the current and voltage distribution along the line, showing location and magnitude of E_{max}, E_{min}, I_{max}, and I_{min}. *Ans.:* $E_{max} = 2{,}000$ volts; $E_{min} = 1{,}000$ volts; $I_{max} = 10$ amp; $I_{min} = 5$ amp.

9. What is the required length of a short-circuited coaxial line to provide an inductive reactance of 250 ohms, if $\lambda = 80$ cm and $R_c = 70$ ohms. ***Ans.:*** 16.5 cm.

10. What must be the characteristic impedance of a quarter-wave section of line to match a generator of $50 + j0$ ohms impedance to a load of $5{,}000 + j0$ ohms. If the frequency is 300 mcps and the wire is No. 10, whose diameter is 0.1 in., what should be the length and spacing of the wires for a two-wire matching section? *Ans.:* 500 ohms; length, 25 cm = 9.84 in.; spacing, 3.23 in., center to center.

11. A line of characteristic resistance $R_c = 60$ ohms is to be matched to a resistive load of 300 ohms by means of a closed stub. The wavelength is 50 cm. Determine the distance between the stub and the load and the length of the stub in centimeters. Solve by means of the formulas and with the aid of a circle diagram. Since R_c is 60 ohms, what type of construction does the line probably have? *Ans.:* Distance from load 9.2 cm; stub length 4.1 cm; coaxial construction.

12. In Prob. 11, (*a*) what are the impedance and admittance at the junction between the stub and the line, looking into the short section of line to which the load is attached? What is the Q of this impedance? (*b*) the admittance and impedance of the closed stub? *Ans.:* (*a*) $Z = 14.3 - j25.8$, $Y = 0.0167 + j0.0298$, $Q = 1.78$; (*b*) $Y = j0.0298$, $Z = +j33.6$.

13. A line whose characteristic impedance is 300 ohms is terminated by an antenna whose impedance is $70 + j0$ ohms. What are the location, length, and termination of a stub to match this antenna to the line? Express the lengths in terms of the wavelength. *Ans.:* Open stub, 0.161λ long, located 0.072λ from antenna.

14. Determine the inside diameter of the outer conductor of a concentric cable that will have no standing waves on it when used as a feeder connected to a load of $60 + j0$ ohms. The inner conductor is to be No. 14 copper wire whose diameter is 64 mils, or 0.064 in. *Ans.:* 174 mils, or 0.174 in.

15. A coaxial cable is operated at a frequency of 100 mcps and is loaded by an impedance equal to its characteristic impedance, 90 ohms. The power loss per meter is 0.1 per cent. Find the radii a and b. *Ans.:* $(b/a) = 4.49$; $b = 2.56$ cm $\doteq 1$ in.; $a = 0.57$ cm $= 0.225$ in.

16. A line of characteristic resistance 469 ohms is 30 cm long and is operated at $\omega = 10^9$. (*a*) What inductance in microhenrys should be connected to the end of the line to make the input impedance infinite (assuming no losses)? (*b*) What capacitance in micromicrofarads should be connected to the end of the line to make the input impedance equal to zero? *Ans.:* (*a*) $Z_R = j300$ ohms, $L = 0.3\ \mu$h; (*b*) $Z_R = -j734$, $C = 1.36\ \mu\mu$f.

17. Power supplied at T, Fig. 17, is to be transmitted alternately to A and A' by alternately closing and opening the line terminations at C and C'. The terminations at A and A' are arranged to be equal to 440 ohms. The lengths BC and $B'C'$ are adjusted to antiresonance when bridged and are resonant when open as viewed from B and B'. For the arrangement in Fig. 17 and with DB and DB' each $\lambda/4$ determine approximately the impedance of the line (*a*) looking to the left, and (*b*) looking to the right at D, and (*c*) looking in at T. Consider the lines to be dissipationless and to have a characteristic resistance of 440 ohms. Discuss the operation of the circuit. *Ans.:* (*a*) 440 ohms; (*b*) ideally infinite, practically very high.

18. A line having losses is terminated in an impedance equal to Z_c. The attenuation is 0.1 db/m, and the wavelength on the line is 1 m. The voltage applied to the line is 1,000 volts, and the line is 10 m long. (*a*) What is the voltage at the load? (*b*) What is the phase shift in degrees per centimeter length of line?

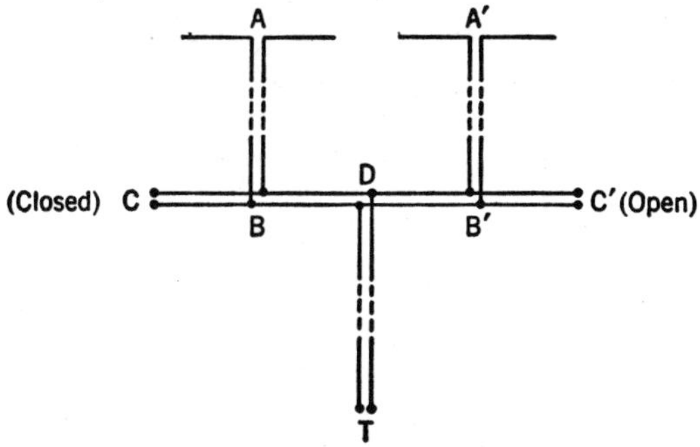

FIG. 17.

19. A dissipationless two-wire line of exposed construction (air dielectric), has a characteristic impedance of 500 ohms, is terminated in a resistance of 500 ohms, and is two wavelengths long. (*a*) At the instant when the input current at *A* in the top wire is at its positive maximum of 5 amp to the right, what is the instantaneous current, in magnitude and direction, at the point *C*? (*b*) What is the rms value of the currents at *A* and *C*, and what is the phase difference between them?

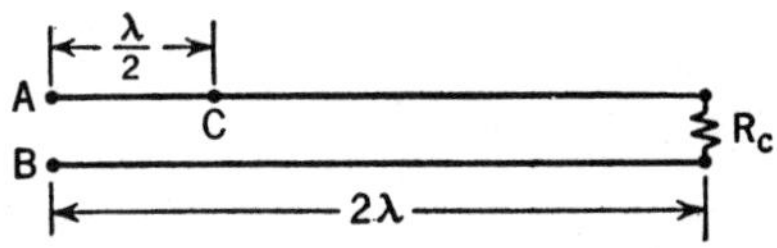

FIG. 19.

20. A certain Army-Navy standardized solid dielectric, coaxial transmission line has the following specifications: outside diameter of dielectric 0.910 in., characteristic resistance 69 ohms, capacitance per foot 22 $\mu\mu$f. (1 m = 3.28 ft) Assuming the cable to have no losses, calculate (*a*) the inductance per meter, (*b*) the diameter of the inside conductor in inches, and (*c*) the relative dielectric constant of the solid dielectric.

21. A coaxial cable whose R_c is 72 ohms is available. It is desired to use a short-circuited length of this cable as an inductance. At a wavelength of 1 m, what length of short-circuited cable is required, if the input impedance is to be 150 ohms of inductive reactance?

22. A short-circuited section of transmission line is used as an inductance in a resonant tank circuit. The capacitor *C* has a reactance of 500 ohms when the wavelength is 1 m. What is the required length *s*, in centimeters, of the short-circuited section? The *Q* of this tank circuit may be considered suffi-

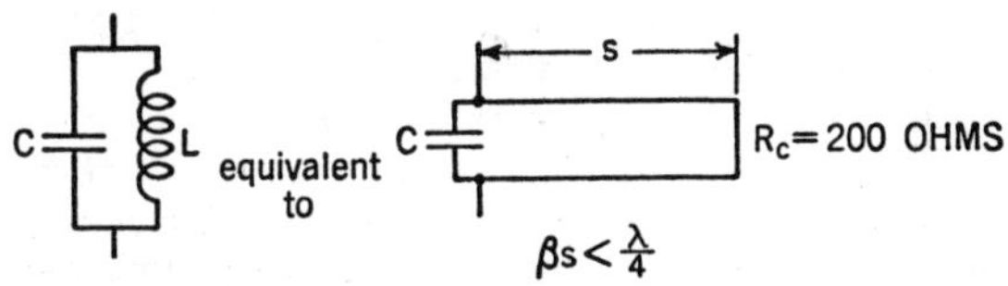

FIG. 22.

ciently high to make $\omega L \doteq 1/\omega C$. The length of the shorting bar may be neglected.

23. A dissipationless transmission line having a characteristic impedance of 600 ohms is feeding a load resistance of 300 ohms. The voltage across the load is 1,200 volts. At a point on the line one-quarter wavelength from the load, what is (*a*) the voltage across the line? (*b*) the current in the line?

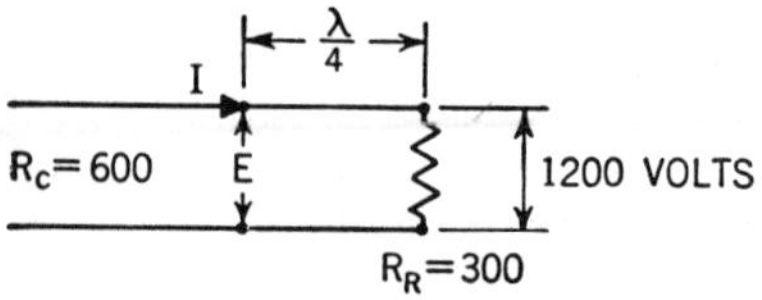

FIG. 23.

24. A certain coaxial transmission line may be assumed to have neither series resistance nor shunt conductance. (*a*) State the nature of Z_c (*i.e.*, a pure resistance, a pure reactance, or a complex impedance). (*b*) Determine the phase shift per meter along the line when it is terminated in Z_c, the frequency of the impressed signal is 300 mcps ($\lambda = 1$ m), and the dielectric is air. (*c*) What is the characteristic impedance of a coaxial cable with air dielectric, so proportioned that the attenuation is a minimum? (*d*) How will the use of a solid dielectric (assumed to have no losses) affect the phase shift per unit length of line? (*e*) How will the use of a solid dielectric affect the phase velocity on the line? (*f*) How will the use of a solid dielectric affect the wavelength on the line?

25. A dissipationless line having a characteristic impedance of 200 ohms is one-eighth wavelength long and is terminated in a resistance of 500 ohms. Determine the complex and polar expressions for the input impedance.

26. (*a*) What is the input impedance of a quarter wavelength of transmission line whose $Z_c = 200 + j0$ ohms and whose losses are negligible, when the distant end is (1) open, (2) connected to a load of $65 + j0$ ohms, and (3) short-circuited? (*b*) Answer the same questions when the line is one-half wavelength long.

27. An open-wire transmission line is steadily feeding power from a transmitter to a load. It is desired to know whether or not the load impedance is equal to the Z_c of the line. A vacuum-tube voltmeter is the only equipment at hand. Explain briefly how one would determine whether or not the load impedance equals Z_c.

28. A dissipationless line is terminated in a resistance. The characteristic impedance of the line is 70 ohms. What is the value of the load resistance when (*a*) the magnitude of the voltage is the same everywhere along the line? (*b*) the standing-wave ratio $|E_{max}|/|E_{min}|$ is 4 and $|E_{max}|$ occurs at the load? (*c*) the standing-wave ratio is 4 and $|E_{max}|$ occurs one-quarter wavelength from the load?

29. (*a*) A transmission line with a characteristic resistance of 500 ohms is delivering 10 kw to a load. The standing-wave ratio is 3. Make a sketch of the standing-wave pattern, marking the *relative* location of E_{max}, I_{max}, E_{min}, I_{min}. (*b*) What are the magnitudes of these four quantities?

30. A dissipationless transmission line having a characteristic impedance of 100 ohms is one wavelength long and feeds a resistive load of 50 ohms. The power in the load is 5 kw. Make a sketch showing the magnitudes of the voltage and current along this line (*i.e.*, draw the standing-wave patterns). Show actual location and numerical values of E_{max}, E_{min}, I_{max}, I_{min}.

31. A dissipationless transmission line whose R_c is 100 ohms is terminated in a load whose impedance is $130 + j85$ ohms. The wavelength is 360 cm. (*a*) Determine the impedance looking toward the load at a point 25 cm from the load. (*b*) What is the standing-wave ratio on this line? (*c*) If E_{max} is 1,000 volts, what is the power delivered to the load?

32. Figure 32 represents a dissipationless transmission line terminated in an impedance $Z_R \neq R_c$. The graph shows the relative magnitude of the

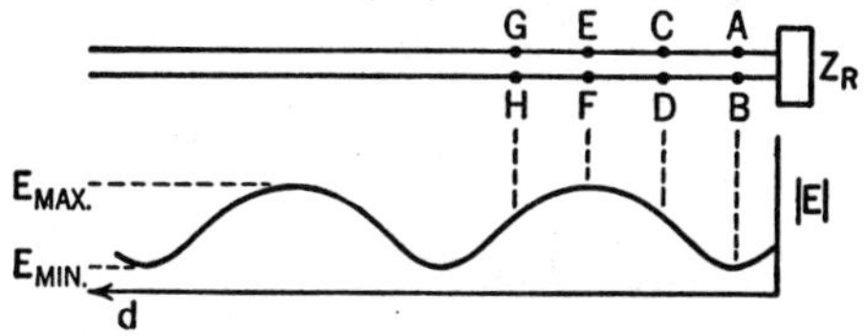

FIG. 32.

voltage at different points on the line. In the following table, put a check mark in that box opposite each of the impedances in the first column, which properly designates the nature of that impedance. The definitions of Z_{AB}, etc., are given below.

	Resistive	Resistive capacitive	Resistive inductive
Z_{AB}			
Z_{CD}			
Z_{EF}			
Z_{GH}			
Z_R			

By Z_{AB} is meant the impedance looking toward the load at the points AB, the remainder of the line being neglected. Z_{CD}, Z_{EF}, and Z_{GH} are similarly defined. Z_R designates the load impedance itself.

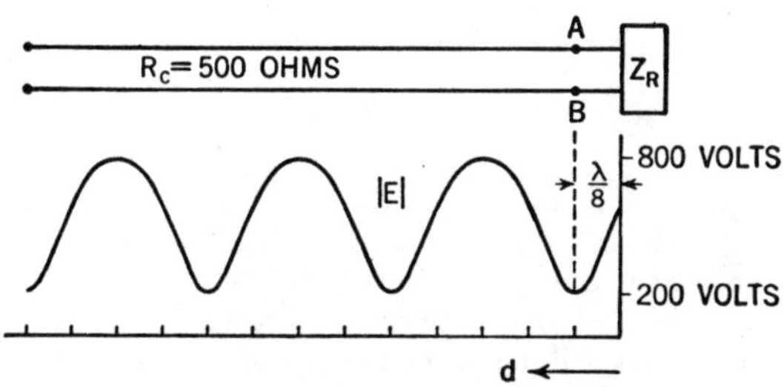

FIG. 33.

33. Figure 33 shows the observed standing-wave distribution of voltage on a line whose R_c is 500 ohms and whose dissipation is negligible. (*a*) How long is the line in wavelengths? (*b*) What is the impedance looking toward the load at AB? (*c*) What is the magnitude of the current in the line at the points AB? (*d*) What is the power delivered to the load? (*e*) What are the resistance and reactance of the load?

34. For the dissipationless line indicated in Fig. 34, $R_c = 500$ ohms. $|E_{max}| = 1{,}200$ volts rms, and $|E_{min}| = 400$ volts rms. (*a*) What is the power delivered to the load Z_R? (*b*) What are $|I_{max}|$ and $|I_{min}|$ in rms magnitude? (*c*) What is the impedance looking toward the load at AB, the portion to the left of AB being disconnected? (*d*) Is the load capacitive or inductive? (*e*) What is the distance in wavelengths between successive maxima of current or voltage?

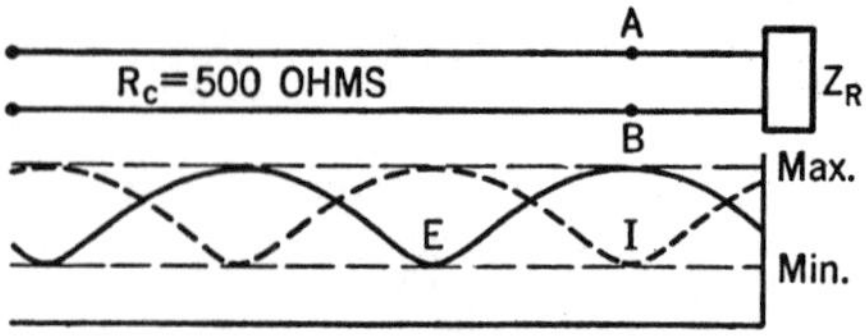

FIG. 34.

35. (*a*) If the maximum instantaneous voltage across the wires of a two-wire transmission line is not to exceed 5,000 volts, what is the maximum average power that this line can transmit? The R_c of the line is 500 ohms. What must be the impedance of the load to secure delivery of maximum power to the load under the above limitations? (*b*) A transmission line is terminated in a load which is a pure resistance. This load is matched to the R_c of the line by a closed stub, less than $\lambda/4$ long, placed on the line at a distance less than $\lambda/4$ from the load. Is the load resistance greater or less than the R_c of the line? Explain your reasoning concisely.

36. A quarter-wave section matches a 66.7-ohm resistive load to a 600-ohm feeder. The power delivered to the load is 1 kw. Neglecting losses in the feeder and in the matching stub, determine the magnitude of (*a*) current and voltage at the terminals AB, (*b*) current and voltage at the terminals CD, and (*c*) current and voltage at the terminals EF, Fig. 36.

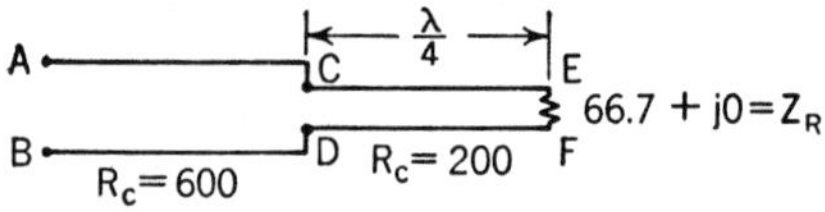

FIG. 36.

37. In Fig. 37, the distance between the stubs is $d = \lambda/4$. Determine the lengths d_1 and d_2 that will make the portion of the feeder to the left of CD nonresonant. The wavelength is 80 cm.

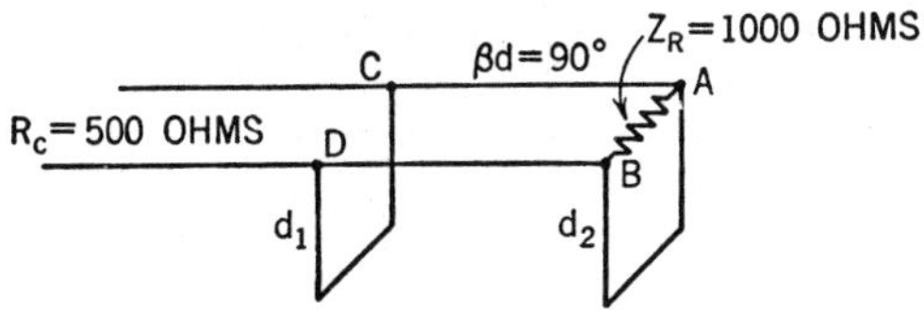

FIG. 37.

38. (*a*) A line having an R_c of 250 ohms is terminated by a load impedance Z_R of $300 + j300$ ohms. Determine the distance d between load and stub and the length d_1 of a stub that will match the load to the R_c of the line. The distance d is to be as small as possible. Should the stub be open or closed ($\lambda = 1$ m)? (*b*) If the power delivered to the load is 5,000 watts, what is the rms voltage between the wires at the point where the stub is shunted across the line?

39. A double-stub tuner employing closed stubs with $3\lambda/8$ distance between them is to be used to match a load whose admittance is $(3 - j3)\cdot 10^{-3}$ mhos to a line whose R_c is 500 ohms. The wavelength is 1 m. Determine (*a*) the length of the stub at the load, (*b*) the length of the stub $3\lambda/8$ from the load.

40. In Fig. 40, a resonant antenna is connected at AB and an open stub at CD. The R_c of the transmission line and stub is 300 ohms. The stub is located

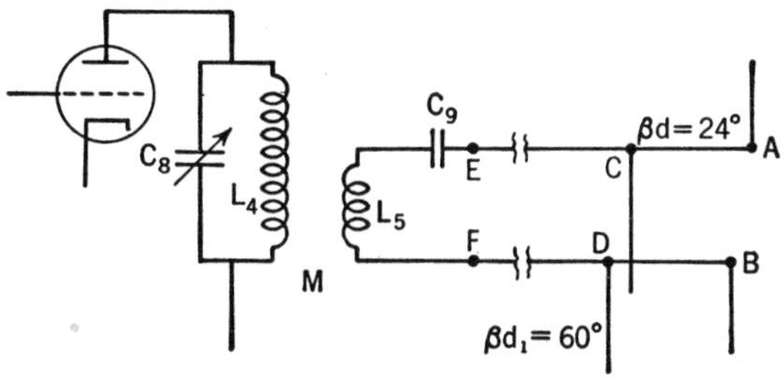

FIG. 40.

24° from the antenna and is 60° long. The stub is adjusted to eliminate standing waves on the line to the left of CD. (*a*) What is the impedance Z_{AB} of the antenna? (*b*) If the power fed into the antenna is 10 kw, what is the rms voltage across the line at the points CD? (*c*) What is the rms voltage across the open end of the stub?

If the line develops a short circuit at a distance $\lambda/4$ from the input end EF, what is the effect on (*d*) the plate dissipation in the tube? (*e*) the plate-circuit efficiency of the tube? (*f*) the r-f power output of the tube? The tube is a conventional Class C power amplifier tube. The input circuit to the tube and the power supplies are not shown in the diagram. C_8 and L_4 are tuned for parallel resonance and C_9 and L_5 are tuned for series resonance. The Q of the tank circuit under normal load conditions may be taken as 15.

41. A transmission line whose R_c is 70 ohms is terminated by a load whose impedance is $70 + j70$. A stub is to be placed in parallel with the line, as close to the load as possible, in order to eliminate standing waves on the line between the generator and the stub. Determine the length, termination, and position of the shortest stub that will accomplish this purpose. The wavelength is 40 cm.

42. The impedance of the antenna connected at AB is $65 + j20$ ohms. The stub at CD eliminates standing waves on the line. The stub at EF is the same

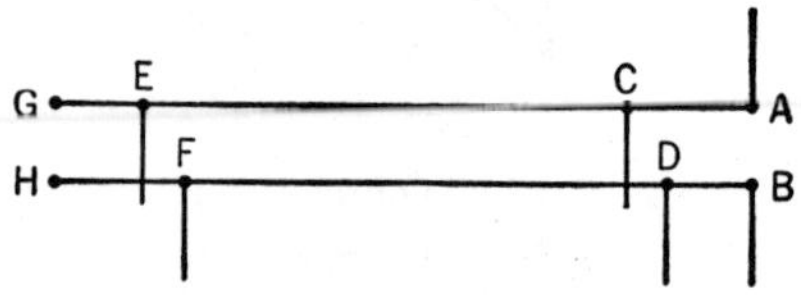

FIG. 42.

as that at CD, and the distance GE is the same as the distance AC. The losses in the line and stubs may be neglected. What is the input impedance at GH?

Hint.—Apply the principle of conjugates to the junctions CD and AB, considering AB and CD to be the terminals of a four-terminal network. Then do the same for terminals EF and GH.

Chapter II

1. Sketch the distribution of current along an antenna of half length $h = 3\lambda/8$, (*a*) if it is center-driven so that (approximately)

$$I_z = I_0 \frac{\sin \beta(h - |z|)}{\sin \beta h}$$

and (*b*) if it is a receiving antenna immersed in an electric field of uniform amplitude parallel to the antenna so that (approximately)

$$I_z = I_0 \frac{\cos \beta z - \cos \beta h}{1 - \cos \beta h}$$

2. Two symmetrical center-driven antennas are oriented in the same plane at right angles to a line joining their centers, as shown in Fig. 2. Antenna 1 has a half length $h_1 = 3.75$ m. Antenna 2 has a half length $h_2 = 1.25$ m. The frequency of the driving potential is 60 mcps for each antenna. (*a*) Calculate the minimum separating distance r required to ensure that the antennas be in the far zone with respect to each other if the condition for the far zone is interpreted to mean $\beta r \geqq 25$. (*b*) Sketch the current and charge distribution on each antenna at the time $t = T/8$ (take $\omega t = 45°$). For this purpose, assume that the radius of the antenna wires is extremely small.

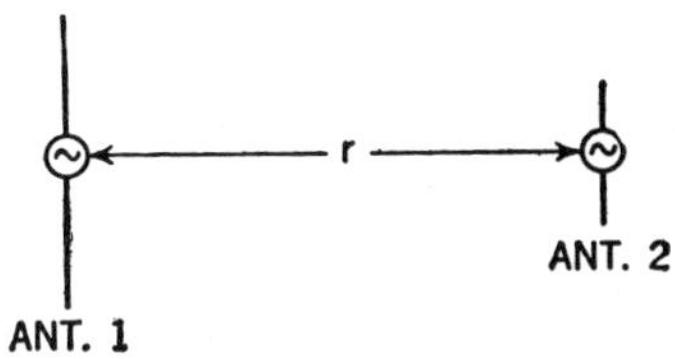

FIG. 2.

3. An antenna is constructed of brass rod of $\frac{2}{3}$ mm diameter to be operated at 50 mcps. What is the input self-impedance of this antenna if it is center-driven and has a half length of exactly one-half wavelength?
Ans.: $1250 - j1300$ ohms.

4. (*a*) What must be the length of an antenna made of $\frac{2}{3}$ mm brass rod if it is to be antiresonant at 50 mcps? (*b*) What is the input self-impedance in this case? *Ans.:* (*a*) $h = 2.83$ m, (*b*) $2400 + j0$ ohms.

5. A generator having an open-circuit voltage of 300 rms volts at 10^6 cps, and having an internal impedance of $200 + j300$ ohms, is connected to one end of a transmission line of characteristic resistance $R_c = 500$ ohms. The length of the line is $7\lambda/8$. A symmetrical center-driven antenna with its loading reactance presents an impedance of $147 + j88$ ohms across the other end of the line. Assuming no line losses, find: (*a*) the current in the antenna at its input terminals; (*b*) the power radiated by the antenna; and (*c*) the rms current at a point $\lambda/4$ up on the antenna from the input terminals, assuming that $h = 3\lambda/8$, and that a sinusoidal distribution of current is a satisfactory approximation. *Ans.:* (*a*) 876 ma, (*b*) 113 watts, (*c*) 876 ma.

6. A vertical tower radiator is used at a broadcasting station operating on a frequency of 1,110 kcps. The tower height is 0.20λ. By experimental means, it has been determined that the equivalent circular radius a of this tower antenna is 6 cm. Assuming no ground losses, and that none of the power supplied to the antenna is dissipated in heating it or the tuning reactance, determine: (*a*) the input impedance of the antenna; (*b*) the *magnitude* and *sign* of the lumped reactance to be connected in series to tune the circuit to resonance; (*c*) the power radiated by the antenna, *assuming that the circuit has been tuned to resonance*, for a driving potential difference of 312 volts rms applied to the circuit terminals; and (*d*) the voltage across the base insulator. *Ans.:* (*a*) $20 - j100$ ohms, (*b*) $X = +100$ ohms, (*c*) 4,870 watts, (*d*) 1,590 volts rms.

7. What is the input impedance (*a*) at the first resonance; (*b*) at the first antiresonance, of a center-driven antenna for which $\Omega = 15$? Which of the two cases is current-fed? Which is voltage-fed? *Ans.:* (*a*) $71.7 + j0$, (*b*) 2,430 $+ j0$ ohms.

8. The power radiated by a relatively thin antenna of half length $h = 3\lambda/8$ is 5 kw for an input current of 3.78 amp rms. Assuming a sinusoidally distributed current, calculate the radiation resistance of this antenna referred to the maximum current. No power is dissipated in heating the antenna, and it may be assumed that the circuit has been tuned to resonance by the use of an appropriate reactance in series with the antenna. *Ans.:* $R_m = 175$ ohms.

9. In Fig. 9(*a*), a wire of half length h is center-driven and isolated. In Fig. 9(*b*), a wire of length h is erected vertically over a perfectly conducting

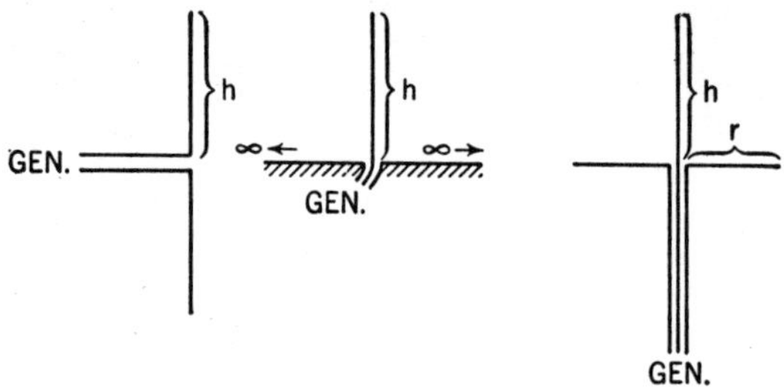

(*a*) (*b*) (*c*)

FIG. 9.

infinite plane. It is driven by a coaxial line connecting to a generator below the plane. In Fig. 9(*c*), a wire of length h is erected vertically over a perfectly conducting circular disk of radius r. It is driven by a coaxial line through the disk connecting to a generator very far below it. *Assume that there is no current on the underside of the disk or on the outside of the coaxial line.*

In all three cases, the wires of length h are identical and of such radius that $h/a = 1.1 \times 10^4$. The lengths h [and in Fig. 9(*c*) the radius r] are so adjusted, near $\lambda/2$, that the impedance loading the transmission line is a pure resistance at antiresonance. (*a*) Where do currents that contribute significantly to the distant field exist in each of the three cases? (*b*) With the aid of diagrams, describe the *approximate* distribution of all these currents when they have a maximum value in the cycle. Assume the wires infinitely thin for this purpose. (*c*) Determine the load across the end of the transmission line in the cases of Fig. 9(*a*) and (*b*). (*d*) Estimate the load across the end of the line in the case of Fig. 9(*c*). Do this by stating whether it will be larger or smaller than the loads in Fig. 9(*a*) and (*b*).

10. Two vertical radiators are used in an array for transmitting. The height of each antenna is $h = \lambda/4$, the spacing is $b/\lambda = 0.30$, and $\Omega = 15$ for each antenna. The input current to antenna 1 is 10 amp rms and the input current to antenna 2 is 5 amp rms. Verify that $Z_S = 78.4 + j40.9$, $Z_{12} = 30.4 - j40.7$. Find the power radiated by this antenna array. (Assume impedanceless generators and a perfectly conducting earth.)

11. (*a*) Describe the principal effect of each coaxial sleeve in each of the four antenna arrangements shown in Fig. 11. (*b*) In each case, state what parts of the arrangement, if any, carry large antenna currents and indicate

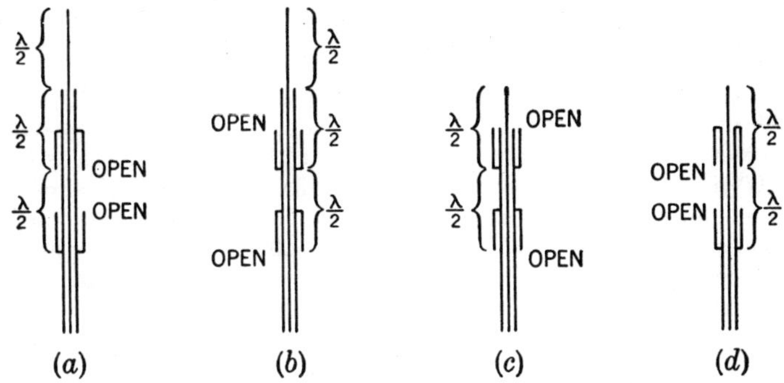

FIG. 11.

their approximate distribution. (*c*) In each case, is the outer surface of the line below the lower sleeve detuned? Explain.

12. With the aid of diagrams, describe briefly the construction and one application of each of the following: (*a*) insulating stub (or sleeve); (*b*) phase-reversing stub (or sleeve); (*c*) detuning stub (or sleeve); (*d*) matching stub.

13. (*a*) Determine the direction and magnitude of the electric field and of the magnetic field at the point *P* in Fig. 13 if the current at the input terminals of the antenna is 2 amp. Assume the antenna infinitely thin in calculating the field. (*b*) The antenna in (*a*) is lengthened so that $h = \lambda/2$ and the *maximum* current in the antenna is kept the same as in (*a*). (1) Does the electric field at *P* increase or decrease, and by how much? (2) Is the direction of the electric field at *P* changed?

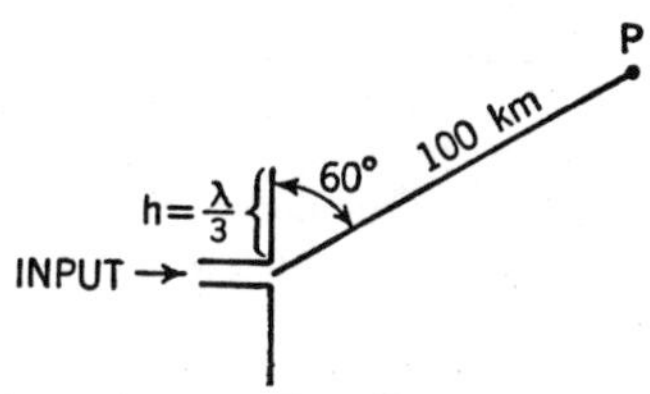

FIG. 13.

14. A center-driven antenna of half length *h* near $\lambda/4$ has an input resistance at resonance of 70 ohms. The voltage applied at the terminals is 100 volts rms at 300 mcps. (*a*) Calculate the rms electric field $\mathcal{E}_\theta$ at a distance of 2 km from the antenna at the angle $\theta = 60°$, as shown in Fig. 14. Assume the actual half length $h = \lambda/4$ for this purpose. (*b*) Determine the power transferred to the *matched load* of a receiving antenna placed parallel to the field calculated in (*a*), *i.e.*, at 2 km from the transmitting antenna with $\theta = 60°$ as before. The receiving antenna is exactly like the transmitting antenna as described. Assume its effective half length $h_e = \lambda/2\pi$. *Ans.:* (*a*) 0.035 volt/m, (*b*) 0.52 μwatt.

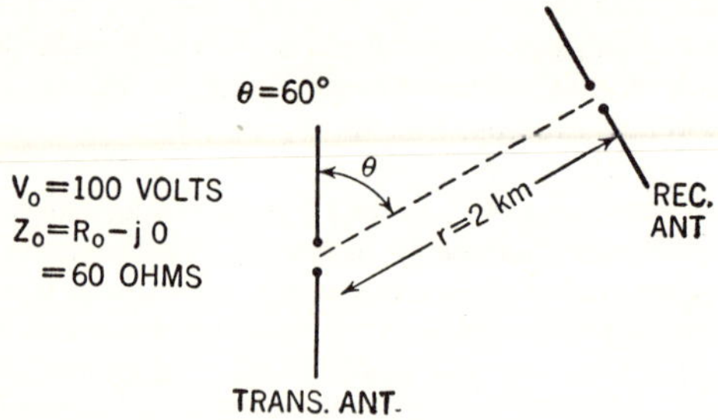

FIG. 14.

15. A symmetrical center-driven antenna has an input impedance at resonance of 72 ohms resistance if $\Omega = 20$. The corresponding h is 0.367 m if $f = 200$ mcps. Check the above values of R_0 and h. It is desired to find the voltage developed across a *conjugate matched load* used in conjunction with this antenna for receiving. The electric field is 500 μvolts/m rms in the vicinity of the receiving antenna and has a direction $\psi = 60°$ and $\theta_1 = 36.9°$. *Ans.:* 41.25 μvolts rms.

16. (*a*) Using the relation $F(\theta) = \dfrac{\cos(\pi/2 \cos\theta)}{\sin\theta}$, calculate sufficient points to plot the "vertical" field pattern for an isolated, extremely thin antenna. (*b*) What is the value of h, the antenna half length? (*c*) What is the shape of the "horizontal" pattern, *i.e.*, the pattern in the plane at right angles to the antenna?

17. Using the formula

$$|\mathcal{E}_\theta| = \frac{60 I_0}{R} \frac{\cos(H\cos\theta) - \cos H}{\sin H \sin\theta}$$

calculate the rms magnitude of the electric field in volts/m in the mid- or horizontal plane, $\theta = 90°$, at a point $r = 5$ km away from the axis of the antenna wire under the following conditions: (*a*) $h = \lambda/4$, $I_{max} = 10.0$ amp rms; (*b*) $h = 3\lambda/8$, $I_{max} = 10.0$ amp rms; and (*c*) $h = \lambda/2$, $I_{max} = 10.0$ amp rms. *Ans.:* (*a*) 0.12, (*b*) 0.204, (*c*) 0.24 volt/m.

18. (*a*) Calculate the *magnitude* of the distant electric field $\mathcal{E}_\theta$ in volts/meter at the point P, which is located in the horizontal plane, for two parallel, vertical

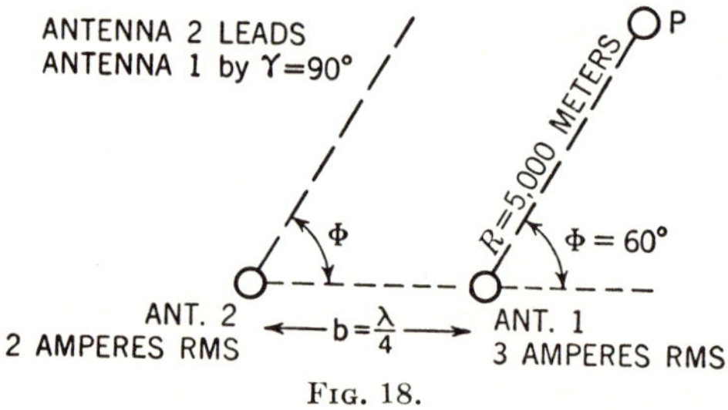

FIG. 18.

antennas of half length $h = \lambda/4$ [$F(\theta) = 1.0$]. Assume that antenna 1 has an input current of 3.0 amp rms and that antenna 2 has an input current of 2.0 amp rms. The phase angle between the currents is $\delta = 90°$, with antenna 2 leading antenna 1. The distance between the antennas is $b = \lambda/4$ for $f = 12$ mcps. The angle Φ is 60°, as shown in Fig. 18. Take $R = 5{,}000$ m. (*b*) Determine the magnitude of the magnetic field $\mathcal{H}_\Phi$ in amperes/meter at the same point. *Ans.:* (*a*) 0.0555 volt/m, (*b*) 1.472×10^{-4} amp/m.

19. (*a*) Design an array (consisting of four identical antennas, each of height $\lambda/2$, erected vertically on a perfectly conducting half-space) which will maintain a vertical electric field of 50 μvolts/m at points near the conducting plane, 1,200 km north and south of the array. The field is to have as sharp maxima as possible in north and south directions with zero field in the east and west directions. Describe the geometrical arrangement of the antennas with distances in fractions of a wavelength. Calculate magnitudes and relative phases of the maximum amplitudes of the currents to be maintained in the four units. The frequency is 100 mcps. (*b*) Sketch the horizontal pattern of the array, and the vertical pattern in the quadrant containing north. (*c*) Show a circuit arrangement using transmission lines for driving the four units from a single generator.

20. For the accompanying Fig. 20, calculate the field in volts/meter at the point P when $\theta = 45°$. The two dipoles are fed in phase, and the input currents are 10 amp rms. *Ans.:* $480/R$ volts/m.

The following is suggested as a suitable approach to the solution of this problem: (*a*) Find the amplitude and direction of the electric field of antenna 1 at P. (*b*) Follow the same procedure for antenna 2. (*c*) Combine these fields noting that they must be computed from the currents in the antennas at appropriately earlier times.

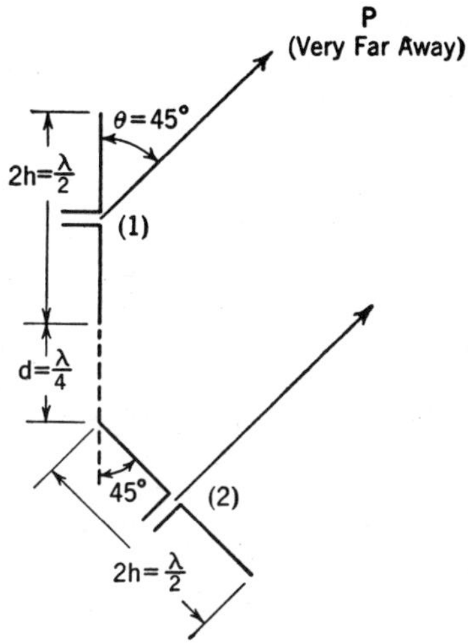

Fig. 20.

21. Repeat Prob. 20, but assume that the input current to antenna 2 leads the input current to antenna 1 by an electrical phase angle of 127.3°. *Ans.:* $976/R$ volts/m.

22. A symmetrical center-driven antenna of half length $h = \lambda/4$ is oriented *parallel* to the surface of the earth, which is to be assumed *perfectly conducting.* The antenna is supported at a height of 15 m by two poles. The current I_0 is 6 amp rms and the frequency is 4 mcps. (*a*) Find the rms magnitude of the electric field at a point in the vertical plane in a direction making an angle $\alpha = 30°$ with the earth's surface. The distance from the point to the driven antenna is 5 km. (*b*) Find the rms magnitude of the magnetic field at the same point.

Note: The vertical plane includes the point of field calculation, the center of the antenna, and the center of the earth. The antenna wires are perpendicular to this plane. Assume that for this purpose a sinusoidal distribution of current in the antenna is a good approximation. *Ans.:* (*a*) 0.0846 volt/m, (*b*) $2.25 \cdot 10^{-4}$ amp/m.

23. (*a*) Determine the equation for the vertical pattern of the isolated collinear array shown in Fig. 23. All elements carry currents of equal magnitude and in the same phase. (*b*) Using the equation determined in (*a*), calculate the distant field for $\theta = 0°, 30°, 60°, 90°$. Plot the result. (*c*) What is the shape of the "horizontal" field pattern? What is the direction of the electric field?

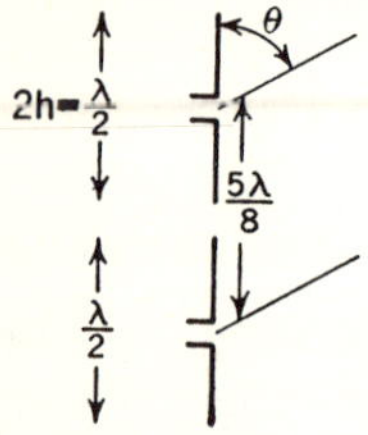

FIG. 23.

24. A center-driven antenna has an input resistance at resonance of 70 ohms. The voltage applied at the terminals is 11 volts at 3,000 mcps. (*a*) Calculate the electric field at a distance of 10 km from the antenna in a direction in which it is a maximum. Specify this direction. Assume the antenna equivalent to an indefinitely thin antenna of half length $h = \lambda/4$ for this purpose. (*b*) Determine the power transferred to the *matched load* of a receiving antenna placed parallel to the field calculated in (*a*), *i.e.*, at 10 km from the transmitter. The receiving antenna is exactly like the transmitting antenna as described. Assume its effective half length $h_e = \lambda/2\pi$.

25. A parasitic array consists of a vertical center-driven antenna and a parallel parasitic antenna, each of half length $h = \lambda/4$, and radius such that $\Omega = 2 \ln \frac{2h}{a} = 20$. They are separated a horizontal distance of 5 cm. A potential difference of 100 volts at $f = 300$ mcps is maintained at the terminals of the driven antenna. The parasitic antenna has a pure reactance of -40 ohms at its center. Determine the ratio of forward to backward electric field in the horizontal mid-plane of the array.

26. The following information applies to the portion of Fig. 26 including the generator and transmitting antenna:

Generator: emf is $V_1 = 55$ volts; impedance is 50 ohms pure resistance; frequency is 200 mcps.

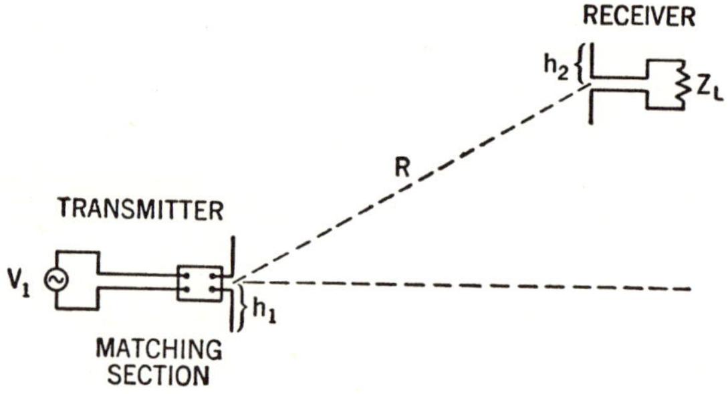

FIG. 26.

Line: Characteristic resistance is 500 ohms; length is 10 wavelengths to matching section; power loss is 10 per cent.

Matching section: Terminates line in R_c; power loss negligible.

Antenna: Radius is $a = 0.083$ cm; length is adjusted so that $h_1 = \lambda/2$; power loss in heating the antenna is negligible.

Determine the following: (*a*) input impedance of the antenna; (*b*) magnitude of the input current to the antenna; and (*c*) total power radiated.

27. The following information applies to the receiving antenna and load in Fig. 26.

Electric field (due to transmitter) at the receiving antenna: It is perpendicular to the line joining the centers of the two antennas; it is in the plane containing the antennas; its magnitude is 4 mv/m; the frequency is 200 mcps.

Receiving antenna: $\Omega = 2 \ln \frac{2h}{a} = 20$; its full length is adjusted near $\lambda/2$ so that the input impedance is a pure resistance; its center is 1.5 km from the center of the transmitting antenna and is 300 m higher.

Transmission line connecting the receiving antenna to the load Z_L: Input impedance is equal to the impedance of the antenna; power losses in the line are negligible because its length is less than a wavelength.

Load impedance: Z_L is $30 + j25$ ohms.

Determine the following: (*a*) impedance, actual length $2h$, and effective length $2h_e$ of the receiving antenna; (*b*) voltage across the input terminals of the line; and (*c*) power received by the load.

28. The following information applies to the portion of Fig. 28 including the generator and transmitting antenna:

Generator: Supplying 100 watts at 300 mcps to the transmission line.

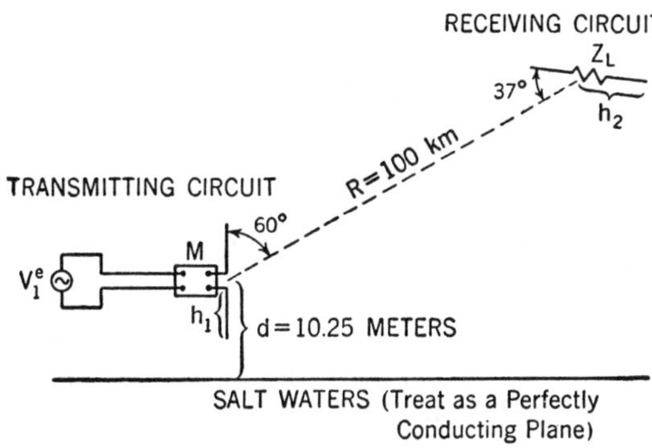

FIG. 28.

Line: Characteristic resistance is 400 ohms; length to matching section M is 8 wavelengths; power loss is 8 per cent.

Matching section (M): Terminates line in R_c; power loss is negligible.

Antenna: Radius $a = 0.1$ cm, length adjusted so that the antenna is resonant near $h_1 = \lambda/4$; power loss in heating the antenna is 2 per cent of that supplied to the antenna.

Determine the following: (*a*) input impedance of the antenna; (*b*) magnitude of input current to antenna; (*c*) half length h_1; and (*d*) total power radiated. (*e*) Determine the current at the center of a resonant, perfectly conducting, and infinitely thin antenna of half length $\lambda/4$ which radiates the power calculated in (*d*). (*f*) Calculate the rms electric field (due to the transmitting antenna) at the center of the receiving antenna, assuming the actual antenna replaced by the infinitely thin one described in (*e*). (*g*) Specify the direction of the electric field (due to the transmitter) at the center of the receiving antenna.

The following data apply to the receiving circuit in Fig. 28: $Z_L = 10^6$ ohms pure resistance. h_2 is adjusted near $\lambda/2$ for *maximum* effective half length h_e (this half length would make the antenna antiresonant if it were center-driven). h_2/a for the antenna is such that $2 \ln \frac{2h_2}{a} = 10$ (a is the radius of the antenna). (*h*) Determine the voltage across the load.

29. The two parallel, center-driven antennas in Fig. 29 have a half length $h = \lambda/2$ and a radius such that $\Omega = 2 \ln \frac{2h}{a} = 20$. They are driven at a frequency of 50 mcps. The section of two-wire line connecting their terminals has a length $\lambda/2$ and a characteristic resistance of 400 ohms, and the driving terminals AB are at the center. The maximum amplitude of the current is the same in each antenna and is 2 amp. (*a*) Assuming a sinusoidal distribution of current, determine the direction and the numerical magnitude of the distant electric field due to the array in the direction or directions in which this is a maximum at a distance of 300 km. (*b*) Make a sketch of the horizontal field pattern and a sketch of the vertical field pattern. In each sketch, show the antennas properly located; also indicate the magnitude of the electric field calculated in (*a*).

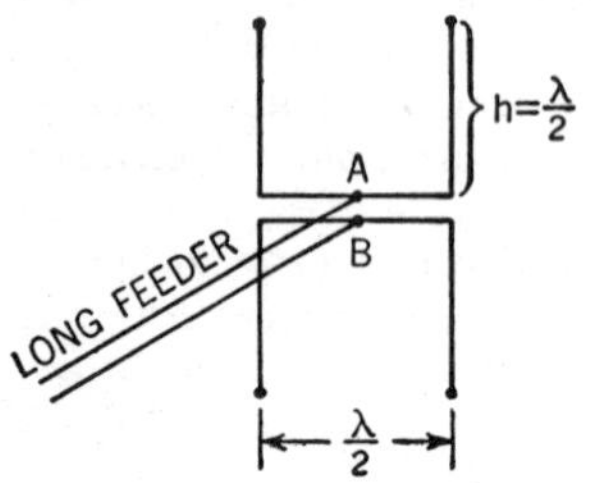

FIG. 29.

30. (*a*) Does the long (horizontal) feeding line carry line currents, antenna currents, or both, if arranged perpendicular to the plane containing the antennas, as in Fig. 29? Justify your answer. (*b*) Repeat (*a*) if the long vertical line feeds the array from below, with each wire of the line in a position equidistant from the parallel arms of the antennas. (*c*) Would it be preferable, from the point of view of minimum antenna currents on the feeder, to use a coaxial feeding line from the generator to the driving point AB instead of the long two-wire line in (*a*) or in (*b*)? Justify your answer.

31. Determine the numerical value of the impedance of the array (at AB) in Fig. 29.

Chapter III

1. It is desired to use a cylindrical wave guide to transmit power at 2,000 mcps from a generator with coaxial output to an antenna fed by a coaxial line. (*a*) Sketch the complete system, including matching devices, needed to make the system efficient for use with the $TE_{1,1}$ mode. (*b*) What should be the inside diameter of the cylindrical wave guide in order to pass the $TE_{1,1}$ mode but not the $TM_{0,1}$ mode at 2,000 mcps? (*c*) Under what conditions should the antenna (used to extract energy from the wave guide) not be placed parallel to the antenna used to excite the wave guide?

2. An u-h-f oscillator generating 4,000 mcps is to be used to drive an electromagnetic horn using a short coaxial line and a long *rectangular* hollow wave guide. (*a*) Show diagrammatically the complete circuit including necessary accessories to ensure optimum performance. Explain why these accessories are necessary. (*b*) What numerical values would you recommend for the cross-sectional dimensions of the rectangular pipe? Why?

3. A *rectangular* hollow wave guide is to be used as a transmission circuit at 3,000 mcps. It is desired to make use of the $TE_{1,0}$ mode. (*a*) What are practical cross-sectional dimensions a and b for the wave guide? (*b*) Show a sketch of a cross section where at a given instant the electric field distribution and the charge distribution are maximum. What are the currents and where? Label the sides a and b. (*c*) With the aid of a sketch of the same cross section describe the charge, current, and field distributions a quarter period later. (*d*) State under what conditions a rectangular pipe may be nonresonant in the axial direction, yet resonant in each transverse plane.

4. A cylindrical hollow wave guide is to be used to transmit a pure 1,000-mcps signal from the generator to the transmitting antenna. The wave-guide diameter is 22 cm. (*a*) Show, by calculation, what mode or modes can be excited at this frequency. (*b*) The locations of two adjacent resonance peaks are determined by means of a standing-wave detector, in "tuning up" this system. How far apart in centimeters are the peaks? (*c*) It is desired to obtain the most efficient transfer of power by the wave guide. Would you adjust for a high or a low standing-wave ratio? Why? (*d*) Where should a slot be cut in the guide in measuring the standing-wave ratio? Why?

5. Compute the following quantities for a hollow conductor of rectangular cross section with sides $a = 6.35$ cm, $b = 3.15$ cm, when filled with air. The wavelength of the driving generator as measured in a coaxial wavemeter is 9.725 cm. $TE_{1,0}$ mode is excited. (*a*) The cutoff frequency f_{cutoff} and wavelength λ_{cutoff}; (*b*) the wavelength λ_g in the pipe; (*c*) phase and group velocities v_{gp} and v_{gg}; (*d*) the characteristic resistance R_{cTE}.

6. Outline, with the aid of a labeled diagram, a procedure for measuring experimentally the phase function Φ_s of a termination for a metal pipe.

7. Describe or show sketches of reactive elements (for use with the $TE_{1,0}$ mode in a rectangular pipe) which behave insofar as their phase functions Φ_s

are concerned, like the following essentially reactive elements shunted across a parallel line: capacitor, inductor, series-resonant *LC* circuit, parallel-resonant *LC* circuit, stub section of line, a piece of straight wire.

8. Summarize the similarities and differences between the following quantity or quantities as defined for conventional parallel and coaxial lines and for circular pipes in the $TM_{0,1}$ and $TE_{1,1}$ modes: (*a*) characteristic impedance, (*b*) open end, (*c*) current and charge distributions, (*d*) wavelength measured along the line or guide and λ_{TEM}, (*e*) cutoff wavelength, and (*f*) attenuation constant.

9. A rectangular wave guide of sides $a = 7$ cm, $b = 3$ cm, and of great length is terminated at the far end in a highly absorbing load. It is driven at 3,000 mcps by an antenna inserted a short distance from a metal piston constituting the termination at the near end of the pipe. State briefly where and in what directions large currents may be expected under the following circumstances: (*a*) the antenna is inserted at the center of side *b* parallel to side *a*; and (*b*) the antenna is inserted at the center of side *a* parallel to side *b*.

10. A circular pipe of inner radius *b* is driven at 3,000 mcps from a short movable antenna which is kept at a fixed distance $\lambda_g/4$ from a movable piston at one end. A movable detector is kept at a distance $2\lambda_g$ from the antenna. The other end of the pipe is terminated in a slit. The dominant mode is excited. Maximum deflection is observed in the detector when the length of the pipe between slit and piston is 2.015 m, and 2.090 m when the dominant mode is excited. The widths of the two resonance curves at these resonant lengths as measured between locations of the piston giving 0.707 of maximum deflection on each side of the maximum are found to be the same within the available precision and equal to 0.7 cm. When the slit is closed, maximum deflection occurs with a distance between closed slit and movable piston of 2.040 m. The width of the resonance curve is 0.2 cm. (*a*) Determine the wavelength and the phase constant in the pipe. (*b*) Show that the over-all attenuation of the pipe is negligible. (*c*) What is the inner radius of the pipe? (*d*) Is the $TM_{0,1}$ mode possible? (*e*) Determine the attenuation factor A_0 for the generator end of the pipe including the detector. (*f*) Determine the attenuation factor A_s for the slit. (*g*) Determine the phase factor Φ_s for the slit. (*h*) Determine the theoretical standing-wave ratio with the slit open. If the detector were moved, would the same value be observed? Explain.

INDEX

D

M

N

O

P

Q

R

CATALOGUE OF DOVER BOOKS

PHYSICS

General physics

FOUNDATIONS OF PHYSICS, R. B. Lindsay & H. Margenau. Excellent bridge between semi-popular works & technical treatises. A discussion of methods of physical description, construction of theory; valuable for physicist with elementary calculus who is interested in ideas that give meaning to data, tools of modern physics. Contents include symbolism, mathematical equations; space & time foundations of mechanics; probability; physics & continua; electron theory; special & general relativity; quantum mechanics; causality. "Thorough and yet not overdetailed. Unreservedly recommended," NATURE (London). Unabridged, corrected edition. List of recommended readings. 35 illustrations. xi + 537pp. 5⅜ x 8.
S377 Paperbound **$2.75**

FUNDAMENTAL FORMULAS OF PHYSICS, ed. by D. H. Menzel. Highly useful, fully inexpensive reference and study text, ranging from simple to highly sophisticated operations. Mathematics integrated into text—each chapter stands as short textbook of field represented. Vol. 1: Statistics, Physical Constants, Special Theory of Relativity, Hydrodynamics, Aerodynamics, Boundary Value Problems in Math. Physics; Viscosity, Electromagnetic Theory, etc. Vol. 2: Sound, Acoustics, Geometrical Optics, Electron Optics, High-Energy Phenomena, Magnetism, Biophysics, much more. Index. Total of 800pp. 5⅜ x 8. Vol. 1 S595 Paperbound **$2.00**
Vol. 2 S596 Paperbound **$2.00**

MATHEMATICAL PHYSICS, D. H. Menzel. Thorough one-volume treatment of the mathematical techniques vital for classic mechanics, electromagnetic theory, quantum theory, and relativity. Written by the Harvard Professor of Astrophysics for junior, senior, and graduate courses, it gives clear explanations of all those aspects of function theory, vectors, matrices, dyadics, tensors, partial differential equations, etc., necessary for the understanding of the various physical theories. Electron theory, relativity, and other topics seldom presented appear here in considerable detail. Scores of definitions, conversion factors, dimensional constants, etc. "More detailed than normal for an advanced text . . . excellent set of sections on Dyadics, Matrices, and Tensors," JOURNAL OF THE FRANKLIN INSTITUTE. Index. 193 problems, with answers. x + 412pp. 5⅜ x 8. S56 Paperbound **$2.00**

THE SCIENTIFIC PAPERS OF J. WILLARD GIBBS. All the published papers of America's outstanding theoretical scientist (except for "Statistical Mechanics" and "Vector Analysis"). Vol I (thermodynamics) contains one of the most brilliant of all 19th-century scientific papers—the 300-page "On the Equilibrium of Heterogeneous Substances," which founded the science of physical chemistry, and clearly stated a number of highly important natural laws for the first time; 8 other papers complete the first volume. Vol II includes 2 papers on dynamics, 8 on vector analysis and multiple algebra, 5 on the electromagnetic theory of light, and 6 miscellaneous papers. Biographical sketch by H. A. Bumstead. Total of xxxvi + 718pp. 5⅝ x 8⅜.
S721 Vol I Paperbound **$2.50**
S722 Vol II Paperbound **$2.00**
The set **$4.50**

BASIC THEORIES OF PHYSICS, Peter Gabriel Bergmann. Two-volume set which presents a critical examination of important topics in the major subdivisions of classical and modern physics. The first volume is concerned with classical mechanics and electrodynamics: mechanics of mass points, analytical mechanics, matter in bulk, electrostatics and magnetostatics, electromagnetic interaction, the field waves, special relativity, and waves. The second volume (Heat and Quanta) contains discussions of the kinetic hypothesis, physics and statistics, stationary ensembles, laws of thermodynamics, early quantum theories, atomic spectra, probability waves, quantization in wave mechanics, approximation methods, and abstract quantum theory. A valuable supplement to any thorough course or text.
Heat and Quanta: Index. 8 figures. x + 300pp. 5⅜ x 8½. S968 Paperbound **$1.75**
Mechanics and Electrodynamics: Index. 14 figures. vii + 280pp. 5⅜ x 8½.
S969 Paperbound **$1.75**

THEORETICAL PHYSICS, A. S. Kompaneyets. One of the very few thorough studies of the subject in this price range. Provides advanced students with a comprehensive theoretical background. Especially strong on recent experimentation and developments in quantum theory. Contents: Mechanics (Generalized Coordinates, Lagrange's Equation, Collision of Particles, etc.), Electrodynamics (Vector Analysis, Maxwell's equations, Transmission of Signals, Theory of Relativity, etc.), Quantum Mechanics (the Inadequacy of Classical Mechanics, the Wave Equation, Motion in a Central Field, Quantum Theory of Radiation, Quantum Theories of Dispersion and Scattering, etc.), and Statistical Physics (Equilibrium Distribution of Molecules in an Ideal Gas, Boltzmann statistics, Bose and Fermi Distribution, Thermodynamic Quantities, etc.). Revised to 1961. Translated by George Yankovsky, authorized by Kompaneyets. 137 exercises. 56 figures. 529pp. 5⅜ x 8½. S972 Paperbound **$2.50**

ANALYTICAL AND CANONICAL FORMALISM IN PHYSICS, André Mercier. A survey, in one volume, of the variational principles (the key principles—in mathematical form—from which the basic laws of any one branch of physics can be derived) of the several branches of physical theory, together with an examination of the relationships among them. Contents: the Lagrangian Formalism, Lagrangian Densities, Canonical Formalism, Canonical Form of Electrodynamics, Hamiltonian Densities, Transformations, and Canonical Form with Vanishing Jacobian Determinant. Numerous examples and exercises. For advanced students, teachers, etc. 6 figures. Index. viii + 222pp. 5⅜ x 8½. S1077 Paperbound **$1.75**

Acoustics, optics, electricity and magnetism, electromagnetics, magneto-hydrodynamics

THE THEORY OF SOUND, Lord Rayleigh. Most vibrating systems likely to be encountered in practice can be tackled successfully by the methods set forth by the great Nobel laureate, Lord Rayleigh. Complete coverage of experimental, mathematical aspects of sound theory. Partial contents: Harmonic motions, vibrating systems in general, lateral vibrations of bars, curved plates or shells, applications of Laplace's functions to acoustical problems, fluid friction, plane vortex-sheet, vibrations of solid bodies, etc. This is the first inexpensive edition of this great reference and study work. Bibliography. Historical introduction by R. B. Lindsay. Total of 1040pp. 97 figures. 5⅜ x 8.
S292, S293, Two volume set, paperbound, **$4.70**

THE DYNAMICAL THEORY OF SOUND, H. Lamb. Comprehensive mathematical treatment of the physical aspects of sound, covering the theory of vibrations, the general theory of sound, and the equations of motion of strings, bars, membranes, pipes, and resonators. Includes chapters on plane, spherical, and simple harmonic waves, and the Helmholtz Theory of Audition. Complete and self-contained development for student and specialist; all fundamental differential equations solved completely. Specific mathematical details for such important phenomena as harmonics, normal modes, forced vibrations of strings, theory of reed pipes, etc. Index. Bibliography. 86 diagrams. viii + 307pp. 5⅜ x 8. S655 Paperbound **$1.50**

WAVE PROPAGATION IN PERIODIC STRUCTURES, L. Brillouin. A general method and application to different problems: pure physics, such as scattering of X-rays of crystals, thermal vibration in crystal lattices, electronic motion in metals; and also problems of electrical engineering. Partial contents: elastic waves in 1-dimensional lattices of point masses. Propagation of waves along 1-dimensional lattices. Energy flow. 2 dimensional, 3 dimensional lattices. Mathieu's equation. Matrices and propagation of waves along an electric line. Continuous electric lines. 131 illustrations. Bibliography. Index. xii + 253pp. 5⅜ x 8.
S34 Paperbound **$2.00**

THEORY OF VIBRATIONS, N. W. McLachlan. Based on an exceptionally successful graduate course given at Brown University, this discusses linear systems having 1 degree of freedom, forced vibrations of simple linear systems, vibration of flexible strings, transverse vibrations of bars and tubes, transverse vibration of circular plate, sound waves of finite amplitude, etc. Index. 99 diagrams. 160pp. 5⅜ x 8. S190 Paperbound **$1.35**

LIGHT: PRINCIPLES AND EXPERIMENTS, George S. Monk. Covers theory, experimentation, and research. Intended for students with some background in general physics and elementary calculus. Three main divisions: 1) Eight chapters on geometrical optics—fundamental concepts (the ray and its optical length, Fermat's principle, etc.), laws of image formation, apertures in optical systems, photometry, optical instruments etc.; 2) 9 chapters on physical optics—interference, diffraction, polarization, spectra, the Rayleigh refractometer, the wave theory of light, etc.; 3) 23 instructive experiments based directly on the theoretical text. "Probably the best intermediate textbook on light in the English language. Certainly, it is the best book which includes both geometrical and physical optics," J. Rud Nielson, PHYSICS FORUM. Revised edition. 102 problems and answers. 12 appendices. 6 tables. Index. 270 illustrations. xi +489pp. 5⅜ x 8½. S341 Paperbound **$2.50**

PHOTOMETRY, John W. T. Walsh. The best treatment of both "bench" and "illumination" photometry in English by one of Britain's foremost experts in the field (President of the International Commission on Illumination). Limited to those matters, theoretical and practical, which affect the measurement of light flux, candlepower, illumination, etc., and excludes treatment of the use to which such measurements may be put after they have been made. Chapters on Radiation, The Eye and Vision, Photo-Electric Cells, The Principles of Photometry, The Measurement of Luminous Intensity, Colorimetry, Spectrophotometry, Stellar Photometry, The Photometric Laboratory, etc. Third revised (1958) edition. 281 illustrations. 10 appendices. xxiv + 544pp. 5½ x 9¼. S319 Clothbound **$10.00**

EXPERIMENTAL SPECTROSCOPY, R. A. Sawyer. Clear discussion of prism and grating spectrographs and the techniques of their use in research, with emphasis on those principles and techniques that are fundamental to practically all uses of spectroscopic equipment. Beginning with a brief history of spectroscopy, the author covers such topics as light sources, spectroscopic apparatus, prism spectroscopes and graphs, diffraction grating, the photographic process, determination of wave length, spectral intensity, infrared spectroscopy, spectrochemical analysis, etc. This revised edition contains new material on the production of replica gratings, solar spectroscopy from rockets, new standard of wave length, etc. Index. Bibliography. 111 illustrations. x + 358pp. 5⅜ x 8½. S1045 Paperbound **$2.25**

FUNDAMENTALS OF ELECTRICITY AND MAGNETISM, L. B. Loeb. For students of physics, chemistry, or engineering who want an introduction to electricity and magnetism on a higher level and in more detail than general elementary physics texts provide. Only elementary differential and integral calculus is assumed. Physical laws developed logically, from magnetism to electric currents, Ohm's law, electrolysis, and on to static electricity, induction, etc. Covers an unusual amount of material; one third of book on modern material: solution of wave equation, photoelectric and thermionic effects, etc. Complete statement of the various electrical systems of units and interrelations. 2 Indexes. 75 pages of problems with answers stated. Over 300 figures and diagrams. xix +669pp. 5⅜ x 8. S745 Paperbound **$2.75**

MATHEMATICAL ANALYSIS OF ELECTRICAL AND OPTICAL WAVE-MOTION, Harry Bateman. Written by one of this century's most distinguished mathematical physicists, this is a practical introduction to those developments of Maxwell's electromagnetic theory which are directly connected with the solution of the partial differential equation of wave motion. Methods of solving wave-equation, polar-cylindrical coordinates, diffraction, transformation of coordinates, homogeneous solutions, electromagnetic fields with moving singularities, etc. Index. 168pp. 5⅜ x 8. S14 Paperbound **$1.75**

PRINCIPLES OF PHYSICAL OPTICS, Ernst Mach. This classical examination of the propagation of light, color, polarization, etc. offers an historical and philosophical treatment that has never been surpassed for breadth and easy readability. Contents: Rectilinear propagation of light. Reflection, refraction. Early knowledge of vision. Dioptrics. Composition of light. Theory of color and dispersion. Periodicity. Theory of interference. Polarization. Mathematical representation of properties of light. Propagation of waves, etc. 279 illustrations, 10 portraits. Appendix. Indexes. 324pp. 5⅜ x 8. S178 Paperbound **$1.75**

THE THEORY OF OPTICS, Paul Drude. One of finest fundamental texts in physical optics, classic offers thorough coverage, complete mathematical treatment of basic ideas. Includes fullest treatment of application of thermodynamics to optics; sine law in formation of images, transparent crystals, magnetically active substances, velocity of light, apertures, effects depending upon them, polarization, optical instruments, etc. Introduction by A. A. Michelson. Index. 110 illus. 567pp. 5⅜ x 8. S532 Paperbound **$2.45**

ELECTRICAL THEORY ON THE GIORGI SYSTEM, P. Cornelius. A new clarification of the fundamental concepts of electricity and magnetism, advocating the convenient m.k.s. system of units that is steadily gaining followers in the sciences. Illustrating the use and effectiveness of his terminology with numerous applications to concrete technical problems, the author here expounds the famous Giorgi system of electrical physics. His lucid presentation and well-reasoned, cogent argument for the universal adoption of this system form one of the finest pieces of scientific exposition in recent years. 28 figures. Index. Conversion tables for translating earlier data into modern units. Translated from 3rd Dutch edition by L. J. Jolley. x + 187pp. 5½ x 8¾. S909 Clothbound **$6.00**

ELECTRIC WAVES: BEING RESEARCHES ON THE PROPAGATION OF ELECTRIC ACTION WITH FINITE VELOCITY THROUGH SPACE, Heinrich Hertz. This classic work brings together the original papers in which Hertz—Helmholtz's protegé and one of the most brilliant figures in 19th-century research—probed the existence of electromagnetic waves and showed experimentally that their velocity equalled that of light, research that helped lay the groundwork for the development of radio, television, telephone, telegraph, and other modern technological marvels. Unabridged republication of original edition. Authorized translation by D. E. Jones. Preface by Lord Kelvin. Index of names. 40 illustrations. xvii + 278pp. 5⅜ x 8½. S57 Paperbound **$1.75**

PIEZOELECTRICITY: AN INTRODUCTION TO THE THEORY AND APPLICATIONS OF ELECTROMECHANICAL PHENOMENA IN CRYSTALS, Walter G. Cady. This is the most complete and systematic coverage of this important field in print—now regarded as something of scientific classic. This republication, revised and corrected by Prof. Cady—one of the foremost contributors in this area—contains a sketch of recent progress and new material on Ferroelectrics. Time Standards, etc. The first 7 chapters deal with fundamental theory of crystal electricity. 5 important chapters cover basic concepts of piezoelectricity, including comparisons of various competing theories in the field. Also discussed: piezoelectric resonators (theory, methods of manufacture, influences of air-gaps, etc.); the piezo oscillator; the properties, history, and observations relating to Rochelle salt; ferroelectric crystals; miscellaneous applications of piezoelectricity; pyroelectricity; etc. "A great work," W. A. Wooster, NATURE. Revised (1963) and corrected edition. New preface by Prof. Cady. 2 Appendices. Indices. Illustrations. 62 tables. Bibliography. Problems. Total of 1 + 822pp. 5⅜ x 8½.
S1094 Vol. I Paperbound **$2.50**
S1095 Vol. II Paperbound **$2.50**
Two volume set Paperbound **$5.00**

MAGNETISM AND VERY LOW TEMPERATURES, H. B. G. Casimir. A basic work in the literature of low temperature physics. Presents a concise survey of fundamental theoretical principles, and also points out promising lines of investigation. Contents: Classical Theory and Experimental Methods, Quantum Theory of Paramagnetism, Experiments on Adiabatic Demagnetization. Theoretical Discussion of Paramagnetism at Very Low Temperatures, Some Experimental Results, Relaxation Phenomena. Index. 89-item bibliography. ix + 95pp. 5⅜ x 8. S943 Paperbound **$1.25**

SELECTED PAPERS ON NEW TECHNIQUES FOR ENERGY CONVERSION: THERMOELECTRIC METHODS; THERMIONIC; PHOTOVOLTAIC AND ELECTRICAL EFFECTS; FUSION, Edited by Sumner N. Levine. Brings together in one volume the most important papers (1954-1961) in modern energy technology. Included among the 37 papers are general and qualitative descriptions of the field as a whole, indicating promising lines of research. Also: 15 papers on thermoelectric methods, 7 on thermionic, 5 on photovoltaic, 4 on electrochemical effect, and 2 on controlled fusion research. Among the contributors are: Joffe, Maria Telkes, Herold, Herring, Douglas, Jaumot, Post, Austin, Wilson, Pfann, Rappaport, Morehouse, Domenicali, Moss, Bowers, Harman, Von Doenhoef. Preface and introduction by the editor. Bibliographies. xxviii + 451pp. 6⅛ x 9¼. S37 Paperbound **$3.00**

SUPERFLUIDS: MACROSCOPIC THEORY OF SUPERCONDUCTIVITY, Vol. I, Fritz London. The major work by one of the founders and great theoreticians of modern quantum physics. Consolidates the researches that led to the present understanding of the nature of superconductivity. Prof. London here reveals that quantum mechanics is operative on the macroscopic plane as well as the submolecular level. Contents: Properties of Superconductors and Their Thermodynamical Correlation; Electrodynamics of the Pure Superconducting State; Relation between Current and Field; Measurements of the Penetration Depth; Non-Viscous Flow vs. Superconductivity; Micro-waves in Superconductors; Reality of the Domain Structure; and many other related topics. A new epilogue by M. J. Buckingham discusses developments in the field up to 1960. Corrected and expanded edition. An appreciation of the author's life and work by L. W. Nordheim. Biography by Edith London. Bibliography of his publications. 45 figures. 2 Indices. xviii + 173pp. 5⅝ x 8⅜. S44 Paperbound **$1.45**

SELECTED PAPERS ON PHYSICAL PROCESSES IN IONIZED PLASMAS, Edited by Donald H. Menzel, Director, Harvard College Observatory. 30 important papers relating to the study of highly ionized gases or plasmas selected by a foremost contributor in the field, with the assistance of Dr. L. H. Aller. The essays include 18 on the physical processes in gaseous nebulae, covering problems of radiation and radiative transfer, the Balmer decrement, electron temperatives, spectrophotometry, etc. 10 papers deal with the interpretation of nebular spectra, by Bohm, Van Vleck, Aller, Minkowski, etc. There is also a discussion of the intensities of "forbidden" spectral lines by George Shortley and a paper concerning the theory of hydrogenic spectra by Menzel and Pekeris. Other contributors: Goldberg, Hebb, Baker, Bowen, Ufford, Liller, etc. viii + 374pp. 6⅛ x 9¼. S60 Paperbound **$2.95**

THE ELECTROMAGNETIC FIELD, Max Mason & Warren Weaver. Used constantly by graduate engineers. Vector methods exclusively: detailed treatment of electrostatics, expansion methods, with tables converting any quantity into absolute electromagnetic, absolute electrostatic, practical units. Discrete charges, ponderable bodies, Maxwell field equations, etc. Introduction. Indexes. 416pp. 5⅜ x 8. S185 Paperbound **$2.00**

THEORY OF ELECTRONS AND ITS APPLICATION TO THE PHENOMENA OF LIGHT AND RADIANT HEAT, H. Lorentz. Lectures delivered at Columbia University by Nobel laureate Lorentz. Unabridged, they form a historical coverage of the theory of free electrons, motion, absorption of heat, Zeeman effect, propagation of light in molecular bodies, inverse Zeeman effect, optical phenomena in moving bodies, etc. 109 pages of notes explain the more advanced sections. Index. 9 figures. 352pp. 5⅜ x 8. S173 Paperbound **$1.85**

FUNDAMENTAL ELECTROMAGNETIC THEORY, Ronold P. King, Professor Applied Physics, Harvard University. Original and valuable introduction to electromagnetic theory and to circuit theory from the standpoint of electromagnetic theory. Contents: Mathematical Description of Matter—stationary and nonstationary states; Mathematical Description of Space and of Simple Media—Field Equations, Integral Forms of Field Equations, Electromagnetic Force, etc.; Transformation of Field and Force Equations; Electromagnetic Waves in Unbounded Regions; Skin Effect and Internal Impedance—in a solid cylindrical conductor, etc.; and Electrical Circuits—Analytical Foundations, Near-zone and quasi-near zone circuits, Balanced two-wire and four-wire transmission lines. Revised and enlarged version. New preface by the author. 5 appendices (Differential operators: Vector Formulas and Identities, etc.). Problems. Indexes. Bibliography. xvi + 580pp. 5⅜ x 8½. S1023 Paperbound **$2.75**

Hydrodynamics

A TREATISE ON HYDRODYNAMICS, A. B. Basset. Favorite text on hydrodynamics for 2 generations of physicists, hydrodynamical engineers, oceanographers, ship designers, etc. Clear enough for the beginning student, and thorough source for graduate students and engineers on the work of d'Alembert, Euler, Laplace, Lagrange, Poisson, Green, Clebsch, Stokes, Cauchy, Helmholtz, J. J. Thomson, Love, Hicks, Greenhill, Besant, Lamb, etc. Great amount of documentation on entire theory of classical hydrodynamics. Vol I: theory of motion of frictionless liquids, vortex, and cyclic irrotational motion, etc. 132 exercises. Bibliography. 3 Appendixes. xii + 264pp. Vol II: motion in viscous liquids, harmonic analysis, theory of tides, etc. 112 exercises, Bibliography. 4 Appendixes. xv + 328pp. Two volume set. 5⅜ x 8.

S724 Vol I Paperbound **$1.75**
S725 Vol II Paperbound **$1.75**
The set **$3.50**

HYDRODYNAMICS, Horace Lamb. Internationally famous complete coverage of standard reference work on dynamics of liquids & gases. Fundamental theorems, equations, methods, solutions, background, for classical hydrodynamics. Chapters include Equations of Motion, Integration of Equations in Special Gases, Irrotational Motion, Motion of Liquid in 2 Dimensions, Motion of Solids through Liquid-Dynamical Theory, Vortex Motion, Tidal Waves, Surface Waves, Waves of Expansion, Viscosity, Rotating Masses of liquids. Excellently planned, arranged; clear, lucid presentation. 6th enlarged, revised edition. Index. Over 900 footnotes, mostly bibliographical. 119 figures. xv + 738pp. 6⅛ x 9¼. S256 Paperbound **$3.25**

HYDRODYNAMICS, H. Dryden, F. Murnaghan, Harry Bateman. Published by the National Research Council in 1932 this enormous volume offers a complete coverage of classical hydrodynamics. Encyclopedic in quality. Partial contents: physics of fluids, motion, turbulent flow, compressible fluids, motion in 1, 2, 3 dimensions; viscous fluids rotating, laminar motion, resistance of motion through viscous fluid, eddy viscosity, hydraulic flow in channels of various shapes, discharge of gases, flow past obstacles, etc. Bibliography of over 2,900 items. Indexes. 23 figures. 634pp. 5⅜ x 8. S303 Paperbound **$2.75**

Mechanics, dynamics, thermodynamics, elasticity

MECHANICS, J. P. Den Hartog. Already a classic among introductory texts, the M.I.T. professor's lively and discursive presentation is equally valuable as a beginner's text, an engineering student's refresher, or a practicing engineer's reference. Emphasis in this highly readable text is on illuminating fundamental principles and showing how they are embodied in a great number of real engineering and design problems: trusses, loaded cables, beams, jacks, hoists, etc. Provides advanced material on relative motion and gyroscopes not usual in introductory texts. "Very thoroughly recommended to all those anxious to improve their real understanding of the principles of mechanics." MECHANICAL WORLD. Index. List of equations. 334 problems, all with answers. Over 550 diagrams and drawings. ix + 462pp. 5⅜ x 8.
S754 Paperbound **$2.00**

THEORETICAL MECHANICS: AN INTRODUCTION TO MATHEMATICAL PHYSICS, J. S. Ames, F. D. Murnaghan. A mathematically rigorous development of theoretical mechanics for the advanced student, with constant practical applications. Used in hundreds of advanced courses. An unusually thorough coverage of gyroscopic and baryscopic material, detailed analyses of the Coriolis acceleration, applications of Lagrange's equations, motion of the double pendulum, Hamilton-Jacobi partial differential equations, group velocity and dispersion, etc. Special relativity is also included. 159 problems. 44 figures. ix + 462pp. 5⅜ x 8.
S461 Paperbound **$2.00**

THEORETICAL MECHANICS: STATICS AND THE DYNAMICS OF A PARTICLE, W. D. MacMillan. Used for over 3 decades as a self-contained and extremely comprehensive advanced undergraduate text in mathematical physics, physics, astronomy, and deeper foundations of engineering. Early sections require only a knowledge of geometry; later, a working knowledge of calculus. Hundreds of basic problems, including projectiles to the moon, escape velocity, harmonic motion, ballistics, falling bodies, transmission of power, stress and strain, elasticity, astronomical problems. 340 practice problems plus many fully worked out examples make it possible to test and extend principles developed in the text. 200 figures. xvii + 430pp. 5⅜ x 8. S467 Paperbound **$2.00**

THEORETICAL MECHANICS: THE THEORY OF THE POTENTIAL, W. D. MacMillan. A comprehensive, well balanced presentation of potential theory, serving both as an introduction and a reference work with regard to specific problems, for physicists and mathematicians. No prior knowledge of integral relations is assumed, and all mathematical material is developed as it becomes necessary. Includes: Attraction of Finite Bodies; Newtonian Potential Function; Vector Fields, Green and Gauss Theorems; Attractions of Surfaces and Lines; Surface Distribution of Matter; Two-Layer Surfaces; Spherical Harmonics; Ellipsoidal Harmonics; etc. "The great number of particular cases . . . should make the book valuable to geophysicists and others actively engaged in practical applications of the potential theory," Review of Scientific Instruments. Index. Bibliography. xiii + 469pp. 5⅜ x 8. S486 Paperbound **$2.25**

THEORETICAL MECHANICS: DYNAMICS OF RIGID BODIES, W. D. MacMillan. Theory of dynamics of a rigid body is developed, using both the geometrical and analytical methods of instruction. Begins with exposition of algebra of vectors, it goes through momentum principles, motion in space, use of differential equations and infinite series to solve more sophisticated dynamics problems. Partial contents: moments of inertia, systems of free particles, motion parallel to a fixed plane, rolling motion, method of periodic solutions, much more. 82 figs. 199 problems. Bibliography. Indexes. xii + 476pp. 5⅜ x 8. S641 Paperbound **$2.00**

MATHEMATICAL FOUNDATIONS OF STATISTICAL MECHANICS, A. I. Khinchin. Offering a precise and rigorous formulation of problems, this book supplies a thorough and up-to-date exposition. It provides analytical tools needed to replace cumbersome concepts, and furnishes for the first time a logical step-by-step introduction to the subject. Partial contents: geometry & kinematics of the phase space, ergodic problem, reduction to theory of probability, application of central limit problem, ideal monatomic gas, foundation of thermo-dynamics, dispersion and distribution of sum functions. Key to notations. Index. viii + 179pp. 5⅜ x 8.
S147 Paperbound **$1.50**

ELEMENTARY PRINCIPLES IN STATISTICAL MECHANICS, J. W. Gibbs. Last work of the great Yale mathematical physicist, still one of the most fundamental treatments available for advanced students and workers in the field. Covers the basic principle of conservation of probability of phase, theory of errors in the calculated phases of a system, the contributions of Clausius, Maxwell, Boltzmann, and Gibbs himself, and much more. Includes valuable comparison of statistical mechanics with thermodynamics: Carnot's cycle, mechanical definitions of entropy, etc. xvi + 208pp. 5⅜ x 8. S707 Paperbound **$1.45**

PRINCIPLES OF MECHANICS AND DYNAMICS, Sir William Thomson (Lord Kelvin) and Peter Guthrie Tait. The principles and theories of fundamental branches of classical physics explained by two of the greatest physicists of all time. A broad survey of mechanics, with material on hydrodynamics, elasticity, potential theory, and what is now standard mechanics Thorough and detailed coverage, with many examples, derivations, and topics not included in more recent studies. Only a knowledge of calculus is needed to work through this book. Vol. I (Preliminary): Kinematics; Dynamical Laws and Principles; Experience (observation, experimentation, formation of hypotheses, scientific method); Measures and Instruments; Continuous Calculating Machines. Vol. II (Abstract Dynamics): Statics of a Particle—Attraction; Statics of Solids and Fluids. Formerly Titled "Treatise on Natural Philosophy." Unabridged reprint of revised edition. Index. 168 diagrams. Total of xlii + 1035pp. 5⅜ x 8½.
Vol. I: S966 Paperbound **$2.35**
Vol. II: S967 Paperbound **$2.35**
Two volume Set Paperbound **$4.70**

INVESTIGATIONS ON THE THEORY OF THE BROWNIAN MOVEMENT, Albert Einstein. Reprints from rare European journals. 5 basic papers, including the Elementary Theory of the Brownian Movement, written at the request of Lorentz to provide a simple explanation. Translated by A. D. Cowper. Annotated, edited by R. Fürth. 33pp. of notes elucidate, give history of previous investigations. Author, subject indexes. 62 footnotes. 124pp. 5⅜ x 8.
S304 Paperbound **$1.25**

MECHANICS VIA THE CALCULUS, P. W. Norris, W. S. Legge. Covers almost everything, from linear motion to vector analysis: equations determining motion, linear methods, compounding of simple harmonic motions, Newton's laws of motion, Hooke's law, the simple pendulum, motion of a particle in 1 plane, centers of gravity, virtual work, friction, kinetic energy of rotating bodies, equilibrium of strings, hydrostatics, sheering stresses, elasticity, etc. 550 problems. 3rd revised edition. xii + 367pp. 6 x 9. S207 Clothbound **$4.95**

THE DYNAMICS OF PARTICLES AND OF RIGID, ELASTIC, AND FLUID BODIES; BEING LECTURES ON MATHEMATICAL PHYSICS, A. G. Webster. The reissuing of this classic fills the need for a comprehensive work on dynamics. A wide range of topics is covered in unusually great depth, applying ordinary and partial differential equations. Part I considers laws of motion and methods applicable to systems of all sorts; oscillation, resonance, cyclic systems, etc. Part 2 is a detailed study of the dynamics of rigid bodies. Part 3 introduces the theory of potential; stress and strain, Newtonian potential functions, gyrostatics, wave and vortex motion, etc. Further contents: Kinematics of a point; Lagrange's equations; Hamilton's principle; Systems of vectors; Statics and dynamics of deformable bodies; much more, not easily found together in one volume. Unabridged reprinting of 2nd edition. 20 pages of notes on differential equations and the higher analysis. 203 illustrations. Selected bibliography. Index. xi + 588pp. 5⅜ x 8. S522 Paperbound **$2.45**

A TREATISE ON DYNAMICS OF A PARTICLE, E. J. Routh. Elementary text on dynamics for beginning mathematics or physics student. Unusually detailed treatment from elementary definitions to motion in 3 dimensions, emphasizing concrete aspects. Much unique material important in recent applications. Covers impulsive forces, rectilinear and constrained motion in 2 dimensions, harmonic and parabolic motion, degrees of freedom, closed orbits, the conical pendulum, the principle of least action, Jacobi's method, and much more. Index. 559 problems, many fully worked out, incorporated into text. xiii + 418pp. 5⅜ x 8.
S696 Paperbound **$2.25**

DYNAMICS OF A SYSTEM OF RIGID BODIES (Elementary Section), E. J. Routh. Revised 7th edition of this standard reference. This volume covers the dynamical principles of the subject, and its more elementary applications: finding moments of inertia by integration, foci of inertia, d'Alembert's principle, impulsive forces, motion in 2 and 3 dimensions, Lagrange's equations, relative indicatrix, Euler's theorem, large tautochronous motions, etc. Index. 55 figures. Scores of problems. xv + 443pp. 5⅜ x 8. S664 Paperbound **$2.50**

DYNAMICS OF A SYSTEM OF RIGID BODIES (Advanced Section), E. J. Routh. Revised 6th edition of a classic reference aid. Much of its material remains unique. Partial contents: moving axes, relative motion, oscillations about equilibrium, motion. Motion of a body under no forces, any forces. Nature of motion given by linear equations and conditions of stability. Free, forced vibrations, constants of integration, calculus of finite differences, variations, precession and nutation, motion of the moon, motion of string, chain, membranes. 64 figures. 498pp. 5⅜ x 8. S229 Paperbound **$2.45**

DYNAMICAL THEORY OF GASES, James Jeans. Divided into mathematical and physical chapters for the convenience of those not expert in mathematics, this volume discusses the mathematical theory of gas in a steady state, thermodynamics, Boltzmann and Maxwell, kinetic theory, quantum theory, exponentials, etc. 4th enlarged edition, with new material on quantum theory, quantum dynamics, etc. Indexes. 28 figures. 444pp. 6⅛ x 9¼.
S136 Paperbound **$2.65**

THE THEORY OF HEAT RADIATION, Max Planck. A pioneering work in thermodynamics, providing basis for most later work, Nobel laureate Planck writes on Deductions from Electrodynamics and Thermodynamics, Entropy and Probability, Irreversible Radiation Processes, etc. Starts with simple experimental laws of optics, advances to problems of spectral distribution of energy and irreversibility. Bibliography. 7 illustrations, xiv + 224pp. 5⅜ x 8.
S546 Paperbound **$1.50**

FOUNDATIONS OF POTENTIAL THEORY, O. D. Kellogg. Based on courses given at Harvard this is suitable for both advanced and beginning mathematicians. Proofs are rigorous, and much material not generally avaliable elsewhere is included. Partial contents: forces of gravity, fields of force, divergence theorem, properties of Newtonian potentials at points of free space, potentials as solutions of Laplace's equations, harmonic functions, electrostatics, electric images, logarithmic potential, etc. One of Grundlehren Series. ix + 384pp. 5⅜ x 8.
S144 Paperbound **$1.98**

THERMODYNAMICS, Enrico Fermi. Unabridged reproduction of 1937 edition. Elementary in treatment; remarkable for clarity, organization. Requires no knowledge of advanced math beyond calculus, only familiarity with fundamentals of thermometry, calorimetry. Partial Contents: Thermodynamic systems; First & Second laws of thermodynamics; Entropy; Thermodynamic potentials: phase rule, reversible electric cell; Gaseous reactions: van't Hoff reaction box, principle of LeChatelier; Thermodynamics of dilute solutions: osmotic & vapor pressures, boiling & freezing points; Entropy constant. Index. 25 problems. 24 illustrations. x + 160pp. 5⅜ x 8.
S361 Paperbound **$1.75**

THE THERMODYNAMICS OF ELECTRICAL PHENOMENA IN METALS and A CONDENSED COLLECTION OF THERMODYNAMIC FORMULAS, P. W. Bridgman. Major work by the Nobel Prizewinner: stimulating conceptual introduction to aspects of the electron theory of metals, giving an intuitive understanding of fundamental relationships concealed by the formal systems of Onsager and others. Elementary mathematical formulations show clearly the fundamental thermodynamical relationships of the electric field, and a complete phenomenological theory of metals is created. This is the work in which Bridgman announced his famous "thermomotive force" and his distinction between "driving" and "working" electromotive force. We have added in this Dover edition the author's long unavailable tables of thermodynamic formulas, extremely valuable for the speed of reference they allow. Two works bound as one. Index. 33 figures. Bibliography. xviii + 256pp. 5⅜ x 8. S723 Paperbound **$1.65**

TREATISE ON THERMODYNAMICS, Max Planck. Based on Planck's original papers this offers a uniform point of view for the entire field and has been used as an introduction for students who have studied elementary chemistry, physics, and calculus. Rejecting the earlier approaches of Helmholtz and Maxwell, the author makes no assumptions regarding the nature of heat, but begins with a few empirical facts, and from these deduces new physical and chemical laws. 3rd English edition of this standard text by a Nobel laureate. xvi + 297pp. 5⅜ x 8.
S219 Paperbound **$1.75**

THE MATHEMATICAL THEORY OF ELASTICITY, A. E. H. Love. A wealth of practical illustration combined with thorough discussion of fundamentals—theory, application, special problems and solutions. Partial Contents: Analysis of Strain & Stress, Elasticity of Solid Bodies, Elasticity of Crystals, Vibration of Spheres, Cylinders, Propagation of Waves in Elastic Solid Media, Torsion, Theory of Continuous Beams, Plates. Rigorous treatment of Volterra's theory of dislocations, 2-dimensional elastic systems, other topics of modern interest. "For years the standard treatise on elasticity," AMERICAN MATHEMATICAL MONTHLY. 4th revised edition. Index. 76 figures. xviii + 643pp. 6⅛ x 9¼.
S174 Paperbound **$3.00**

STRESS WAVES IN SOLIDS, H. Kolsky, Professor of Applied Physics, Brown University. The most readable survey of the theoretical core of current knowledge about the propagation of waves in solids, fully correlated with experimental research. Contents: Part I—Elastic Waves: propagation in an extended plastic medium, propagation in bounded elastic media, experimental investigations with elastic materials. Part II—Stress Waves in Imperfectly Elastic Media: internal friction, experimental investigations of dynamic elastic properties, plastic waves and shock waves, fractures produced by stress waves. List of symbols. Appendix. Supplemented bibliography. 3 full-page plates. 46 figures. x + 213pp. 5⅜ x 8½.
S1098 Paperbound **$1.55**

Relativity, quantum theory, atomic and nuclear physics

SPACE TIME MATTER, Hermann Weyl. "The standard treatise on the general theory of relativity" (Nature), written by a world-renowned scientist, provides a deep clear discussion of the logical coherence of the general theory, with introduction to all the mathematical tools needed: Maxwell, analytical geometry, non-Euclidean geometry, tensor calculus, etc. Basis is classical space-time, before absorption of relativity. Partial contents: Euclidean space, mathematical form, metrical continuum, relativity of time and space, general theory. 15 diagrams. Bibliography. New preface for this edition. xviii + 330pp. 5⅜ x 8.
S267 Paperbound **$2.00**

ATOMIC SPECTRA AND ATOMIC STRUCTURE, G. Herzberg. Excellent general survey for chemists, physicists specializing in other fields. Partial contents: simplest line spectra and elements of atomic theory, building-up principle and periodic system of elements, hyperfine structure of spectral lines, some experiments and applications. Bibilography. 80 figures. Index. xii + 257pp. 5⅜ x 8.
S115 Paperbound **$2.00**

PHYSICS, HISTORIES AND CLASSICS

A HISTORY OF PHYSICS: IN ITS ELEMENTARY BRANCHES (THROUGH 1925), INCLUDING THE EVOLUTION OF PHYSICAL LABORATORIES, Florian Cajori. Revised and enlarged edition. The only first-rate brief history of physics. Still the best entry for a student or teacher into the antecedents of modern theories of physics. A clear, non-mathematical, handy reference work which traces in critical fashion the developments of ideas, theories, techniques, and apparatus from the Greeks to the 1920's. Within each period he analyzes the basic topics of mechanics, light, electricity and magnetism, sound, atomic theory and structure of matter, radioactivity, etc. A chapter on modern research: Curie, Kelvin, Planck's quantum theory, thermodynamics, Fitzgerald and Lorentz, special and general relativity, J. J. Thomson's model of an atom, Bohr's discoveries and later results, wave mechanics, and many other matters. Much bibliographic detail in footnotes. Index. 16 figures. xv + 424pp. 5⅜ x 8. T970 Paperbound **$2.00**

A HISTORY OF THE MATHEMATICAL THEORIES OF ATTRACTION AND THE FIGURE OF THE EARTH: FROM THE TIME OF NEWTON TO THAT OF LAPLACE, I. Todhunter. A technical and detailed review of the theories concerning the shape of the earth and its gravitational pull, from the earliest investigations in the seventeenth century up to the middle of the nineteenth. Some of the greatest mathematicians and scientists in history applied themselves to these questions: Newton ("Principia Mathematica"), Huygens, Maupertuis, Simpson, d'Alembert, etc. Others discussed are Poisson, Gauss, Plana, Lagrange, Boit, and many more. Particular emphasis is placed on the theories of Laplace and Legendre, several chapters being devoted to Laplace's "Mécanique Céleste" and his memoirs, and several others to the memoirs of Legendre. Important to historians of science and mathematics and to the specialist who desires background information in the field. 2 volumes bound as 1. Index. xxxvi + 984pp. 5⅜ x 8.
S148 Clothbound **$7.50**

OPTICKS, Sir Isaac Newton. In its discussions of light, reflection, color, refraction, theories of wave and corpuscular theories of light, this work is packed with scores of insights and discoveries. In its precise and practical discussion of construction of optical apparatus, contemporary understandings of phenomena it is truly fascinating to modern physicists, astronomers, mathematicians. Foreword by Albert Einstein. Preface by I. B. Cohen of Harvard University. 7 pages of portraits, facsimile pages, letters, etc. cxvi + 414pp. 5⅜ x 8.
S205 Paperbound **$2.25**

TREATISE ON LIGHT, Christiaan Huygens. The famous original formulation of the wave theory of light, this readable book is one of the two decisive and definitive works in the field of light (Newton's "Optics" is the other). A scientific giant whose researches ranged over mathematics, astronomy, and physics, Huygens, in this historic work, covers such topics as rays propagated in straight lines, reflection and refraction, the spreading and velocity of light, the nature of opaque bodies, the non-spherical nature of light in the atmosphere, properties of Iceland Crystal, and other related matters. Unabridged republication of original (1912) English edition. Translated and introduced by Silvanus P. Thompson. 52 illustrations. xii + 129pp. 5⅜ x 8. S179 Paperbound **$1.35**

FARADAY'S EXPERIMENTAL RESEARCHES IN ELECTRICITY. Faraday's historic series of papers containing the fruits of years of original experimentation in electrical theory and electrochemistry. Covers his findings in a variety of areas: Induction of electric currents, Evolution of electricity from magnetism, New electrical state or condition of matter, Explication of Arago's magnetic phenomena, New law of electric conduction, Electro-chemical decomposition, Electricity of the Voltaic Pile, Static Induction, Nature of the electric force or forces, Nature of electric current, The character and direction of the electric force of the Gymnotus, Magneto-electric spark, The magnetization of light and the illumination of magnetic lines of force, The possible relation of gravity to electricity, Sub-terraneous electro-telegraph wires, Some points of magnetic philosophy, The diamagnetic conditions of flame and gases, and many other matters. Complete and unabridged republication. 3 vols. bound as 2. Originally reprinted from the Philosophical Transactions of 1831-8. Indices. Illustrations. Total of 1463pp. 5⅜ x 8. S783-4, Clothbound **$17.50** (tentative)

REFLECTIONS ON THE MOTIVE POWER OF FIRE, Sadi Carnot, and other papers on the 2nd law of thermodynamics by E. Clapeyron and R. Clausius. Carnot's "Reflections" laid the groundwork of modern thermodynamics. Its non-technical, mostly verbal statements examine the relations between heat and the work done by heat in engines, establishing conditions for the economical working of these engines. The papers by Clapeyron and Clausius here reprinted added further refinements to Carnot's work, and led to its final acceptance by physicists. Selections from posthumous manuscripts of Carnot are also included. All papers in English. New introduction by E. Mendoza. 12 illustrations. xxii + 152pp. 5⅜ x 8.
S661 Paperbound **$1.50**

DIALOGUES CONCERNING TWO NEW SCIENCES, Galileo Galilei. This classic of experimental science, mechanics, engineering, is as enjoyable as it is important. A great historical document giving insights into one of the world's most original thinkers, it is based on 30 years' experimentation. It offers a lively exposition of dynamics, elasticity, sound, ballistics, strength of materials, the scientific method. "Superior to everything else of mine," Galileo. Trans. by H. Crew, A. Salvio. 126 diagrams. Index. xxi + 288pp. 5⅜ x 8.
S99 Paperbound **$1.75**

TREATISE ON ELECTRICITY AND MAGNETISM, James Clerk Maxwell. For more than 80 years a seemingly inexhaustible source of leads for physicists, mathematicians, engineers. Total of 1082pp. on such topics as Measurement of Quantities, Electrostatics, Elementary Mathematical Theory of Electricity, Electrical Work and Energy in a System of Conductors, General Theorems, Theory of Electrical Images, Electrolysis, Conduction, Polarization, Dielectrics, Resistance, etc. "The greatest mathematical physicist since Newton," Sir James Jeans. 3rd edition. 107 figures, 21 plates. 1082pp. 5⅜ x 8. S636-7, 2 volume set, paperbound **$4.00**

A HISTORY OF THE THEORY OF ELASTICITY AND THE STRENGTH OF MATERIALS, I. Todhunter and K. Pearson. For over 60 years a basic reference, unsurpassed in scope or authority. Both a history of the mathematical theory of elasticity from Galileo, Hooke, and Mariotte to Saint Venant, Kirchhoff, Clebsch, and Lord Kelvin and a detailed presentation of every important mathematical contribution during this period. Presents proofs of thousands of theorems and laws, summarizes every relevant treatise, many unavailable elsewhere. Practically a book apiece is devoted to modern founders: Saint Venant, Lamé, Boussinesq, Rankine, Lord Kelvin, F. Neumann, Kirchhoff, Clebsch. Hundreds of pages of technical and physical treatises on specific applications of elasticity to particular materials. Indispensable for the mathematician, physicist, or engineer working with elasticity. Unabridged, corrected reprint of original 3-volume 1886-1893 edition. Three volume set. Two indexes. Appendix to Vol. I. Total of 2344pp. 5⅜ x 8⅜. S914–916 The set, Clothbound **$15.00**

DE MAGNETE, William Gilbert. This classic work on magnetism founded a new science. Gilbert was the first to use the word "electricity", to recognize mass as distinct from weight, to discover the effect of heat on magnetic bodies; invent an electroscope, differentiate between static electricity and magnetism, conceive of the earth as a magnet. Written by the first great experimental scientist, this lively work is valuable not only as an historical landmark, but as the delightfully easy to follow record of a perpetually searching, ingenious mind. Translated by P. F. Mottelay. 25-page biographical memoir. 90 figures. lix +368pp. 5⅜ x 8. S470 Paperbound **$2.00**

ASTRONOMY

THE INTERNAL CONSTITUTION OF THE STARS, Sir A. S. Eddington. Influence of this has been enormous; first detailed exposition of theory of radiative equilibrium for stellar interiors, of all available evidence for existence of diffuse matter in interstellar space. Studies quantum theory, polytropic gas spheres, mass-luminosity relations, variable stars, etc. Discussions of equations paralleled with informal exposition of intimate relationship of astrophysics with great discoveries in atomic physics, radiation. Introduction. Appendix. Index. 421pp. 5⅜ x 8. S563 Paperbound **$2.25**

PLANETARY THEORY, E. W. Brown and C. A. Shook. Provides a clear presentation of basic methods for calculating planetary orbits for today's astronomer. Begins with a careful exposition of specialized mathematical topics essential for handling perturbation theory and then goes on to indicate how most of the previous methods reduce ultimately to two general calculation methods: obtaining expressions either for the coordinates of planetary positions or for the elements which determine the perturbed paths. An example of each is given and worked in detail. Corrected edition. Preface. Appendix. Index. xii + 302pp. 5⅜ x 8½. S1133 Paperbound **$2.25**

CANON OF ECLIPSES (CANON DER FINSTERNISSE), Prof. Theodor Ritter von Oppolzer. Since its original publication in 1887, this has been the standard reference and the most extensive single volume of data on the calculation of solar and lunar eclipses, past and future. A comprehensive introduction gives a full explanation of the use of the tables for the calculations of the exact dates of eclipses, etc. Data furnished for the calculation of 8,000 solar and 5,200 lunar eclipses, going back as far as 1200 B.C. and giving predictions up to the year 2161. Information is also given for partial and ring eclipses. All calculations based on Universal (Greenwich) Time. An unsurpassed reference work for astronomers, scientists engaged in space research and developments, historians, etc. Unabridged republication, with corrections. Preface to this edition by Donald Menzel and Owen Gingerich of the Harvard College Observatory. Translated by Owen Gingerich. 160 charts. lxx + 538pp. 8⅜ x 11¼. S114 Clothbound **$10.00**

THEORY OF THE MOTION OF THE HEAVENLY BODIES MOVING ABOUT THE SUN IN CONIC SECTIONS, Karl Friedrich Gauss. A landmark of theoretical astronomy by the great German scientist. Still authoritative and invaluable to the practicing astronomer. Part I develops the relations between the quantities on which the motion about the sun of the heavenly bodies depends—relations pertaining simply to position in the orbit, simply to position in space, between several places in orbit, and between several places in space. The calculation methods of Part II based on the groundwork of Part I include: determination of an orbit from 3 complete observations, from 4 observations (of which only two are complete), determination of an orbit satisfying as nearly as possible any number of observations whatever, and determination of orbits, taking into account the perturbations. Translation of "Theoria Motus" and with an appendix by C. H. Davis. Unabridged republication. Appendices and tables. 13 figures. xviii + 376pp. 6½ x 9¼. S1056 Paperbound **$2.95**

THE GALACTIC NOVAE, C. Payne-Gaposchkin, Prof. of Astronomy, Harvard Univ. A work that will be the standard reference source for years to come. Gathers together all the pertinent data, results recorded by countless observers of galactic novae over the centuries, in order to formulate a valid starting point for an interpretation of the nova process. Covers information and statistics on known novae, their variations in luminosity, distribution in the sky, spectral changes, etc.; symbiotic novae; frequently-recurring variables of the U Geminorum and Z Camelopardis class; supernovae; comparison of spectral changes; theories and interpretations of these phenomena, etc. "A comprehensive summary of everything that is now known about these stars," SCIENCE. Bibliographical references. Preface. Indices. 49 figures. 6 plates. 101 tables. x + 336pp. 5⅜ x 8⅜. S1170 Paperbound **$2.45**

BINARY STARS, R. G. Aitken. Still the definitive work in the field of double star astronomy. Written by the director of the Lick Observatory (considered the father of the modern study of binary star systems), this book sums up the results of 40 years of experience in the field, plus the work of centuries of research. Includes historical survey of major discoveries and contributions of the past, observational methods for visual binary stars, the radial velocity of a star (by Dr. J. H. Moore), eclipsing binary stars, known orbits of binary stars, some binary systems of special interest, the origin of binary stars. Much information on methods of spectrum analysis, orbit plotting, use of the telescope, and other practical matters. Useful for classroom study and advanced hobbyists, etc. Revised edition, corrected and with additional notes by Prof. J. T. Kent. New preface. 50 tables, 13 figures, 4 full-page plates. Bibliographies. Appendix. Indices. xii + 309pp. 5⅜ x 8½. S1102 Paperbound **$2.00**

THE NATURE OF COMETS, N. B. Richter. An authority on comets presents a concise, but thorough survey of the state of our present-day knowledge of comets and cometary activity. Based on over 20 years of research, this is a middle-level account that even the layman can appreciate, providing a fund of information on historical theories (from 1700 to the present); statistical research on total number of comets, orbital forms, perturbations caused by Jupiter, comet groups, etc.; the structure of a comet; comets as processes of cosmic decay; origin and formation of comets; etc. Also: a lengthy introduction on modern theories by Dr. R. A. Lyttleton, much technical data and observational material of specific comets, supplementary tables, and the like. Revised (1963) edition. Translated and revised by Arthur Beer. 69 illustrations, including 54 photographs of comets, tails, spectra. 41 tables. Bibliography. Index. xli + 221pp. S1111 Clothbound **$10.00**

CELESTIAL OBJECTS FOR COMMON TELESCOPES, Rev. T. W. Webb. Classic handbook for the use and pleasure of the amateur astronomer. Of inestimable aid in locating and identifying thousands of celestial objects. Vol. I, The Solar System: discussions of the principle and operation of the telescope, procedures of observations and telescope-photography, spectroscopy, etc., precise location information of sun, moon, planets, meteors. Vol. II, The Stars: alphabetical listing of constellations, information on double stars, clusters, stars with unusual spectra, variables, and nebulae, etc. Nearly 4,000 objects noted. Edited and extensively revised by Margaret W. Mayall, director of the American Assn. of Variable Star Observers. New Index by Mrs. Mayall giving the location of all objects mentioned in the text for Epoch 2000. New Precession Table added. New appendices on the planetary satellites, constellation names and abbreviations, and solar system data. Total of 46 illustrations. Total of xxxix + 606pp. 5⅜ x 8.
Vol. I: T917 Paperbound **$2.25**
Vol. II: T918 Paperbound **$2.25**
Two Volume Set Paperbound **$4.50**

ASTRONOMY AND COSMOGONY, Sir James Jeans. A modern classic which is still of enormous value to everyone in astronomy, etc., this is Jean's last and most famous exposition. The summation of a lifetime's devotion to science, it presents his final conclusions on a host of problems ranging over the whole of descriptive astronomy, astrophysics, stellar dynamics, and cosmology. Contents: The Light from the Stars, Gaseous Stars, the Source of Stellar Energy, Liquid Stars, The Evolution of the Stars, The Configuration of Rotating Masses, The Evolution of Binary Systems, The Ages of the Stars, The Great Nebulae, The Galactic Systems, Variable Stars, etc. New preface by L. Motz, Columbia U. 16 full-page photographic illustrations. xv + 428pp. 5⅝ x 8⅜. S923 Paperbound **$2.45**

ASTRONOMY OF STELLAR ENERGY AND DECAY, Martin Johnson. Middle level treatment of astronomy as interpreted by modern atomic physics. Part One is non-technical, examines physical properties, source of energy, spectroscopy, fluctuating stars, various models and theories, etc. Part Two parallels these topics, providing their mathematical foundation. "Clear, concise, and readily understandable," American Library Assoc. Bibliography. 3 indexes. 29 illustrations. 216pp. 5⅜ x 8. S537 Paperbound **$1.50**

MATHEMATICAL THEORIES OF PLANETARY MOTIONS, Otto Dziobek. Translated by Mark W. Harrington and William J. Hussey. Lucid account of the principles of mathematical astronomy. It examines that part of celestial mechanics which deals with the motions of heavenly bodies considered as material points. Contents: Solution of the Problem of Two Bodies; Formation of the General Integrals for Problem of **n** Bodies . . . including discussions of elliptic, parabolic, and hyperbolic orbits, the solution of Kepler's equation, etc.; and sections headed The General Properties of the Integrals and The Theory of Perturbations . . . which deals with the theory of absolute perturbations, analytical development of the perturbing function, the variation of the elements, the secular variation of the mean longitude, etc. vi + 294pp. 5⅜ x 8½. S129 Paperbound **$2.00**

A COMPENDIUM OF SPHERICAL ASTRONOMY, S. Newcomb. Long a standard collection of basic methods and formulas most useful to the working astronomer, and clear full text for students. Includes the most important common approximations; 40 pages on the method of least squares; general theory of spherical coordinates; parallax; aberration; astronomical refraction; theory of precession; proper motion of the stars; methods of deriving positions of stars; and much more. Index. 9 Appendices of tables, formulas, etc. 36 figures. xviii + 444pp. 5⅜ x 8.
S690 Paperbound **$2.25**

PRINCIPLES OF STELLAR DYNAMICS, S. Chandrasekhar. A leading astrophysicist here presents the theory of stellar dynamics as a branch of classical dynamics, clarifying the fundamental issues and the underlying motivations of the theory. He analyzes the effects of stellar encounters in terms of the classical 2-body problem, and investigates problems centering about Liouville's theorem and the solutions of the equations of continuity. This edition also includes 4 important papers by the author published since "Stellar Dynamics," and equally indispensable for all workers in the field: "New Methods in Stellar Dynamics" and "Dynamical Friction," Parts I, II, and III. Index. 3 Appendixes. Bibliography. 50 illustrations. x + 313pp. 5⅜ x8.
S659 Paperbound **$2.00**

AN INTRODUCTION TO THE STUDY OF STELLAR STRUCTURE, Subrahmanyan Chandrasekhar. Outstanding treatise on stellar dynamics by one of world's greatest astrophysicists. Uses classical & modern math methods to examine relationship between loss of energy, the mass, and radius of stars in a steady state. Discusses thermodynamic laws from Carathéodory's axiomatic standpoint; adiabatic, polytropic laws; work of Ritter, Emden, Kelvin, others; Stroemgren envelopes as starter for theory of gaseous stars; Gibbs statistical mechanics (quantum); degenerate stellar configuration & theory of white dwarfs, etc. "Highest level of scientific merit," BULLETIN, AMER. MATH. SOC. Bibliography. Appendixes. Index. 33 figures. 509pp. 5⅜ x 8. S413 Paperbound **$2.75**

STATISTICAL ASTRONOMY, Robert J. Trumpler and Harold F. Weaver, University of California. Standard introduction to the principles and techniques of statistical astronomy, a field of rapidly growing importance in this space age. An extensive section, "Elements of Statistical Theory," provides the astronomer with the tools for solving problems of descriptive astronomy, observational errors, constitution of extra-galactic nebulae, etc. Procedures used in statistical astronomy are related to basic mathematical principles of statistics such as unvariate distribution, integral equations, general theory of samples, etc. Other sections deal with: Statistical Description of the Galactic System; Stellar Motions in the Vicinity of the Sun; Luminosity—Spectral Type Distribution; Space Distribution of Stars; and Galactic Rotation. List of symbols. Appendix (10 tables). 2 Indexes. Extensive bibliography. 31 tables. 97 figures. xxi + 644pp. 5⅜ x 8½. S301 Paperbound **$3.00**

AN INTRODUCTORY TREATISE ON DYNAMICAL ASTRONOMY, H. C. Plummer. Unusually wide connected and concise coverage of nearly every significant branch of dynamical astronomy, stressing basic principles throughout: determination of orbits, planetary theory, lunar theory, precession and nutation, and many of their applications. Hundreds of formulas and theorems worked out completely, important methods thoroughly explained. Covers motion under a central attraction, orbits of double stars and spectroscopic binaries, the libration of the moon, and much more. Index. 8 diagrams. xxi + 343pp. 5⅝ x 8⅜. S689 Paperbound **$2.35**

AN INTRODUCTORY TREATISE ON THE LUNAR THEORY, E. W. Brown. Indispensable for all scientists and engineers interested in orbital calculation, satellites, or navigation of space. Only work in English to explain in detail 5 major mathematical approaches to the problem of 3 bodies, those of Laplace, de Pontécoulant, Hansen, Delaunay, and Hill. Covers expressions for mutual attraction, equations of motion, forms of solution, variations of the elements in disturbed motion, the constants and their interpretations, planetary and other disturbing influences, etc. Index. Bibliography. Tables. xvi + 292pp. 5⅝ x 8⅜.
S666 Paperbound **$2.00**

SPHERICAL AND PRACTICAL ASTRONOMY, W. Chauvenet. First book in English to apply mathematical techniques to astronomical problems is still standard work. Covers almost entire field, rigorously, with over 300 examples worked out. Vol. 1, spherical astronomy, applications to nautical astronomy; determination of hour angles, parallactic angle for known stars; interpolation; parallax; laws of refraction; predicting eclipses; precession, nutation of fixed stars; etc. Vol. 2, theory, use, of instruments; telescope; measurement of arcs, angles in general; electro-chronograph; sextant, reflecting circles; zenith telescope; etc. 100-page appendix of detailed proof of Gauss' method of least squares. 5th revised edition. Index. 15 plates, 20 tables. 1340pp. 5⅜ x 8. Vol. 1 S618 Paperbound **$2.75**
Vol. 2 S619 Paperbound **$2.75**
The set **$5.50**

RADIATIVE TRANSFER, S. Chandrasekhar. Definitive work in field provides foundation for analysis of stellar atmospheres, planetary illumination, sky radiation; to physicists, a study of problems analogous to those in theory of diffusion of neutrons. Partial contents: equation of transfer, isotropic scattering, H-functions, diffuse reflection and transmission, Rayleigh scattering, X, Y functions, radiative equilibrium of stellar atmospheres. Extensive bibliography. 3 appendices. 35 tables. 35 figures. 407pp. 5⅝ x 8⅜. S599 Paperbound **$2.25**

ENGINEERING AND TECHNOLOGY

General and mathematical

ENGINEERING MATHEMATICS, Kenneth S. Miller. A text for graduate students of engineering to strengthen their mathematical background in differential equations, etc. Mathematical steps very explicitly indicated. Contents: Determinants and Matrices, Integrals, Linear Differential Equations, Fourier Series and Integrals, Laplace Transform, Network Theory, Random Function . . . all vital requisites for advanced modern engineering studies. Unabridged republication. Appendices: Borel Sets; Riemann-Stieltjes Integral; Fourier Series and Integrals. Index. References at Chapter Ends. xii + 417pp. 6 x 8½. S1121 Paperbound **$2.00**

MATHEMATICAL ENGINEERING ANALYSIS, Rufus Oldenburger. A book designed to assist the research engineer and scientist in making the transition from physical engineering situations to the corresponding mathematics. Scores of common practical situations found in all major fields of physics are supplied with their correct mathematical formulations—applications to automobile springs and shock absorbers, clocks, throttle torque of diesel engines, resistance networks, capacitors, transmission lines, microphones, neon tubes, gasoline engines, refrigeration cycles, etc. Each section reviews basic principles of underlying various fields: mechanics of rigid bodies, electricity and magnetism, heat, elasticity, fluid mechanics, and aerodynamics. Comprehensive and eminently useful. Index. 169 problems, answers. 200 photos and diagrams. xiv + 426pp. 5⅜ x 8½. S919 Paperbound **$2.00**

MATHEMATICS OF MODERN ENGINEERING, E. G. Keller and R. E. Doherty. Written for the Advanced Course in Engineering of the General Electric Corporation, deals with the engineering use of determinants, tensors, the Heaviside operational calculus, dyadics, the calculus of variations, etc. Presents underlying principles fully, but purpose is to teach engineers to deal with modern engineering problems, and emphasis is on the perennial engineering attack of set-up and solve. Indexes. Over 185 figures and tables. Hundreds of exercises, problems, and worked-out examples. References. Two volume set. Total of xxxiii + 623pp. 5⅜ x 8.
S734 Vol I Paperbound **$1.85**
S735 Vol II Paperbound **$1.85**
The set **$3.70**

MATHEMATICAL METHODS FOR SCIENTISTS AND ENGINEERS, L. P. Smith. For scientists and engineers, as well as advanced math students. Full investigation of methods and practical description of conditions under which each should be used. Elements of real functions, differential and integral calculus, space geometry, theory of residues, vector and tensor analysis, series of Bessel functions, etc. Each method illustrated by completely-worked-out examples, mostly from scientific literature. 368 graded unsolved problems. 100 diagrams. x + 453pp. 5⅝ x 8⅜. S220 Paperbound **$2.00**

THEORY OF FUNCTIONS AS APPLIED TO ENGINEERING PROBLEMS, edited by R. Rothe, F. Ollendorff, and K. Pohlhausen. A series of lectures given at the Berlin Institute of Technology that shows the specific applications of function theory in electrical and allied fields of engineering. Six lectures provide the elements of function theory in a simple and practical form, covering complex quantities and variables, integration in the complex plane, residue theorems, etc. Then 5 lectures show the exact uses of this powerful mathematical tool, with full discussions of problem methods. Index. Bibliography. 108 figures. x + 189pp. 5⅜ x 8.
S733 Paperbound **$1.35**

Aerodynamics and hydrodynamics

AIRPLANE STRUCTURAL ANALYSIS AND DESIGN, E. E. Sechler and L. G. Dunn. Systematic authoritative book which summarizes a large amount of theoretical and experimental work on structural analysis and design. Strong on classical subsonic material still basic to much aeronautic design . . . remains a highly useful source of information. Covers such areas as layout of the airplane, applied and design loads, stress-strain relationships for stable structures, truss and frame analysis, the problem of instability, the ultimate strength of stiffened flat sheet, analysis of cylindrical structures, wings and control surfaces, fuselage analysis, engine mounts, landing gears, etc. Originally published as part of the CALCIT Aeronautical Series. 256 Illustrations. 47 study problems. Indexes. xi + 420pp. 5⅜ x 8½.
S1043 Paperbound **$2.25**

FUNDAMENTALS OF HYDRO- AND AEROMECHANICS, L. Prandtl and O. G. Tietjens. The well-known standard work based upon Prandtl's lectures at Goettingen. Wherever possible hydrodynamics theory is referred to practical considerations in hydraulics, with the view of unifying theory and experience. Presentation is extremely clear and though primarily physical, mathematical proofs are rigorous and use vector analysis to a considerable extent. An Engineering Society Monograph, 1934. 186 figures. Index. xvi + 270pp. 5⅜ x 8.
S374 Paperbound **$1.85**

FLUID MECHANICS FOR HYDRAULIC ENGINEERS, H. Rouse. Standard work that gives a coherent picture of fluid mechanics from the point of view of the hydraulic engineer. Based on courses given to civil and mechanical engineering students at Columbia and the California Institute of Technology, this work covers every basic principle, method, equation, or theory of interest to the hydraulic engineer. Much of the material, diagrams, charts, etc., in this self-contained text are not duplicated elsewhere. Covers irrotational motion, conformal mapping, problems in laminar motion, fluid turbulence, flow around immersed bodies, transportation of sediment, general charcteristics of wave phenomena, gravity waves in open channels, etc. Index. Appendix of physical properties of common fluids. Frontispiece + 245 figures and photographs. xvi + 422pp. 5⅜ x 8. S729 Paperbound **$2.25**

WATERHAMMER ANALYSIS, John Parmakian. Valuable exposition of the graphical method of solving waterhammer problems by Assistant Chief Designing Engineer, U.S. Bureau of Reclamation. Discussions of rigid and elastic water column theory, velocity of waterhammer waves, theory of graphical waterhammer analysis for gate operation, closings, openings, rapid and slow movements, etc., waterhammer in pump discharge caused by power failure, waterhammer analysis for compound pipes, and numerous related problems. "With a concise and lucid style, clear printing, adequate bibliography and graphs for approximate solutions at the project stage, it fills a vacant place in waterhammer literature," WATER POWER. 43 problems. Bibliography. Index. 113 illustrations. xiv + 161pp. 5⅜ x 8½.
S1061 Paperbound **$1.65**

AERODYNAMIC THEORY: A GENERAL REVIEW OF PROGRESS, William F. Durand, editor-in-chief. A monumental joint effort by the world's leading authorities prepared under a grant of the Guggenheim Fund for the Promotion of Aeronautics. Intended to provide the student and aeronautic designer with the theoretical and experimental background of aeronautics. Never equalled for breadth, depth, reliability. Contains discussions of special mathematical topics not usually taught in the engineering or technical courses. Also: an extended two-part treatise on Fluid Mechanics, discussions of aerodynamics of perfect fluids, analyses of experiments with wind tunnels, applied airfoil theory, the non-lifting system of the airplane, the air propeller, hydrodynamics of boats and floats, the aerodynamics of cooling, etc. Contributing experts include Munk, Giacomelli, Prandtl, Toussaint, Von Karman, Klemperer, among others. Unabridged republication. 6 volumes bound as 3. Total of 1,012 figures, 12 plates. Total of 2,186pp. Bibliographies. Notes. Indices. 5⅜ x 8.
S328-S330 Clothbound, The Set **$17.50**

APPLIED HYDRO- AND AEROMECHANICS, L. Prandtl and O. G. Tietjens. Presents, for the most part, methods which will be valuable to engineers. Covers flow in pipes, boundary layers, airfoil theory, entry conditions, turbulent flow in pipes, and the boundary layer, determining drag from measurements of pressure and velocity, etc. "Will be welcomed by all students of aerodynamics," NATURE. Unabridged, unaltered. An Engineering Society Monograph, 1934. Index. 226 figures, 28 photographic plates illustrating flow patterns. xvi + 311pp. 5⅜ x 8.
S375 Paperbound **$1.85**

SUPERSONIC AERODYNAMICS, E. R. C. Miles. Valuable theoretical introduction to the supersonic domain, with emphasis on mathematical tools and principles, for practicing aerodynamicists and advanced students in aeronautical engineering. Covers fundamental theory, divergence theorem and principles of circulation, compressible flow and Helmholtz laws, the Prandtl-Busemann graphic method for 2-dimensional flow, oblique shock waves, the Taylor-Maccoll method for cones in supersonic flow, the Chaplygin method for 2-dimensional flow, etc. Problems range from practical engineering problems to development of theoretical results. "Rendered outstanding by the unprecedented scope of its contents . . . has undoubtedly filled a vital gap," AERONAUTICAL ENGINEERING REVIEW. Index. 173 problems, answers. 106 diagrams. 7 tables. xii + 255pp. 5⅜ x 8. S214 Paperbound **$1.45**

HYDRAULIC TRANSIENTS, G. R. Rich. The best text in hydraulics ever printed in English . . . by one of America's foremost engineers (former Chief Design Engineer for T.V.A.). Provides a transition from the basic differential equations of hydraulic transient theory to the arithmetic intergration computation required by practicing engineers. Sections cover Water Hammer, Turbine Speed Regulation, Stability of Governing, Water-Hammer Pressures in Pump Discharge Lines, The Differential and Restricted Orifice Surge Tanks, The Normalized Surge Tank Charts of Calame and Gaden, Navigation Locks, Surges in Power Canals—Tidal Harmonics, etc. Revised and enlarged. Author's prefaces. Index. xiv + 409pp. 5⅜ x 8½.
S116 Paperbound **$2.50**

HYDRAULICS AND ITS APPLICATIONS, A. H. Gibson. Excellent comprehensive textbook for the student and thorough practical manual for the professional worker, a work of great stature in its area. Half the book is devoted to theory and half to applications and practical problems met in the field. Covers modes of motion of a fluid, critical velocity, viscous flow, eddy formation, Bernoulli's theorem, flow in converging passages, vortex motion, form of effluent streams, notches and weirs, skin friction, losses at valves and elbows, siphons, erosion of channels, jet propulsion, waves of oscillation, and over 100 similar topics. Final chapters (nearly 400 pages) cover more than 100 kinds of hydraulic machinery: Pelton wheel, speed regulators, the hydraulic ram, surge tanks, the scoop wheel, the Venturi meter, etc. A special chapter treats methods of testing theoretical hypotheses: scale models of rivers, tidal estuaries, siphon spillways, etc. 5th revised and enlarged (1952) edition. Index. Appendix. 427 photographs and diagrams. 95 examples, answers. xv + 813pp. 6 x 9.
S791 Clothbound **$8.00**

FLUID MECHANICS THROUGH WORKED EXAMPLES, D. R. L. Smith and J. Houghton. Advanced text covering principles and applications to practical situations. Each chapter begins with concise summaries of fundamental ideas. 163 fully worked out examples applying principles outlined in the text. 275 other problems, with answers. Contents; The Pressure of Liquids on Surfaces; Floating Bodies; Flow Under Constant Head in Pipes; Circulation; Vorticity; The Potential Function; Laminar Flow and Lubrication; Impact of Jets; Hydraulic Turbines; Centrifugal and Reciprocating Pumps; Compressible Fluids; and many other items. Total of 438 examples. 250 line illustrations. 340pp. Index. 6 x 8⅞. S981 Clothbound **$6.00**

THEORY OF SHIP MOTIONS, S. N. Blagoveshchensky. The only detailed text in English in a rapidly developing branch of engineering and physics, it is the work of one of the world's foremost authorities—Blagoveshchensky of Leningrad Shipbuilding Institute. A senior-level treatment written primarily for engineering students, but also of great importance to naval architects, designers, contractors, researchers in hydrodynamics, and other students. No mathematics beyond ordinary differential equations is required for understanding the text. Translated by T. & L. Strelkoff, under editorship of Louis Landweber, Iowa Institute of Hydraulic Research, under auspices of Office of Naval Research. Bibliography. Index. 231 diagrams and illustrations. Total of 649pp. 5⅜ x 8½. Vol. I: S234 Paperbound **$2.00**
Vol. II: S235 Paperbound **$2.00**

THEORY OF FLIGHT, Richard von Mises. Remains almost unsurpassed as balanced, well-written account of fundamental fluid dynamics, and situations in which air compressibility effects are unimportant. Stressing equally theory and practice, avoiding formidable mathematical structure, it conveys a full understanding of physical phenomena and mathematical concepts. Contains perhaps the best introduction to general theory of stability. "Outstanding," Scientific, Medical, and Technical Books. New introduction by K. H. Hohenemser. Bibliographical, historical notes. Index. 408 illustrations. xvi + 620pp. 5⅜ x 8⅜. S541 Paperbound **$2.95**

THEORY OF WING SECTIONS, I. H. Abbott, A. E. von Doenhoff. Concise compilation of subsonic aerodynamic characteristics of modern NASA wing sections, with description of their geometry, associated theory. Primarily reference work for engineers, students, it gives methods, data for using wing-section data to predict characteristics. Particularly valuable: chapters on thin wings, airfoils; complete summary of NACA's experimental observations, system of construction families of airfoils. 350pp. of tables on Basic Thickness Forms, Mean Lines, Airfoil Ordinates, Aerodynamic Characteristics of Wing Sections. Index. Bibliography. 191 illustrations. Appendix. 705pp. 5⅜ x 8. S558 Paperbound **$3.25**

WEIGHT-STRENGTH ANALYSIS OF AIRCRAFT STRUCTURES, F. R. Shanley. Scientifically sound methods of analyzing and predicting the structural weight of aircraft and missiles. Deals directly with forces and the distances over which they must be transmitted, making it possible to develop methods by which the minimum structural weight can be determined for any material and conditions of loading. Weight equations for wing and fuselage structures. Includes author's original papers on inelastic buckling and creep buckling. "Particularly successful in presenting his analytical methods for investigating various optimum design principles," AERONAUTICAL ENGINEERING REVIEW. Enlarged bibliography. Index. 199 figures. xiv + 404pp. 5⅝ x 8⅜. S660 Paperbound **$2.45**

Electricity

TWO-DIMENSIONAL FIELDS IN ELECTRICAL ENGINEERING, L. V. Bewley. A useful selection of typical engineering problems of interest to practicing electrical engineers. Introduces senior students to the methods and procedures of mathematical physics. Discusses theory of functions of a complex variable, two-dimensional fields of flow, general theorems of mathematical physics and their applications, conformal mapping or transformation, method of images, freehand flux plotting, etc. New preface by the author. Appendix by W. F. Kiltner. Index. Bibliography at chapter ends. xiv + 204pp. 5⅜ x 8½. S1118 Paperbound **$1.50**

FLUX LINKAGES AND ELECTROMAGNETIC INDUCTION, L. V. Bewley. A brief, clear book which shows proper uses and corrects misconceptions of Faraday's law of electromagnetic induction in specific problems. Contents: Circuits, Turns, and Flux Linkages; Substitution of Circuits; Electromagnetic Induction; General Criteria for Electromagnetic Induction; Applications and Paradoxes; Theorem of Constant Flux Linkages. New Section: Rectangular Coil in a Varying Uniform Medium. Valuable supplement to class texts for engineering students. Corrected, enlarged edition. New preface. Bibliography in notes. 49 figures. xi + 106pp. 5⅜ x 8. S1103 Paperbound **$1.25**

INDUCTANCE CALCULATIONS: WORKING FORMULAS AND TABLES, Frederick W. Grover. An invaluable book to everyone in electrical engineering. Provides simple single formulas to cover all the more important cases of inductance. The approach involves only those parameters that naturally enter into each situation, while extensive tables are given to permit easy interpolations. Will save the engineer and student countless hours and enable them to obtain accurate answers with minimal effort. Corrected republication of 1946 edition. 58 tables. 97 completely worked out examples. 66 figures. xiv + 286pp. 5⅜ x 8½. S974 Paperbound **$1.85**

Technological, historical

A DIDEROT PICTORIAL ENCYCLOPEDIA OF TRADES AND INDUSTRY, Manufacturing and the Technical Arts in Plates Selected from "L'Encyclopédie ou Dictionnaire Raisonné des Sciences, des Arts, et des Métiers" of Denis Diderot. Edited with text by C. Gillispie. This first modern selection of plates from the high point of 18th century French engraving is a storehouse of valuable technological information to the historian of arts and science. Over 2000 illustrations on 485 full-page plates, most of them original size, show the trades and industries of a fascinating era in such great detail that the processes and shops might very well be reconstructed from them. The plates teem with life, with men, women, and children performing all of the thousands of operations necessary to the trades before and during the early stages of the industrial revolution. Plates are in sequence, and show general operations, closeups of difficult operations, and details of complex machinery. Such important and interesting trades and industries are illustrated as sowing, harvesting, bee-keeping, cheesemaking, operating windmills, milling flour, charcoal burning, tobacco processing, indigo, fishing, arts of war, salt extraction, mining, smelting, casting iron, steel, extracting mercury, zinc, sulphur, copper, etc., slating, tinning, silverplating, gilding, making gunpowder, cannons, bells, shoeing horses, tanning, papermaking, printing, dyeing, and more than 40 other categories. Professor Gillispie, of Princeton, supplies a full commentary on all the plates, identifying operations, tools, processes, etc. This material, presented in a lively and lucid fashion, is of great interest to the reader interested in history of science and technology. Heavy library cloth. 920pp. 9 x 12. T421 Two volume set **$18.50**

CHARLES BABBAGE AND HIS CALCULATING ENGINES, edited by P. Morrison and E. Morrison. Babbage, leading 19th century pioneer in mathematical machines and herald of modern operational research, was the true father of Harvard's relay computer Mark I. His Difference Engine and Analytical Engine were the first machines in the field. This volume contains a valuable introduction on his life and work; major excerpts from his autobiography, revealing his eccentric and unusual personality; and extensive selections from "Babbage's Calculating Engines," a compilation of hard-to-find journal articles by Babbage, the Countess of Lovelace, L. F. Menabrea, and Dionysius Lardner. 8 illustrations, Appendix of miscellaneous papers. Index. Bibliography. xxxviii + 400pp. 5⅜ x 8. T12 Paperbound **$2.00**

HISTORY OF HYDRAULICS, Hunter Rouse and Simon Ince. First history of hydraulics and hydrodynamics available in English. Presented in readable, non-mathematical form, the text is made especially easy to follow by the many supplementary photographs, diagrams, drawings, etc. Covers the great discoveries and developments from Archimedes and Galileo to modern giants—von Mises, Prandtl, von Karman, etc. Interesting browsing for the specialist; excellent introduction for teachers and students. Discusses such milestones as the two-piston pump of Ctesibius, the aqueducts of Frontius, the anticipations of da Vinci, Stevin and the first book on hydrodynamics, experimental hydraulics of the 18th century, the 19th-century expansion of practical hydraulics and classical and applied hydrodynamics, the rise of fluid mechanics in our time, etc. 200 illustrations. Bibliographies. Index. xii + 270pp. 5¾ x 8.
S1131 Paperbound **$2.00**

BRIDGES AND THEIR BUILDERS, David Steinman and Sara Ruth Watson. Engineers, historians, everyone who has ever been fascinated by great spans will find this book an endless source of information and interest. Dr. Steinman, recipient of the Louis Levy medal, was one of the great bridge architects and engineers of all time, and his analysis of the great bridges of history is both authoritative and easily followed. Greek and Roman bridges, medieval bridges, Oriental bridges, modern works such as the Brooklyn Bridge and the Golden Gate Bridge, and many others are described in terms of history, constructional principles, artistry, and function. All in all this book is the most comprehensive and accurate semipopular history of bridges in print in English. New, greatly revised, enlarged edition. 23 photographs, 26 line drawings. Index. xvii + 401pp. 5⅜ x 8. T431 Paperbound **$2.00**

Prices subject to change without notice.

Dover publishes books on art, music, philosophy, literature, languages, history, social sciences, psychology, handcrafts, orientalia, puzzles and entertainments, chess, pets and gardens, books explaining science, intermediate and higher mathematics, mathematical physics, engineering, biological sciences, earth sciences, classics of science, etc. Write to:

Dept. catrr.
Dover Publications, Inc.
180 Varick Street, N.Y. 14, N.Y.

(continued from front flap)

Mathematical Tables of Elementary and Some Higher Mathematical Functions, Herbert B. Dwight. $1.75

The Theory and Operation of the Slide Rule, John P. Ellis. $1.50

Differential Equations for Engineers, Philip Franklin. $1.65

Hydraulics and Its Applications, A. H. Gibson. Clothbound $8.00

A Treatise on Gyrostatics and Rotational Motion, Andrew Gray. $2.75

Inductance Calculations: Working Formulas and Tables, Frederick W. Grover. $1.85

The Principles of Electromagnetism Applied to Electrical Machines, Bernard Hague. $2.25

Elasticity, Plasticity and Structure of Matter, Roelof Houwink. $2.45

Applied Mechanics for Engineers, Sir Charles Inglis. $2.00

Tables of Functions with Formulae and Curves, Eugene Jahnke and Fritz Emde. $2.00

Photoelasticity: Principles and Methods, H. T. Jessop and F. C. Harris. $2.00

Optics and Optical Instruments, B. K. Johnson. $1.65

Mathematics of Modern Engineering, Ernest G. Keller and Robert E. Doherty. Two volume set $3.70

Fundamental Electromagnetic Theory, Ronold P. King. $2.75

Calculus Refresher for Technical Men, A. Albert Klaf. $2.00

Trigonometry Refresher for Technical Men, A. Albert Klaf. $2.00

Stress Waves in Solids, H. Kolsky. $1.55

Tensors for Circuits, Gabriel Kron. $2.00

The Dynamical Theory of Sound, Horace Lamb. $1.50

Hydrodynamics, Horace Lamb. $3.75

Selected Papers on New Techniques for Energy Conversion, edited by Sumner N. Levine. $3.00

Selected Papers on Semiconductor Microwave Electronics, edited by Sumner N. Levine and Richard M. Kurzrok. $2.25

Fundamentals of Electricity and Magnetism, Leonard B. Loeb. $2.75

The Principles of Electrochemistry, Duncan A. MacInnes. $2.45

Loud Speakers: Theory, Performance, Testing, and Design, Norman W. McLachlan. $2.25

The Electromagnetic Field, Max Mason and Warren Weaver. $2.00

Higher Mathematics for Students of Chemistry and Physics, Joseph W. Mellor. $2.25